Beyond Apoptosis

Beyond Apoptosis

Cellular Outcomes of Cancer Therapy

Edited by

Igor B. Roninson
Ordway Research Institute
Albany, New York, USA

J. Martin Brown
Stanford University
Stanford, California, USA

Dale E. Bredesen
Buck Institute for Age Research
Novato, California, USA

informa
healthcare

New York London

Informa Healthcare USA, Inc.
52 Vanderbilt Avenue
New York, NY 10017

Printed in the United States of America on acid-free paper
10 9 8 7 6 5 4 3 2 1

International Standard Book Number-10: 0-8493-9192-X (Hardcover)
International Standard Book Number-13: 978-0-8493-9192-7 (Hardcover)

Library of Congress Cataloging-in-Publication Data

Beyond apoptosis : cellular outcomes of cancer therapy / edited by Igor B. Roninson, J. Martin Brown, Dale E. Bredesen.
p. ; cm.
Includes bibliographical references and index.
ISBN-13: 978-0-8493-9192-7 (hardcover : alk. paper)
ISBN-10: 0-8493-9192-X (hardcover : alk. paper)

1. Cancer—Chemotherapy. 2. Apoptosis. I. Roninson, Igor B. II. Brown, Martin J. III. Bredesen, Dale E.
[DNLM: 1. Neoplasms—drug therapy. 2. Antineoplastic Agents. 3. Apoptosis—drug effects. 4. Cell Aging—drug effects. 5. Mitosis—drug effects. QZ 267 B573 2008]
RC271.C5B49 2008
616.99′4061—dc22

2008025103

For Corporate Sales and Reprint Permissions call 212-520-2700 or write to: Sales Department, 52 Vanderbilt Avenue, 7th floor, New York, NY 10017.

Visit the Informa Web site at
www.informa.com

and the Informa Healthcare Web site at
www.informahealthcare.com

Preface

BEYOND APOPTOSIS

Over the past decade, much of the effort in understanding how chemotherapy or radiation therapy stops tumor growth has been concentrated on apoptosis, a physiological program of cell death. Although the existing anticancer agents are capable of inducing apoptosis in tumor cells, the accumulating body of evidence indicates that apoptosis plays only a limited role in determining the success of cancer therapy. In recent years, this realization prompted the examination of alternative mechanisms of the antiproliferative action of anticancer agents. These alternative mechanisms include non-apoptotic forms of cell death, terminal cell cycle arrest through the program of cell senescence, and the process of mitotic catastrophe, a general term that designates different forms of abnormal mitosis that lead to eventual cell death or cessation of cell division. Some of these non-apoptotic responses have been already shown to be significant players in stopping the growth of tumor cells, while some other responses (in particular non-apoptotic forms of programmed cell death) have not yet been sufficiently investigated in the setting of cancer therapy. The goal of this book is to examine the relative contribution of apoptosis to cancer treatment response and to acquaint the readers with new concepts in non-apoptotic cell death, senescence, and mitotic catastrophe.

Toward this goal, the first section of the book will include a general overview of apoptosis and critical reviews of its role in the toxicity of anticancer agents to both tumor cells and normal tissues. The second section will provide a description of different non-apoptotic forms of cell death (such as necrosis, paraptosis, autophagic cell death, and others) that were defined primarily in non-cancer context. The third section will deal with the relevance and mechanisms of treatment-induced senescence in tumor and normal cells and with both tumor-suppressing and tumor-promoting effects of senescent cells on their environment.

The fourth section will address the forms and consequences of mitotic catastrophe as a critical event preceding tumor cell death or senescence, as well as other factors that determine different forms of cell death or senescence in tumor cells treated with anticancer agents. To help the readers understand the morphological and kinetic differences between different forms of cell death, senescence, or mitotic catastrophe, the book will include not only the necessary morphological illustrations but also a DVD containing additional color images and video clips from time-lapse microscopic studies of different forms of cell death and mitotic catastrophe.

Igor B. Roninson
J. Martin Brown
Dale E. Bredesen

Contents

Contributors

Ian de Belle Centre de Recherche du CHUL, Université Laval, Québec, Canada

Dale E. Bredesen Buck Institute for Age Research, Novato, and University of California, San Francisco, California, U.S.A.

Eugenia V. Broude Cancer Center, Ordway Research Institute, Albany, New York, U.S.A.

J. Martin Brown Division of Radiation and Cancer Biology, Stanford University Medical Center, Stanford, California, U.S.A.

Judith Campisi Lawrence Berkeley National Laboratory, Berkeley; Buck Institute for Age Research, Novato, California, U.S.A.

Kelly Cole Cancer Center, Ordway Research Institute, Albany, New York, U.S.A.

Valina L. Dawson Institute for Cell Engineering, Johns Hopkins University School of Medicine, Baltimore, Maryland, U.S.A.

Ted M. Dawson Institute for Cell Engineering, Johns Hopkins University School of Medicine, Baltimore, Maryland, U.S.A.

William C. Dewey Department of Radiation Oncology, University of California San Francisco, San Francisco, California, U.S.A.

Daniel DiMaio Department of Genetics, Yale University School of Medicine, New Haven, Connecticut, U.S.A.

Lotti Egger Buck Institute for Age Research, Novato, California, U.S.A.

Yongjun Fan Department of Molecular Genetics & Microbiology, Stony Brook University, Stony Brook, New York, U.S.A.

Edward C. Goodwin Department of Genetics, Yale University School of Medicine, New Haven, Connecticut, U.S.A.

Christine Hanko Cancer Center, Ordway Research Institute, Albany, New York, U.S.A.

Stacy M. Horner Department of Genetics, Yale University School of Medicine, New Haven, Connecticut, U.S.A.

Kimberly Johung Department of Genetics, Yale University School of Medicine, New Haven, Connecticut, U.S.A.

John F.R. Kerr Formerly in the Department of Pathology, University of Queensland, Brisbane, Queensland, Australia

Mehtap Kilic Department of Hematology/Oncology, Charité - Universitätsmedizin Berlin, Berlin, Germany

Richard A. Lockshin Department of Biological Sciences, St. John's University, Jamaica, New York, U.S.A.

Jadranka Loncarek Cancer Center, Ordway Research Institute, Albany, New York, U.S.A.

David T. Madden Buck Institute for Age Research, Novato, California, U.S.A.

Brian J. Nickoloff Department of Pathology, Loyola University Medical Center, Maywood, Illinois, U.S.A.

Abdelhadi Rebbaa Children's Memorial Research Center, Department of Pediatrics, Feinberg School of Medicine, Northwestern University, Chicago, Illinois, U.S.A.

Igor B. Roninson Cancer Center, Ordway Research Institute, Albany, New York, U.S.A.

Clemens A. Schmitt Department of Hematology/Oncology, Charité - Universitätsmedizin Berlin, and Max-Delbrück-Center for Molecular Medicine Berlin-Buch, Berlin, Germany

Sabina Sperandio Centre de Recherche du CHUL, Université Laval, Québec, Canada

Mari Swift Department of Molecular Genetics, University of Illinois at Chicago, Chicago, Illinois, U.S.A.

Erica Ullman Graduate Program of Molecular & Cellular Biology, Stony Brook University, Stony Brook, New York, U.S.A.

Hubert J. Vesselle Department of Nuclear Medicine, University of Washington, Seattle, Washington, U.S.A.

Ikuo Wada Cancer Center, Ordway Research Institute, Albany, New York, U.S.A.

Hui Wang Department of Medicine, VA Puget Sound Health Care System; Department of Medicine, Division of Oncology, University of Washington; and Fred Hutchinson Cancer Research Center, Seattle, Washington, U.S.A.

Qin Wang Department of Medicine, VA Puget Sound Health Care System, Seattle, Washington, U.S.A.

George D. Wilson Department of Radiation Oncology, William Beaumont Hospital, Royal Oak, Michigan, U.S.A.

Daniel Y. Wu Department of Medicine, VA Puget Sound Health Care System; Department of Medicine, Division of Oncology, University of Washington; and Fred Hutchinson Cancer Research Center, Seattle, Washington, U.S.A.

Peter C. Wu Department of Surgery, University of Washington, Seattle, Washington, U.S.A.

Kristin E. Yates Department of Genetics, Yale University School of Medicine, New Haven, Connecticut, U.S.A.

Zahra Zakeri Department of Biology, Queens College and Graduate Center of City University of New York, Flushing, New York, U.S.A.

Wei-Xing Zong Department of Molecular Genetics & Microbiology, Stony Brook University, Stony Brook, New York, U.S.A.

1

A Personal History of the Development of the Apoptosis Concept

John F.R. Kerr

Formerly in the Department of Pathology, University of Queensland, Brisbane, Queensland, Australia

INTRODUCTION

In this chapter, I will trace the sequence of observations that led to the formulation of the apoptosis concept and will briefly describe its aftermath.

DIFFERENTIATION OF TWO MORPHOLOGICALLY DISTINCT TYPES OF CELL DEATH IN THE LIVER

My interest in cell death began in 1962 during my PhD studies at University College Hospital Medical School in London. My adviser, Sir Roy Cameron, suggested that I examine the cellular processes involved in the shrinkage of liver tissue that was known to follow obstruction of its portal venous blood supply. To this end, I ligated the portal vein branches supplying the left and median lobes of the liver in rats (1). These lobes shrank to about one-sixth of their original weight during the first eight days after operation, the rest of the liver undergoing compensatory enlargement. Within six hours of operation, necrosis developed in the ischemic lobes in discrete foci around the terminal hepatic venules, i.e., downstream in the blood flow to the liver tissue. Typically, the necrosis affected groups of adjoining hepatocytes and involved degeneration with loss of their normal histological staining pattern. The necrotic cells were rapidly removed by mononuclear phagocytes, the abundance of the phagocytes suggesting that many of them were derived from circulating monocytes that had left the microvasculature during an inflammatory response. The remaining parenchyma in the ischemic lobes remained essentially viable, being sustained by blood from the hepatic artery. However, as the lobes shrank, scattered single hepatocytes were converted into small round or oval cytoplasmic masses, some of which contained one or more specks of markedly condensed nuclear chromatin. These bodies were taken up by Kupffer cells, the resident mononuclear phagocytes of the liver, or occasionally by epithelial hepatocytes. Formation of the bodies clearly represented a distinctive form of cell death, which differed from necrosis in its microscopic appearance, in affecting only scattered individual cells and in not evoking inflammation. Importantly, very small numbers of the bodies could be found in the livers of healthy rats.

At the time I was performing these experiments, it had been suggested that rupture of lysosomes with release of their digestive enzymes might play an essential role in causing cell death following various types of injuries (2). Joe Smith, Cameron's deputy, introduced me to histochemical techniques for demonstrating lysosomal acid phosphatase and esterase and I applied these to frozen sections of the experimental rat livers (1). In histochemical preparations of normal liver, hepatocyte lysosomes were evident as discretely stained granules in the cytoplasm bordering bile canaliculi. In the foci of confluent necrosis, there was diffuse staining in these areas, suggesting that lysosomes had indeed ruptured. This was not, however, an early change, and it seemed likely that lysosome rupture occurred along with degeneration of the other cytoplasmic organelles during the evolution of necrosis rather than it being the initiating event causing the necrosis. But by far the most interesting finding was that lysosomes in the small round cytoplasmic bodies stained discretely, suggesting that they were still intact. Further, histochemical techniques for RNA and succinic dehydrogenase suggested that ribosomes and mitochondria in the bodies were also intact. It seemed unlikely that the process involved in the formation of the bodies was degenerative in nature. It was proposed that their small size might be a result of progressive removal of the cytoplasm by autophagy, a hypothesis that was subsequently shown by electron microscopy to be incorrect. The term shrinkage necrosis was suggested for the process, necrosis being synonymous with cell death in those days.

At the beginning of 1965, I returned to my home city of Brisbane, Australia, and joined the University of Queensland Pathology Department. My prime objective was to study the evolution of shrinkage necrosis with the electron microscope. However, the Department's first electron microscope had not yet been installed, and I undertook a histochemical study of liver injury produced in rats by the pyrrolizidine alkaloid heliotrine (3). This was known to produce liver necrosis with a similar distribution to that which I had observed after portal vein branch ligation. As in the ischemic liver injury, cells undergoing shrinkage necrosis were present in moderate numbers in the predominantly viable parenchyma. Histochemical staining patterns for lysosomal enzymes in the two types of cell death were the same as I had seen in London.

DEFINITION OF THE SEQUENCE OF ULTRASTRUCTURAL CHANGES INVOLVED IN SHRINKAGE NECROSIS

With the availability of an electron microscope, I embarked on a study of the evolution of shrinkage necrosis occurring in the rat liver with the technical help of David Collins and later Brian Harmon (4–6). The basic sequence that emerged was subsequently confirmed and refined in studies of many other tissues (7,8).

Small round bodies that still lay free in the extracellular space were found to comprise condensed masses of hepatocyte cytoplasm surrounded by an intact plasma membrane. Their closely packed organelles appeared well preserved. Nuclear fragments present in some of them were only occasionally surrounded by a nuclear envelope; the lack of an envelope was later shown to be an artifact resulting from the electron microscopic preparative techniques used at that time. Nuclear budding and fragmentation in apoptosis are associated with preservation of the nuclear envelope (8). The earliest recognizable nuclear change involved compaction of chromatin into uniformly dense masses with sharply defined edges that abutted on the nuclear envelope. The chromatin condensation persisted in the nuclear fragments, condensed chromatin either being present as peripherally located crescents or filling the fragments. The ultrastructural appearances of the small round bodies were noted to be the same as those described in so-called acidophilic or Councilman bodies occurring in the liver in certain diseases (9–11).

Some of the rounded bodies were extremely small, being comparable in size to mitochondria. Further, the bodies were often present in clusters. These findings, taken in conjunction with the marked crowding of their organelles, clearly suggested that they arose by separation of protuberances forming on the surface of condensing cells with sealing of the plasma membrane. The actual process of cellular budding was, however, rarely observed. I was able to produce only rather pathetic electron micrographs showing budding at the time. Much more satisfactory illustrations were produced later (7,8). It was correctly inferred that the cellular condensation and budding are effected very quickly (4). This was noted to be consistent with the classic phase-contrast microscopic studies of cell death occurring in vitro carried out by Marcel Bessis (12). Electron microscopy confirmed that the cellular fragments resulting from shrinkage necrosis in the liver were taken up by epithelial hepatocytes as well as by mononuclear phagocytes (6). The progressive degradation of the phagocytosed bodies within phagolysosomes was also followed.

The electron microscopic features of necrosis occurring in the liver were totally different from those of shrinkage necrosis (4,5). In necrosis, there was swelling of the cytoplasm rather than condensation, and mitochondria became swollen and developed matrix densities. Internal and plasma membranes progressively broke down and this was followed by dissolution of all cytoplasmic components. While nuclear chromatin often underwent irregular clumping at an early stage of necrosis, the chromatin clumps were not uniformly dense, their edges were not sharply delineated, and the condensed chromatin was not relocated to abut on the nuclear envelope. The nuclei never budded to form discrete fragments. Eventually, the chromatin disappeared. Debris of necrotic cells was taken up by mononuclear phagocytes but not by epithelial hepatocytes.

At this stage of my research (6) I concluded that while severe injury to cells by toxins and ischemia causes necrosis, mild injury by the same agents can enhance the extent of shrinkage necrosis in a tissue. Secondly, evidence was provided that cells undergoing shrinkage necrosis are still capable of some metabolic activity, and it was proposed that the rapid cellular condensation and budding are the result of inherent activity of the cells. Thirdly, it was stated that shrinkage necrosis constitutes at least one type of cell death occurring in normal tissues.

RECOGNITION OF SHRINKAGE NECROSIS IN MALIGNANT TUMORS

About the middle of 1970, I attended a seminar on tumors given by a student. He referred to the well-recognized slow growth rate of basal cell carcinomas of human skin, which was surprising in view of the large number of mitotic figures evident within them. I recalled that Jeffrey Searle, who was undertaking postgraduate training as an anatomical pathologist in Brisbane at the time, had mentioned that cells showing the light microscopic features of shrinkage necrosis were numerous in basal cell carcinomas. We decided to study them with the electron microscope (13,14).

We found that ultrastructurally typical shrinkage necrosis was indeed present in basal cell carcinomas and that the resulting cellular fragments were taken up and digested by the carcinoma cells. The extent of the process in parts of some tumors was comparable to that seen in rat liver lobes following obstruction of their portal venous blood supply; it will be recalled that these lobes decreased to about one-sixth of their original weight during the first eight days after operation. We concluded that the effect of the shrinkage necrosis on the growth kinetics of the tumors must be considerable. When we consulted the literature, we found that several groups of investigators had recently shown that there is a discrepancy between the rate of enlargement of a number of types of malignant tumors and the rate of proliferation of cells within them as measured by tritiated thymidine labeling of their nuclei. While it was considered by these investigators that cell death was the most likely explanation of the discrepancy, focal necrosis due to ischemia was often not evident. We found that shrinkage necrosis could be detected by light microscopy in a variety of malignant tumors and we proposed that the spontaneous occurrence of the process might be an important general parameter in the kinetics of neoplastic growth.

We also speculated about possible causes of shrinkage necrosis in tumors. While it was known that the blood supply to tumor masses is often precarious, we found shrinkage necrosis in very thin tumor trabeculae, where significant ischemia was unlikely. By this time we had seen shrinkage necrosis in many healthy adult mammalian tissues, where it was clearly involved in cellular turnover. We quoted the seminal suggestion by Anna Laird that death of both normal and neoplastic cells is a preordained, genetically determined phenomenon (15). In retrospect, it is apparent that Laird's suggestion was inspired by her knowledge of cell death occurring during normal development. We did not understand the implications of this at the time, the paper quoted by Laird not being available in Brisbane. I was to catch up with developmental cell death a little later in Scotland.

Finally, we were fortunate in obtaining a human squamous cell carcinoma of skin that had been excised surgically during a course of radiotherapy. It showed an apparent increase in shrinkage necrosis (14). We stated that our preliminary observations suggested that radiotherapy might enhance shrinkage necrosis in tumors. In support of this suggestion, we raised the possibility that some of the unusually large "autophagic vacuoles" reported by others in mouse breast carcinomas following irradiation and chemotherapy might really be phagocytosed cellular fragments, since the published electron micrographs showed them to be surrounded by narrow, membrane-enclosed spaces. I had already realized that the distinction between phagocytosed cellular fragments resulting from shrinkage necrosis and autophagic vacuoles could be difficult if the former lacked a nuclear remnant (16).

FORMULATION OF THE APOPTOSIS CONCEPT

In late 1970, Alastair Currie, at that time Head of the Department of Pathology in the University of Aberdeen, Scotland, came to Brisbane as a guest professor in the University of Queensland. I showed him my electron micrographs of shrinkage necrosis. He was quite excited. He told me that he had seen cell death with the same light microscopic appearances in two situations. First, as a side issue in the induction of breast carcinomas in rats by 9,10 dimethyl-1, 2-benzanthracene (DMBA), he had noticed death of scattered single cells in the inner layers of the adrenal cortices following low doses of the carcinogen; higher doses produced massive adrenal cortical necrosis. Secondly, he and Andrew Wyllie, who had recently started PhD studies under his guidance, had seen similar dying cells in the adrenal cortices of glucocorticoid-treated rats, in which the secretion of adrenocorticotrophic hormone (ACTH) would, of course, have been suppressed. This suggested that in endocrine-dependent tissues, shrinkage necrosis might be under trophic hormonal control. I was due for sabbatical leave the following year; Currie invited me to spend it in Aberdeen.

Before leaving for Aberdeen, I carried out, at Currie's suggestion, an electron microscopic study of the DMBA-induced adrenal cortical lesions (17).

As expected, the single cell death was found to have the ultrastructural features of shrinkage necrosis.

On arriving in Aberdeen in September 1971, I was delighted to discover that shrinkage necrosis had been observed by light microscopy in several sets of experiments being undertaken in the Pathology Department. In each case, I confirmed the nature of the cell death by electron microscopy.

Wyllie had shown that reduction in circulating ACTH levels effected in adult rats by prednisolone injection and in fetal rats either by prednisolone administration to the mothers or by intrauterine decapitation resulted in a decrease in the size of adrenal cortical cells and deletion of many scattered cells in the inner adrenal cortex by shrinkage necrosis; both the decrease in cell size and the cell deletion were prevented by coincident injection of ACTH (18). In a second study, he had shown that shrinkage necrosis is enhanced in the rat adrenal cortex during the neonatal period, when there is a physiological fall in ACTH secretion (19).

Currie had previously studied the regression of DMBA-induced rat breast carcinomas that often follows removal of the ovaries, i.e., following hormone ablation therapy. Electron microscopy of material from these experiments showed that shrinkage necrosis was extensive during the phase of active tumor regression (20).

By good fortune, Allison Crawford, a developmental biologist, was working in the Aberdeen Pathology Department at the time of my visit. She was studying the teratogenic effects of 7-hydroxymethyl-12-methylbenz(a)anthracene, one of the metabolites of DMBA, in Sprague Dawley rats (21). A single intravenous injection of this substance into pregnant rats on days 11 to 14 of pregnancy produced spina bifida with meningomyelocele in all of the fetuses. She had found that these developmental defects could be explained by the occurrence of massive shrinkage necrosis in the presumptive vertebral arches 24 hours after administration of the teratogen. Once again, my electron microscopic studies confirmed the nature of the cell death.

Of greater importance for the formulation of the apoptosis concept, however, was the fact that Allison told us about normal developmental cell death. Since the classic reviews of Glücksmann (22) and Saunders (23), this process had been known to a small coterie of developmental biologists; most people outside this field were unaware of its existence. Our examination of published electron micrographs of normal developmental cell death (24–26) indicated that it is morphologically identical to shrinkage necrosis occurring in adult life.

A serendipitous confluence of observations and ideas had now set the stage for our proposal of the apoptosis concept (20). Here we had a morphologically distinctive type of cell death that was fundamentally different from necrosis. Its ultrastructural features suggested that it was active and inherently programmed rather than being degenerative in nature. It was not accompanied by inflammation. It accounted for focal deletion of tissue during normal development, where it had been shown to be precisely controlled. It occurred in many healthy adult tissues, and here it was likely to be implicated in cellular turnover. In endocrine-dependent tissues, it could be initiated or inhibited by modulation of circulating levels of trophic hormones. It was involved in both normal involution and pathological atrophy of tissues. It occurred spontaneously in growing malignant tumors and was enhanced by some forms of nonsurgical treatment of tumors. We suggested that this hitherto little-recognized phenomenon plays a complementary but opposite role to mitosis in the regulation of animal cell populations and proposed that it be called apoptosis. The term was suggested by Professor James Cormack of the Department of Greek, University of Aberdeen. He told us that it was used in classical Greek to describe the "dropping off" of petals from flowers or leaves from trees. It seemed to embody the biological functions of the process. Cormack said that the second p in the word apoptosis should be silent. This appears to have been incorrect (27).

In view of our contention that apoptosis is a regulated phenomenon, it was puzzling that it could be induced by mild cellular injury by agents such as toxins and ischemia that cause necrosis when the injury is more severe. This apparent paradox may now have been resolved with the discovery that mitochondria are often involved in initiating apoptosis; gross mitochondrical dysfunction has long been known to cause necrosis. It has been suggested (28) that mild cellular injury leads to transient opening of the mitochondrial permeability transition pores with release of pro-apoptotic factors such as cytochrome c, but without ATP depletion, resulting in the induction of apoptosis. Severe cellular injury, on the other hand, probably leads to prolonged opening of the pores, swelling, and decreased functioning of mitochondria, ATP depletion, and collapse of cellular homeostasis, resulting in degeneration and necrosis.

At the time of proposing the apoptosis concept we were not aware of the fact that the light microscopic features of the process had been described and figured as early as 1885. As Majno and Joris (29) have pointed out, Walther Flemming published in that year beautiful camera lucida drawings of apoptosis occurring during normal involution of ovarian follicles in adult rabbits. He appreciated that the morphology of the cell death was distinctive, accurately depicting typical apoptotic chromatin changes, and he clearly understood the role of cell death in the normal regulation of tissue size during adult life. In 1914, Ludwig Gräper proposed that a mechanism must exist to counterbalance cell proliferation in tissues and suggested that the process described by Flemming provided the answer (29). It is extraordinary that Flemming's seminal discovery was ignored, studies of normal cell death in the first six decades of the 20th century being virtually confined to that occurring during development (29–31). This neglect is particularly surprising, since it was the same Walther Flemming who coined the term mitosis (29).

It is interesting to speculate about the reasons for the prolonged neglect. In most adult mammalian tissues apoptosis is quantitatively very sparse. Further, the rapid progression of the early cellular condensation and budding, the small size of many apoptotic bodies, and their phagocytosis and degradation by nearby cells without the occurrence of inflammation make the process inconspicuous in such tissues. It was undoubtedly the frequently massive extent of apoptosis in local areas during normal development that led to its early recognition in this situation. Even here, however, emphasis was placed on the biological implications of the cell death rather than on its morphology (22,23). Indeed, cell death occurring during development was sometimes referred to as necrosis (23). Apoptosis was, nevertheless, detected fairly early in certain specific circumstances in adult mammals where it is unusually extensive. Here the resulting apoptotic bodies were given names unique to the individual circumstances of occurrence, such as tingible body in lymphoid germinal centers, Councilman or acidophilic body in the human liver in viral hepatitis, Civatte body associated with lymphoid infiltration of the epidermis in the human skin disease lichen planus, sunburn cell in the epidermis following exposure to ultraviolet radiation, and karyolytic body occurring in intestinal crypts after X irradiation (7). In these situations, the widely ranging occurrence of the process involved was not appreciated. An important factor militating against early recognition of the morphologically distinct nature of apoptosis was that some condensation of nuclear chromatin often occurs during the development of necrosis (8). Chromatin condensation, termed nuclear pyknosis, had long been correctly recognized as a light microscopic hallmark of cell death. It required electron microscopy, however, to show that the detailed patterns of chromatin condensation in apoptosis and necrosis are quite different. Even at the level of electron microscopy, apoptosis was sometimes misidentified, phagocytosed apoptotic bodies being regarded as autophagic vacuoles despite the presence of nuclear remnants in some of them (7). Finally, it is perhaps a human attribute to favor positive phenomena such as cellular proliferation over negative phenomena such as cellular death. Early studies of the effects of radiation and anticancer drugs on tumors focused on inhibition of the former, whereas cellular deletion was hardly ever mentioned.

SUBSEQUENT PROGRESS IN APOPTOSIS RESEARCH

In 1979, I spent a second period of sabbatical leave with Wyllie and Currie in Scotland, this time in Edinburgh, and we reviewed the modest progress that had been made (7). Andrew Wyllie had shown that glucocorticoid-induced thymocyte death, which was known to display the morphology of apoptosis, was associated with double strand cleavage of nuclear DNA at the linker regions between nucleosomes (32). The resulting oligonucleosomal fragments could be readily demonstrated by agarose gel electrophoresis, a characteristic "ladder" developing. In contrast, DNA from necrotic cells produces a diffuse smear on electrophoresis. Wyllie's work provided the first molecular evidence that apoptosis and necrosis are fundamentally different. It also enabled biochemists to recognize the process without recourse to electron microscopy. Meanwhile, my colleagues and I in Brisbane had extended the apoptosis concept to include bioprotective functions. In some of the circumstances in which apoptosis occurs under pathological conditions, it can be looked upon as serving such a function in that it brings about the elimination of cells whose survival might be harmful to the animal as a whole. For example, we showed that cell death induced by cytotoxic T lymphocytes displayed the ultrastructural features of apoptosis (33,34). This cell death plays a vital role in containing viral infections. It was postulated that the cellular immune system had made opportunistic use during evolution of a preexisting mechanism for deleting cells that enabled it to get rid of cells showing antigenic change with minimal tissue disruption. Secondly, it was shown by electron microscopy that the cell death already known to be induced by DNA-damaging agents such as radiation and cytotoxic anticancer drugs in normal rapidly proliferating cell populations and lymphoid tissues was apoptotic in type, as was much of the cell death induced by these agents in susceptible malignant tumors (35,36). It was argued that the apoptosis induced in normal tissues was effecting the elimination of potentially dangerous cells with unrepaired DNA damage (35,36).

There was prolonged delay in the recognition of apoptosis in the years following its description. This delay has been quantified by Eugene Garfield of the Institute of Scientific Information in Philadelphia (37). Only after 1990 did research on the subject begin to grow rapidly. The surge of activity was undoubtedly a result of the discovery of genes involved in regulation of the process (30,31,37,38). It is probable that the initial lack of interest stemmed from the fact that the definition of a biological concept on the basis of morphology was out of step with the science of the time, which was increasingly focusing on molecular biology. Apoptosis had really been defined in 1885, along with mitosis. It was anomalous that it was then overlooked during the first six decades of the 20th century. This can probably be explained by the need for electron microscopy to appreciate its distinctive features.

ACKNOWLEDGMENT

I have previously written a chapter on the development of the apoptosis concept. The reference is: Kerr JFR. A personal account of events leading to the definition of the apoptosis concept. In Kumar S, ed. Apoptosis: Biology and Mechanisms. Berlin Heidelberg: Springer-Verlag, 1999: 1-10. The current chapter is published with kind permission of Springer Science and Business Media.

REFERENCES

1. Kerr JFR. A histochemical study of hypertrophy and ischaemic injury of rat liver with special reference to changes in lysosomes. J Pathol Bacteriol 1965; 90:419–435.
2. de Reuck AVS, Cameron MP, eds. Ciba Foundation Symposium on Lysosomes. London: Churchill, 1963.

3. Kerr JFR. Lysosome changes in acute liver injury due to heliotrine. J Pathol Bacteriol 1967; 93:167–174.
4. Kerr JFR. An electron-microscope study of liver cell necrosis due to heliotrine. J Pathol 1969; 97:557–562.
5. Kerr JFR. An electron microscopic study of liver cell necrosis due to albitocin. Pathology 1970; 2:251–259.
6. Kerr JFR. Shrinkage necrosis: a distinct mode of cellular death. J Pathol 1971; 105:13–20.
7. Wyllie AH, Kerr JFR, Currie AR. Cell death: the significance of apoptosis. Int Rev Cytol 1980; 68:251–306.
8. Kerr JFR, Gobé GC, Winterford CM, et al. Anatomical methods in cell death. In: Schwartz LM, Osborne BA, eds. Cell Death. Methods in Cell Biology. Vol 46. San Diego: Academic Press, 1995:1–27.
9. Biava C, Mukhlova-Montiel M. Electron microscopic observations on Councilman-like acidophilic bodies and other forms of acidophilic changes in human liver cells. Am J Pathol 1965; 46:775–802.
10. Klion FM, Schaffner F. The ultrastructure of acidophilic "Councilman-like" bodies in the liver. Am J Pathol 1966; 48:755–767.
11. Moppert J, v Ekesparre D, Bianchi L. Zur Morphogenese der eosinophilen Einzelzellnekrose im Leberparenchym des Menschen. Eine licht- und elektronenoptisch korrelierte Untersuchung. Virchows Arch Pathol Anat 1967; 342:210–220.
12. Bessis M. Studies on cell agony and death: an attempt at classification. In: de Reuck AVS, Knight J, eds. Ciba Foundation Symposium on Cellular Injury. London, UK: Churchill, 1964:287–316.
13. Kerr JFR, Searle J. A suggested explanation for the paradoxically slow growth rate of basal-cell carcinomas that contain numerous mitotic figures. J Pathol 1972; 107:41–44.
14. Kerr JFR, Searle J. The digestion of cellular fragments within phagolysosomes in carcinoma cells. J Pathol 1972; 108:55–58.
15. Laird AK. Dynamics of growth in tumors and in normal organisms. In: Perry S, ed. Human Tumor Cell Kinetics. National Cancer Institute Monograph, no. 30. Bethesda: National Cancer Institute, 1969:15–28.
16. Kerr JFR. Some lysosome functions in liver cells reacting to sublethal injury. In: Dingle JT, ed. Lysosomes in Biology and Pathology 3. Frontiers of Biology. Vol 29. Amsterdam: North-Holland, 1973:365–394.
17. Kerr JFR. Shrinkage necrosis of adrenal cortical cells. J Pathol 1972; 107:217–219.
18. Wyllie AH, Kerr JFR, Macaskill IAM, et al. Adrenocortical cell deletion: the role of ACTH. J Pathol 1973; 111:85–94.
19. Wyllie AH, Kerr JFR, Currie AR. Cell death in the normal neonatal rat adrenal cortex. J Pathol 1973; 111:255–261.
20. Kerr JFR, Wyllie AH, Currie AR. Apoptosis: a basic biological phenomenon with wide-ranging implications in tissue kinetics. Br J Cancer 1972; 26:239–257.
21. Crawford AM, Kerr JFR, Currie AR. The relationship of acute mesodermal cell death to the teratogenic effects of 7-OHM-12 MBA in the foetal rat. Br J Cancer 1972; 26:498–503.
22. Glücksmann A. Cell deaths in normal vertebrate ontogeny. Biol Rev 1951; 26:59–86.
23. Saunders JW. Death in embryonic systems. Science 1966; 154:604–612.
24. Saunders JW, Fallon JF. Cell death in morphogenesis. In: Locke M, ed. Major Problems in Developmental Biology. New York: Academic Press, 1966:289–314.

25. Farbman AI. Electron microscope study of palate fusion in mouse embryos. Dev Biol 1968; 18:93–116.
26. Webster DA, Gross J. Studies on possible mechanisms of programmed cell death in the chick embryo. Dev Biol 1970; 22:157–184.
27. Georgatsos JG. The s(p)elling of apo(p)tosis. Nature 1995; 375, 100.
28. Halestrap A. A pore way to die. Nature 2005; 434:578–579.
29. Majno G, Joris I. Apoptosis, oncosis and necrosis. An overview of cell death. Am J Pathol 1995; 146:3–15.
30. Lockshin RA. The early modern period in cell death. Cell Death Differ 1997; 4:347–351.
31. Vaux DL. Apoptosis timeline. Cell Death Differ 2002; 9:349–354.
32. Wyllie AH. Glucocorticoid-induced thymocyte apoptosis is associated with endogenous endonuclease activation. Nature 1980; 284:555–556.
33. Searle J, Kerr JFR, Battersby C, et al. An electron microscopic study of the mode of donor cell death in unmodified rejection of pig liver allografts. Aust J Exp Biol Med Sci 1977; 55:401–406.
34. Don MM, Ablett G, Bishop CJ, et al. Death of cells by apoptosis following attachment of specifically allergized hymphocytes *in vitro*. Aust J Exp Biol Med Sci 1977; 55:407–417.
35. Searle J, Lawson TA, Abbott PJ, et al. An electron-microscope study of the mode of cell death induced by cancer-chemotherapeutic agents in populations of proliferating normal and neoplastic cells. J Pathol 1975; 116:129–138.
36. Kerr JFR, Searle J. Apoptosis: its nature and kinetic role. In: Meyn RE, Withers HR, eds. Radiation Biology in Cancer Research. New York: Raven Press, 1980:367–384.
37. Garfield E, Melino G. The growth of the cell death field: an analysis from the ISI-Science citation index. Cell Death Differ 1997; 4:352–361.
38. Melino G, Knight RA, Green DR. Publications in cell death: the golden age. Cell Death Differ 2001; 8:1–3.

2

What Do the Clinical Data Tell Us About the Role of Apoptosis in Sensitivity to Cancer Therapy?

George D. Wilson
Department of Radiation Oncology, William Beaumont Hospital, Royal Oak, Michigan, U.S.A.

INTRODUCTION

There is no doubt that disruption of apoptosis plays a prominent role in the development of cancer and the progression of the malignant phenotype. Indeed it is widely accepted that evasion of apoptosis is one of the "hallmarks of cancer" (1). There is also no doubt that apoptosis, and proteins involved in this complex process, can convey prognostic information in a variety of different cancers (2–9). There is also no doubt that apoptosis can be invoked by DNA-damaging agents including radiation and chemotherapeutic drugs. However there is considerable doubt whether apoptosis plays a central role in determining the sensitivity of solid tumors to cancer therapy and particularly radiotherapy (RT) (10–12). In this Chapter we will reveal what the clinical data has told us about the significance of apoptosis in determining both the response and clinical outcome of patients treated with cytotoxic therapy. The data will show us that the picture is clear, i.e., there is no clear picture. The clinical literature surrounding apoptosis is plagued by contradictory results and underpowered studies with arbitrary cutoffs, ill-defined patients groups, and overinterpretation of the data. However, there are many well-conducted studies that cast doubt on the significance of apoptosis, or its related proteins, as determinants of cancer therapy in solid tumors.

ASSESSMENT OF APOPTOSIS IN CLINICAL RESEARCH

Apoptotic cells can be recognized by stereotypical morphological changes that are readily identifiable through the electron or light microscope. To the trained eye, there are several characteristic features that distinguish them from non-apoptotic cells. These include a condensed, shrunken cytoplasm, and prominent nuclear changes including homogeneously condensed chromatin either around the nuclear membrane or as solid masses. Usually they are documented as a small number of individual cells surrounded by surviving tumor cells but detached from them in a vacuolar structure. In the final stages of the process, cells fragment into compact membrane-enclosed structures called "apoptotic bodies," which contain cytosol, the condensed chromatin, and organelles. Morphological identification is the mainstay of apoptosis counting in tissue slides but problems can arise in areas close to necrosis where debris can be mistakenly identified as "apoptotic bodies" and apoptotic cells in the stroma are also problematic as they may be of lymphocyte origin. The apoptotic bodies are engulfed by macrophages and thus are removed from the tissue without causing an inflammatory response. Those morphological changes are a consequence of characteristic molecular and biochemical events occurring in an apoptotic cell, most notably the activation of proteolytic enzymes that eventually mediate the cleavage of DNA into oligonucleosomal fragments as well as the cleavage of a multitude of specific protein substrates, which usually determine the integrity and shape of the cytoplasm or organelle. Many proteins are involved in apoptotic processes and it might be expected that they could provide useful clinical information regarding the mode of cell death, the integrity of the apoptotic pathways, and prognostic power.

Apoptotic markers are involved at various levels. Receptors on the surface accept the death signal and transmit it across the cellular membrane. Specific intracellular enzymes are activated in order to coordinate the well-orchestrated breakdown of cellular structures. Sets of inhibitors and enhancers modulate the apoptotic signal. Finally, specific death substrates are degraded and externalized in a defined way. At each of these steps in the apoptotic pathway, various proteins have been investigated for their power to convey prognostic information. Apart from the number of apoptotic cells directly, perhaps the most commonly analyzed proteins that are directly associated with apoptosis have been bcl-2, bax, and more recently survivin. p53 is also studied in the context of apoptosis but the complex nature and function of this tumor suppressor gene and its common deregulation in cancer render the link between its expression and apoptosis tenuous in the majority of solid cancers.

DIRECT MEASURES OF APOPTOSIS

A wide variety of techniques have now been utilized to assess apoptotic cells in tissue sections (13). Electron microscopy is the optimal technique to identify apoptotic cells but is impractical in pathological specimens; the majority of studies have counted apoptotic bodies using light microscopy. On hematoxylin and eosin (H&E)-stained sections, apoptotic bodies are identified as scattered structures with compact segregated masses of chromatin often associated with strongly eosinophilic condensed cytoplasm. The technique is not straightforward and there is some subjectivity in the recognition of apoptotic cells and in the selection of areas to be counted. Often investigators will concentrate on the advancing front in invasive carcinomas and avoid regions with necrosis, inflammation, and fibrosis.

The apoptotic index (AI) is typically reported as apoptotic cells per 100 nonapoptotic tumor cells, per unit area, per grid, per field, or as a grade. The AI is often a small number and thus a significant number of cells need to be counted to minimize errors and take into account heterogeneity; most reports count 10 high-power fields. Several studies had undertaken a more quantitative approach and counted up to 2000 cells.

To overcome the difficulties of the morphological method, many studies have used the in situ end labeling or the terminal deoxynucleotidyl transferase (TdT)–mediated dUTP biotin nick end-labeling (TUNEL) techniques. Both procedures depend on labeling breaks in DNA strands caused by fragmentation. However, fragmentation of DNA is not obligatory for apoptosis and artifacts may occur as DNA cleavage can occur in necrosis. Although some authors have reported a good correlation between histological assessment of apoptosis and end-labeling techniques, a recent study demonstrated a lack of correlation and an overestimated AI derived from TUNEL (mean 8.1%) as opposed to morphology (mean 1.17%) (14). A further development has been the use of antibodies against single-stranded DNA (15) that detect early changes in apoptotic cells. These techniques have been applied to many different tumor types with different treatments.

In Fig. 1, the data for AI and prognosis has been summarized for several of the major cancers in which several independent studies have been undertaken and where radio- and chemotherapy have played a prominent role in the treatment of patient population. Due to the diversity of patient numbers, methodology, cutoffs, and clinical endpoints, the data have been simplified such that the size of the symbol represents the number of cases and prognosis has been classified as better, worse, or noninformative. Studies have been omitted if the number of patients studied were less than 30. In the case of AI, the data are summarized for the prognostic significance of a high AI.

In cervix cancer, where radiation is the prime modality several studies have addressed the issue of pretreatment AI and clinical outcome (14,16–26). Three of the studies involved adenocarcinomas (14,23,25). Although many of

	better	non-informative	worse
Cervix	Chung 1999 Wheeler 1995 Zaghloul 2000 Sheridan 1999	Paxton 2000 Gasinska 2002 Tsang 1999 Jain 2003 Leung 2004	Liu 2001 Levine 1995 Kim 2001
H&N	Grabenbuaer 2003 Xie 1999 Smilek 2005	Stoll 2000 Pulkinnen 1999 Tschuiya 2001 Mandel 2001	Hirikoski 1999 Lera 1998 Sikorska 2000
NSCLC	Tanaka 1999 Hwang 2001 Macluskey 2000 Puglisi 2002 Matturri 1999	Hanaoka 2002	Langendijk 2000 Dworakowska 2005
Bladder	Weiss 2005 Moonen 2001 Rodel 2000 Harada 2000	Lavezzi 2001 Chyle 1999 Matsumoto 2004 Giannopolous 2003	Korkolopolou 2000 Lara 1999
Colorectal	Sugamura 1998 Sinicrope 1999 Marijnen 2003 Adell 2000 Sakakura 1998 Smith 2006 Rodel 2002 Abe 2001	Garrity 2004 Schwander 2000 Michael-Robinson 2001 Noguchi 2003 Paradiso 2001	Benandarf 2003

Prognosis

Figure 1 Summary of prognostic studies assessing apoptotic index. The data are classified as better prognosis if a high apoptotic index was significantly associated with a more favorable clinical outcome or worse if the outcome was significantly shorter. Noninformative denotes no significant correlation with clinical outcome. The studies are arranged in descending order of the number of patients investigated with apoptosis, (●) representing studies with more than 100 patients, (●) 50–99 patients, and (•) less than 50 patients.

these patients received external beam RT or brachytherapy, the data exemplify the recurring theme in this chapter, i.e., some studies finding a positive correlation between AI and outcome while others conclude the opposite and some find no relationship. In two of the largest studies with 146 (22) and 130 (17) patients, no significant association was found between apoptosis and survival specifically in patients receiving RT. Several groups have studied the induction of apoptosis at the end of the first week of RT in cervix cancer (27–31) usually after a dose of

9 Gy, and have noted a significant increase in cells dying from this mode of cell death. However, this is not proof of a direct induction of apoptosis as the indices rose only from 0.2% to 0.8% before treatment to 1.2% to 1.7% after 9 Gy, while the expected cell kill would be greater than 80%. More recently, Bhosle and colleagues (32) have shown a strong correlation between responders to radiation and induction of apoptosis after the first 2 Gy of treatment. Measurements showed that AI increased by two- to threefold in 95% of the patients in complete response group whereas 92% of the partial responders showed increases in AI by a factor less than two. Again, this is an interesting observation but has yet to be correlated with patient outcome.

Another site where radical RT plays a key role in curative treatment is head and neck (H&N) cancer, which again has been the subject of several studies utilizing pretreatment measurement of apoptotic cells (33–42). H&N cancers recapitulate the findings in cervix cancer with no consistent association between AI and clinical outcome (Fig. 1).

In non–small cell lung cancer (NSCLC), the majority of studies suggest that a high level of spontaneous apoptosis before treatment is associated with a more favorable clinical outcome (43–47), while other equally valid studies show the opposite result (48,49) or no clinical significance (50).

A similar pattern of disparity emerges for bladder cancer in Figure 1 where the majority of patients received either RT alone or chemoradiation after transurethral resection. The two largest studies (51,52), failed to find any connection between AI and clinical outcome.

A final group of tumors in which several independent studies have been undertaken and where RT is often employed preoperatively are colorectal adenocarcinomas. Unlike the previous tumor sites, the prevalent finding in these studies is a positive correlation with higher AI's favoring a better clinical outcome (53–66), although several studies showed no correlation [including the largest study of 412 patients (56)].

In summary, of 54 studies where radiation played a key role in treatment, 24 studies found a positive correlation between AI and clinical outcome, 11 showed a significant negative association, and 19 studies found no significant impact of AI on local control, disease-free survival (DFS) or overall survival.

Contradictory data have also been reported in a variety of other tumor types. In five studies involving brain tumors, four showed a lack of clinical significance in glioblastoma (67–71) while one (69) found a positive correlation between high apoptosis and improved outcome. Other notable results include breast cancer where Kato (72) found no significance of AI in 422 breast cancer patients while de Jong (73) noted that a high AI was associated with worse survival in a series of 172 grade I and II invasive breast cancers.

These data underscore the complexity of tumor biology, response to treatment and prognosis. Arguments have been invoked that can reasonably explain why either a high or a low AI could be a good or bad prognostic indicator. An association between poor outcome and a high AI, and vice versa, may be a consequence of the relationship between proliferation and apoptosis. It has been suggested that cells with a diminished apoptotic response have an increased development of aneuploidy and other genetic abnormalities commonly associated with progression and propensity for metastatic survival. On the other hand, if the effects of cytotoxic therapy were mediated through the induction of apoptosis in tumor cells, then tumors that exhibit apoptosis may be more sensitive to treatment and have a better prognosis. The significance of apoptosis may differ from one tumor type to another. In a single study of 363 patients with primary colon or rectal cancer of Dukes Stages A to D, a high AI correlated with poor prognosis in colon cancer but the trend was the opposite in rectal cancer (74) suggesting different biological behavior in these related cancers.

If apoptosis itself is unable to provide a clear relationship with treatment outcome and prognosis, will the status of proteins regulating this process fare better? This is explored in the next sections.

Bcl-2 EXPRESSION

As alluded to earlier, apoptosis is controlled by a complex interplay between regulatory proteins. bcl-2, a 26 kDa integral membrane oncoprotein was the first antiapoptosis gene product discovered. The bcl-2 (B-cell lymphoma gene-2) gene was first cloned from the breakpoint of the t(14;18) translocation in B-cell follicular lymphoma, where it is juxtaposed to the immunoglobulin heavy-chain gene. This translocation results in the constitutive expression of bcl-2 that promotes cell survival by inhibiting the induction of apoptosis. Abnormal expression of bcl-2 by this and other mechanisms is involved in the pathogenesis of other hematopoietic malignancies including B-cell chronic lymphocytic leukemia, acute myeloid leukemia and multiple myeloma. In addition, solid tumors of the breast, bowel, H&N, lung, prostate, cervix, and others also exhibit upregulated bcl-2 protein levels, making it a widespread aberration in many types of human cancer (75).

As an antiapoptotic protein, bcl-2 has been shown in experimental systems to protect cells from growth factor withdrawal, irradiation, cytotoxic drugs, and activated oncogenes. Considering that overexpression of the protein may be prevalent in over 50% of cancers, it was expected that bcl-2 would correlate with poor prognosis. Bcl-2 is easily assessed in fresh and archival histological material using immunohistochemistry and exhibits a cytoplasmic and perinuclear staining pattern. Various methods have been used to assess bcl-2 staining from quantitative cell counting to less formal semiquantitative scoring systems. Bcl-2 expression data is usually dichotomized as positive or negative with variable cutoffs depending on tumor type and individual investigator. However, as with AI, the correlation of bcl-2 with treatment outcome has proven unpredictable and paradoxical. The data presented in Figure 2 summarizes the data in some major tumor types in a similar format to the AI data.

	better	non-informative	worse
Cervix	Uehara 1995 Tjalma 1998 Dimitrakakis 2000 Crawford 1998	Graflund 2002 Jain 2003 Harima 1998	Wootipoom 2004 Pillai 1999 Rajkumar 1198
H&N	Wilson 2001 Fouret 2002 Homma 1999 Gasparini 1995 Lomuzio 2005 Nagy 2003	Stoll 2000 Spafford 1996 Friedman 2001 Veneroni 1997 Pulkkinen 1999 Dijkema 2000 Kunz 2003 Klatka 2001 Casado 2002 Kuropkat 2002	Jackel 2001 Gallo 1999
Bladder	Uchida 2002	Lipponen 1996 Theodoropolous 2005 Nakanishi 2002 Nakapolou 2000 Hussain 2003 Wu 2000 Plastiras 1999 Rodel 2000 Matsumoto 2004 Stavropolous 2002	Pollack 1997 Kong 1998 Ong 2001 Cooke 2000 Wolf 2001
Colorectal	Manne 2000 Kaklamanis 1998 Mottolese 2003 Bugliono 1999 Sun 2003 Krajewska 2005 Ofner 1995 Baretton 1996 Leahy 1999 Zavrides 2005 Elkablawy 2001	Watson 2005 Garrity 2004 Hilska 2005 Giatromanolaki 2001 Schneider 1998 Bosari 1995 Schwander 2000 Lustosa 2005 Zhao 2005 Paradiso 2001 Bendardaf 2003 Rodel 2002 Sarela 2001	Bhatavdekar 1997

Prognosis

Figure 2 Summary of prognostic studies assessing bcl-2 expression. The data are classified as better prognosis if a high level of bcl-2 was significantly associated with a more favorable clinical outcome or worse if the outcome was significantly shorter. Noninformative denotes no significant correlation with clinical outcome. The studies are arranged in descending order of the number of patients investigated with apoptosis, (●) representing studies with more than 100 patients, (●) 50–99 patients, and (•) less than 50 patients.

In cervix cancer, there are several reports (18,76–84) that have addressed the significance of bcl-2 overexpression with respect to treatment outcome with similar disparity (Fig. 2) as reported for AI and without a clear picture emerging. Again, some investigators have studied the influence of a week of RT on the expression of bcl-2 using biopsies but, in contrast to AI, bcl-2 failed to show any consistent changes and none were associated with response to treatment (31,85,86). Interestingly, the proapoptotic protein, and bax, was also studied in these investigations and induction of this protein provided significant clinical outcome correlation. Intrinsic bax expression prior to treatment has also been associated with good prognosis (84).

In H&N cancer, bcl-2 overexpression has been more often associated with good prognosis (87–92) than bad prognosis (93,94). However, the majority of studies have failed to find any significant correlation with clinical outcome (36,39,95–102) (Fig. 2).

In contrast to H&N cancer, bcl-2 overexpression in bladder cancer is rarely associated with better clinical outcome (103). Several studies have reported an adverse influence of bcl-2 expression on clinical outcome (104–108) but the majority fail to find prognostic information when assessing bcl-2 (109–118).

In colorectal cancer, the findings resemble H&N cancer where only one study has reported an unfavorable effect of bcl-2 overexpression on prognosis (119), while many studies report that overexpression of this antiapoptotic protein is associated with a favorable prognosis (120–130). However, over half of the cited references fail to find any relationship between bcl-2 expression and colorectal cancer prognosis (55,56,60,74,131–140). It should be noted that many of the studies in colorectal cancer have studied an impressive number of patients, with nine of the studies addressing 200 patients or more; six of these studies found bcl-2 expression uninformative as an independent prognostic indicator, while three reported a favorable outcome.

Consideration of some of the other tumor sites augments the above results. Of six studies with brain tumors, two found a positive association between bcl-2 and survival (6,141) while four others found no significance (142–145). In prostate cancer treated by RT, most studies have demonstrated that overexpression of bcl-2 is associated with poorer survival or impaired response to treatment (146–149) although the opposite result has also been reported (150) and the most recent study using material from the RTOG 86-10 trial found no correlation (151). In NSCLC, the overwhelming conclusion is that bcl-2 expression does not correlate with response to radiation alone (49), combined modality (152–155), platinum-based treatment (156,157), vinorelbine (158,159), or photodynamic therapy (160). One study suggests that bcl-2 expression is linked with poor response to RT (43). Interestingly two large surgical series suggest that bcl-2 expression predicts patients with better long-term survival (161,162). In breast cancer, high bcl-2 expression has been shown to associate with a number of favorable prognostic factors including ER positivity, PgR positivity, low histological grade, well-differentiated tumor, absence of c-erbB-2 and p53 (34). Numerous studies have shown that tumors with high bcl-2 expression are more responsive to hormone therapy and have more favorable disease-free and overall survival. However, overexpression of bcl-2 has failed to demonstrate a significant influence on the response to post-mastectomy RT (163–165) or to chemotherapy (166–168). Although bcl-2 expression tends to be associated with better prognosis, some studies have indeed suggested that patients with little or no bcl-2 expression respond better to chemotherapy (169).

The literature presents a mystifying series of publications associated with clinical significance of bcl-2 overexpression. Intuitively, it would be expected that tumors that acquire resistance to apoptosis might be expected to have a more aggressive phenotype and reduced sensitivity to treatment. In some tumor types, such as prostate, this appears to be the case, where bcl-2 is usually associated with poor outcome and increasing grade. In others, such as breast, this is not the case and the reverse association with outcome and grade has been observed.

The paradoxical role of bcl-2 in tumor biology has been explored in some detail (170) with the hypothesis that Darwinian selection, driven by hypoxia within a developing tumor mass, selects cells with a reduced sensitivity to apoptosis. This can either be by activation of an antiapoptotic gene, such as bcl-2, or loss of a proapoptotic gene such as p53. Gurova and Gudkov (170) hypothesize that as the appearance of either of these mutations is likely to be a rare event, the expansion of the mutated cell clone will progress rapidly and the event that came first will dominate in the tumor and prevent the expansion of other mutations resulting in the same phenotypic alteration through an alternative mechanism. In this scenario, if p53 mutation occurs first, the tumor will have a high level of genetic instability, more rapid progression, propensity to invade and metastasize and poor prognosis. On the other hand if the bcl-2 overexpressing clone appears first, it leads to development of a tumor in which the control of genomic stability is unaltered (as bcl-2 has been shown to have no influence on this process) and which no longer favors selection of p53-deficient cells. The developing tumor would be characterized by low rates of progression and favorable prognosis. This hypothesis would imply that p53 deficiency and bcl-2 overexpression would be rarely expressed together in one tumor. Indeed, an inverse relationship between these two apoptotic modulators has been observed in several cancer types (139,171,172).

THE ROLE OF P53

To summarize the data on p53 from individual studies is beyond the scope of this chapter as there have been literally hundreds of publications examining the potential role of mutated p53 on the outcome of patients treated for a wide variety of solid tumors. Instead, the implications of recent meta-analyses will be discussed.

In non–small-cell lung cancer (NSCLC) treated with surgery alone, Mitsudomi and colleagues (173) performed a meta-analysis on 43 publications examining the influence of p53 alterations (either by overexpression of p53 protein by immunohistochemistry or by sequencing) on patient outcome. These data, therefore, are relevant to the issue of the influence of p53 on prognosis, but not on response to therapy. They found a highly significant and negative prognostic effect of p53 on survival following surgery for patients with adenocarcinoma, but not for patients with squamous cell carcinoma. They reported similar conclusions for studies using overexpressed p53 as an indication of mutation as with studies in which p53 had been directly sequenced. More recently, Steels and colleagues (174) undertook a systematic review of the literature to assess the prognostic value of p53 abnormalities for the survival of patients presenting with lung cancer (both small and NSCLC). From 74 eligible papers, combined hazard ratios (HR) suggested that an abnormal p53 status had an unfavorable impact on survival in any stage NSCLC (HR = 1.44), in stages I–II NSCLC (HR = 1.68), in stages I–IIIB NSCLC (HR = 1.68), in stages III–IV NSCLC (HR = 1.48), in surgically resected NSCLC (HR = 1.37), in squamous cell carcinoma (HR = 2.24), and in adenocarcinoma (HR = 1.57). However, the data were unable to determine the prognostic value of p53 in small cell lung cancer. The meta-analysis failed to specifically address the influence of p53 alterations on the response to cytotoxic therapy even though a significant proportion of the cited studies involved chemotherapy and radiation treatment.

In bladder cancer, Malats and colleagues (175) analyzed 117 suitable studies comprising of over 10,000 patients. Unfortunately, they restricted their endpoints to the studies concerned with recurrence, progression and mortality. Even though 10 studies addressed the response to chemotherapy and 4 to RT, these were not considered further. The results showed that changes in p53 were only weakly predictive of recurrence, progression, and mortality in bladder cancer.

The effect of mutations in p53 on the prognosis of women with breast cancer was studied in a meta-analysis by Pharoah and colleagues (176). In sixteen eligible studies, 18% of 2993 cases were found to have a mutated p53 (or more accurately contained mutations in exons 5–8 as most studies limited their analyses to this region because it contains 90% of p53 mutations). The authors found a significant influence of p53 mutations on overall survival in unselected patients and those who were either node negative or positive, had a hazard ratio of approximately 2.0. Thus patients with mutations in p53 were twice as likely to die of their disease than those without mutations in p53. No treatment details were given in this publication, so it is likely that these results were relevant to prognosis rather than to the treatment sensitivity.

The preceding meta-analyses are disappointing as therapy was not considered in the analysis. However, Thames and colleagues (177) applied stringent selection rules to identify comparison groups where the same outcome was reported for the same treatment applied to the same tumor with results corrected for important prognostic factors. They restricted themselves to studies of p53 overexpression in patients where chemotherapy or RT was used alone or in combination with surgery and identified 45 reports from a total of 469 in the literature that met their selection criteria. They reported that for Stage I–III breast cancer, the effect of p53 overexpression was marginally significant in predicting a worse outcome for patients treated with surgery and chemotherapy both for DFS and overall survival. For Stage II–IV H&N cancer treated with RT and chemotherapy, the effect of p53 overexpression on treatment outcome was insignificant. However, for ovarian cancer (FIGO I–III) treated with surgery and chemotherapy, the results for overall survival were marginally significant. As the authors corrected for known prognostic factors and used time to failure as their endpoint, they attempted to differentiate the effect on treatment outcome from general prognosis. However, it is doubtful that one can remove the influence of p53 on prognosis from such an analysis as factors such as the rate of cell proliferation could be influenced by p53, which in turn could affect time to failure without specifically affecting tumor cell sensitivity.

The overall conclusion from these meta-analyses is that mutations in p53, detected either by immunohistochemistry or by sequencing, have an effect on the prognosis of some tumors but not others. Whether or not p53 mutations have an additional effect on the sensitivity to RT or chemotherapy, is unclear from the published studies, but if it has an effect, it is small and is not manifested in all tumors. A final caveat to these studies is the presence of publication bias, whereby negative studies go unreported. Although this might reduce the overall hazard of mutated p53 in some tumors, it is unlikely to be responsible for all the negative effect seen in some tumors.

SURVIVIN AND PROGNOSIS

Survivin is a unique member of the inhibitors of apoptosis (IAP) family wherein it shows markedly different tissue expression from other IAPs, being expressed in fetal but not in adult differentiated tissues. It also plays a dual role in suppressing apoptosis and regulating cell division (178). Survivin is regulated in a highly cell cycle-dependent manner, with a marked increase in the G2M phase (179). Several reports have demonstrated survivin expression in a wide variety of human tumors including lung, breast, colon, gastric, esophagus, pancreas, liver, bladder, uterine and ovarian cancers, large-cell non-Hodgkin's lymphomas, leukemias, neuroblastomas, gliomas, soft-tissue sarcomas, melanomas, and other skin cancers (180).

As with other apoptosis-related proteins, a general role for survivin in determining the chemo- and radiosensitivity of tumors has been suggested. Rodel and colleagues (181) used short interfering RNA (siRNA) to decrease survivin in radioresistant SW480 and intermediately radioresistant HCT-15 in colorectal cancer cells. In response to irradiation, both cell lines demonstrated an increase in caspase 3/7 activity and the percentage of apoptotic cells in parallel with decreased cell viability and reduced clonogenic survival. These effects were more pronounced in the radioresistant SW480 cell line. In addition, siRNA treatment caused G2M arrest and increased levels of DNA double-strand breaks in irradiated cells. In a series of 59 patients with rectal cancer treated by combined RT and chemotherapy, an increased survivin expression associated with a significantly higher risk of a local tumor recurrence was noticed (181). Using the human epidermoid carcinoma cell line, A431, Sah and colleagues (182) demonstrated a correlation between inhibition of survivin (using adenoviral-mediated wild-type p53, antisense to survivin, the mitogen-activated protein kinase inhibitor PD98059, the cyclin-dependent kinase inhibitor purvalanol A, or the histone deacetylase inhibitor trichostatin A) and radiosensitization.

The assessment of survivin in clinical specimens has been complicated due to the observation that the protein exists in both a cytoplasmic and nuclear subcellular pool (183) consistent with its function in the regulation of both cell viability and cell division. It has been suggested that the nuclear pool of survivin is involved in promoting cell proliferation whereas the cytoplasmic pool of survivin may participate in controlling cell survival but not cell proliferation. Alternatively, the subcellular localization may be associated with the known splicing variants of survivin, which may differ in their functions with respect to cell survival and cell division (184).

In breast cancer, high survivin levels have been shown to have no significant correlation with clinical outcome (185,186), to be associated with poor prognosis (187,188), with good prognosis (189), and to predict a poor response to endocrine therapy, but a good response to chemotherapy (190). This confusing picture further exemplifies the complexity of apoptosis. In contrast to the antiapoptotic protein bcl-2, which is generally correlated to good prognostic features in breast cancer, survivin expression correlates with adverse features such as high nuclear grade, negative hormone receptor status, HER-2 overexpression, and vascular endothelial growth factor (VEGF) expression (187).

In contrast, in colorectal cancer, high levels of expression of survivin have been universally associated with poor prognosis (136,181,191–193). However, in a study of 98 rectal cancer patients in which 57 underwent surgery alone and 41 underwent RT before surgery, although survivin positivity was related to worse survival in all patients independent of Dukes' stage, local and distant recurrence, differentiation, gender, age, apoptosis, and p53 expression, it was not associated with survival in the patients without ($p = 0.08$) or with ($p = 0.19$) RT (194).

In esophageal cancer, no clear picture has emerged. High survivin has been associated with poor prognosis and response to chemotherapy (195–198), good prognosis and response to chemotherapy (199,200) and of no prognostic value (201).

In NSCLC, the majority of studies have shown high survivin levels to predict poor prognosis in surgically resected cases (202–205). In unresectable cases who received chemotherapy, Karczmarek-Borowska et al. (206) also observed both poor prognosis and response to treatment, but Vischioni et al. (207) found that nuclear survivin levels predicted longer overall and relapse-free survival in both univariate and multivariate analyses but did not correlate with response to chemotherapy. Ling and colleagues (208), highlighted the issue of splice variants by the observation that high levels of expression of survivin-2B was significantly associated with better outcome while high levels of survivin-DeltaEx3 was highly associated with worse clinical outcome.

In a small study of 44 cervical cancer patients treated by RT, all tumors were shown to express survivin (209). The authors classified seven tumors as having a strong expression of survivin while 37 had moderate expression. The five -year overall survival of patients with moderate expression was 66% but only one of the seven patients with strong survivin expression was alive 45 months after treatment; despite the small numbers, this proved to be significant in a Cox regression analysis ($p = 0.02$, RR = 0.3). However, survivin was significantly correlated to anemia, which was an even stronger prognostic indicator in the univariate analysis of this study casting some doubt on the independent significance of survivin determining the response to RT. In another small study of 41 patients in which 43% of the patients received RT and the remainder received RT and chemotherapy, no correlation was found between survivin expression and clinical stage, histology or menopausal state or disease relapse, or death (210).

Preusser and colleagues (211) studied a series of 104 consecutive patients who underwent surgery for glioblastoma and in whom 76% received adjuvant chemo- and RT, 3% received RT alone and 1% received only chemotherapy. Survivin was variably expressed in all tumors but showed no association with overall survival.

The data from survivin studies generally show that increased expression of this antiapoptotic gene is usually associated with poor clinical outcome. However, many of the studies address the role of survivin on the outcome of surgery and the impact of overexpression on patients receiving cytotoxic therapy is less clear. This illustrates a general problem with the interpretation of clinical studies; it is very difficult, if not impossible, in most studies to separate treatment sensitivity from overall prognosis.

SUMMARY

The clinical data for AI, p53, bcl-2, and for survivin do not provide definitive evidence for a prime role of the apoptotic pathway in the response of tumors to cancer therapy. A plethora of confounding factors in human tumors makes the straightforward extrapolation of experimental knowledge to the clinic fraught with difficulty. It is unlikely that any simplistic measurement of apoptosis or apoptosis-related proteins will have a clear-cut influence on the outcome of complex solid tumors treated by cytotoxic therapy. Bcl-2 is a prime example of this difficulty; its clinical significance is variable within and between tumor types (Fig. 1). In breast and H&N cancer, bcl-2 overexpression is more often than not associated with better prognosis, yet in breast cancer it is associated with a more indolent phenotype while in our own studies in H&N cancer involving 400 patients, we demonstrated that overexpression was associated with a more aggressive phenotype (dedifferentiation and increasing N-stage), yet was still associated with better survival (92).

This complexity reflects the "hallmarks of cancer" that are postulated to be six essential alterations in cell physiology that are commonly shared in tumors consisting of self-sufficiency in growth signals, insensitivity to growth-inhibitory signals, evasion of apoptosis, limitless replicative potential, sustained angiogenesis, and tissue invasion and metastasis (1). Apoptosis evasion is only one facet of tumor development and each of these processes, and the alterations which caused them, are likely to affect the response of tumors to cytotoxic therapy. Even the commonly held notion that carcinogenesis is the temporal accumulation of particular somatic mutations may also be an under simplification of cancer development. There is a multitude of genetic and epigenetic changes that occur during carcinogenesis resulting in the observation that no two cancers are genetically or phenotypically the same. Indeed, individual cancer cells within the same tumor express a high degree of genetic and phenotypic heterogeneity and may be the prime reason why cancer therapeutic interventions fail. Chaos theory has been invoked to explain how small perturbations of simple initial conditions can generate highly complex systems that nonetheless behave in a probabilistic manner (212). Unfortunately, with chaos theory it is much easier to predict how small perturbations of simple systems (e.g., the disruption of cell cycle checkpoints or apoptosis in a nontumorigenic cell) can generate complex behavior (e.g., cancer) than it is to work backward and deduce how the perturbations caused cancer and which factors were important.

The demise of a tumor cell is not a simple matter; there are multiple modes of cell death, as outlined in other chapters of this treatise that depend on both intrinsic and extrinsic factors. The complexity of the apoptotic process is still being unraveled (213) and a variety of effector mechanisms have been identified that don't require the activation of caspases, which is the focus of the majority of apoptosis-related prognostic studies. The lack of definitive clinical data does not mean that the apoptosis model is wrong; it is a piece of the puzzle. In the future, we will be seeking biological information on an individual patient basis using genome-wide gene expression profiles and using mathematical models to accurately predict prognosis and response to treatment. However, until that day, we need to ensure that prognostic studies concerned with apoptosis and other biomarkers are carried using the REMARK guidelines (214) to maximize robust, transparent, and complete reporting of information so that clinical usefulness and relevance can be easily ascertained and false negative or positive studies minimized.

REFERENCES

1. Hanahan D, Weinberg RA. The hallmarks of cancer. Cell 2000; 100(1):57–70.
2. Garcea G, Neal CP, Pattenden CJ, et al. Molecular prognostic markers in pancreatic cancer: a systematic review. Eur J Cancer 2005; 41(15):2213–2236.

3. Kahlenberg MS, Sullivan JM, Witmer DD, et al. Molecular prognostics in colorectal cancer. Surg Oncol 2003; 12(3):173–186.
4. Kren L, Brazdil J, Hermanova M, et al. Prognostic significance of anti-apoptosis proteins survivin and bcl-2 in non-small cell lung carcinomas: a clinicopathologic study of 102 cases. Appl Immunohistochem Mol Morphol 2004; 12(1):44–49.
5. Manne U, Myers RB, Moron C, et al. Prognostic significance of Bcl-2 expression and p53 nuclear accumulation in colorectal adenocarcinoma. Int J Cancer 1997; 74(3):346–358.
6. McDonald FE, Ironside JW, Gregor A, et al. The prognostic influence of bcl-2 in malignant glioma. Br J Cancer 2002; 86(12):1899–1904.
7. Quinn DI, Henshall SM, Sutherland RL. Molecular markers of prostate cancer outcome. Eur J Cancer 2005; 41(6):858–887.
8. Schliephake H. Prognostic relevance of molecular markers of oral cancer—a review. Int J Oral Maxillofac Surg 2003; 32(3):233–245.
9. Singhal S, Vachani A, Antin-Ozerkis D, et al. Prognostic implications of cell cycle, apoptosis, and angiogenesis biomarkers in non-small cell lung cancer: a review. Clin Cancer Res 2005; 11(11):3974–3986.
10. Brown JM, Wilson G. Apoptosis genes and resistance to cancer therapy: what does the experimental and clinical data tell us? Cancer Biol Ther 2003; 2(5):477–490.
11. Brown JM, Wouters BG. Apoptosis, p53, and tumor cell sensitivity to anticancer agents. Cancer Res 1999; 59(7):1391–1399.
12. Brown JM, Wouters BG. Apoptosis: mediator or mode of cell killing by anticancer agents? Drug Resist Updat 2001; 4(2):135–136.
13. Stadelmann C, Lassmann H. Detection of apoptosis in tissue sections. Cell Tissue Res 2000; 301(1):19–31.
14. Leung TW, Xue WC, Cheung AN, et al. Proliferation to apoptosis ratio as a prognostic marker in adenocarcinoma of uterine cervix. Gynecol Oncol 2004; 92(3):866–872.
15. Frankfurt OS, Robb JA, Sugarbaker EV, et al. Apoptosis in breast carcinomas detected with monoclonal antibody to single-stranded DNA: relation to bcl-2 expression, hormone receptors, and lymph node metastases. Clin Cancer Res 1997; 3(3):465–471.
16. Chung EJ, Seong J, Yang WI, et al. Spontaneous apoptosis as a predictor of radiotherapy in patients with stage IIB squamous cell carcinoma of the uterine cervix. Acta Oncol 1999; 38(4):449–454.
17. Gasinska A, Urbanski K, Gruchala A, et al. A ratio of apoptosis to mitosis, proliferation pattern and prediction of radiotherapy response in cervical carcinoma. Neoplasma 2002; 49(6):379–386.
18. Jain D, Srinivasan R, Patel FD, et al. Evaluation of p53 and Bcl-2 expression as prognostic markers in invasive cervical carcinoma stage IIb/III patients treated by radiotherapy. Gynecol Oncol 2003; 88(1):22–28.
19. Kim JY, Cho HY, Lee KC, et al. Tumor apoptosis in cervical cancer: its role as a prognostic factor in 42 radiotherapy patients. Int J Cancer 2001; 96(5):305–312.
20. Levine EL, Davidson SE, Roberts SA, et al. Apoptosis as predictor of response to radiotherapy in cervical carcinoma. Lancet 1994; 344(8920):472.
21. Liu SS, Tsang BK, Cheung AN, et al. Anti-apoptotic proteins, apoptotic and proliferative parameters and their prognostic significance in cervical carcinoma. Eur J Cancer 2001; 37(9):1104–1110.

22. Paxton JR, Bolger BS, Armour A, et al. Apoptosis in cervical squamous carcinoma: predictive value for survival following radiotherapy. J Clin Pathol 2000; 53(3): 197–200.
23. Sheridan MT, Cooper RA, West CM. A high ratio of apoptosis to proliferation correlates with improved survival after radiotherapy for cervical adenocarcinoma. Int J Radiat Oncol Biol Phys 1999; 44(3):507–512.
24. Tsang RW, Fyles AW, Li Y, et al. Tumor proliferation and apoptosis in human uterine cervix carcinoma I: correlations between variables. Radiother Oncol 1999; 50(1):85–92.
25. Wheeler JA, Stephens LC, Tornos C, et al. ASTRO Research Fellowship: apoptosis as a predictor of tumor response to radiation in stage IB cervical carcinoma. American Society for Therapeutic Radiology and Oncology. Int J Radiat Oncol Biol Phys 1995; 32(5):1487–1493.
26. Zaghloul MS, El Naggar M, El Deeb A, et al. Prognostic implication of apoptosis and angiogenesis in cervical uteri cancer. Int J Radiat Oncol Biol Phys 2000; 48(5):1409–1415.
27. Kokawa K, Shikone T, Otani T, et al. Transient increases of apoptosis and Bax expression occurring during radiotherapy in patients with invasive cervical carcinoma. Cancer 1999; 86(1):79–87.
28. Lyng H, Sundfor K, Rofstad EK. Changes in tumor oxygen tension during radiotherapy of uterine cervical cancer: relationships to changes in vascular density, cell density, and frequency of mitosis and apoptosis. Int J Radiat Oncol Biol Phys 2000; 46(4):935–946.
29. Niibe Y, Nakano T, Ohno T, et al. Relationship between p21/WAF-1/CIP-1 and apoptosis in cervical cancer during radiation therapy. Int J Radiat Oncol Biol Phys 1999; 44(2):297–303.
30. Ohno T, Nakano T, Niibe Y, et al. Bax protein expression correlates with radiation-induced apoptosis in radiation therapy for cervical carcinoma. Cancer 1998; 83(1):103–110.
31. Yuki H, Fujimura M, Yamakawa Y, et al. Detection of apoptosis and expression of apoptosis-associated proteins as early predictors of prognosis after irradiation therapy in stage IIIb uterine cervical cancer. Jpn J Cancer Res 2000; 91(1):127–134.
32. Bhosle SM, Huilgol NG, Mishra KP. Apoptotic index as predictive marker for radiosensitivity of cervical carcinoma: evaluation of membrane fluidity, biochemical parameters and apoptosis after the first dose of fractionated radiotherapy to patients. Cancer Detect Prev 2005; 29(4):369–375.
33. Grabenbauer GG, Suckorada O, Niedobitek G, et al. Imbalance between proliferation and apoptosis may be responsible for treatment failure after postoperative radiotherapy in squamous cell carcinoma of the oropharynx. Oral Oncol 2003; 39(5):459–469.
34. Zhang GJ, Kimijima I, Tsuchiya A, et al. The role of bcl-2 expression in breast carcinomas (Review). Oncol Rep 1998; 5(5):1211–1216.
35. Mandal AK, Verma D, Mohanta PK, et al. Prognostic significance of apoptosis in squamous cell carcinoma of oral cavity with special reference to TNM stage, histological grade and survival. Indian J Pathol Microbiol 2001; 44(3):257–259.
36. Stoll C, Baretton G, Ahrens C, et al. Prognostic significance of apoptosis and associated factors in oral squamous cell carcinoma. Virchows Arch 2000; 436(2): 102–108.

37. Sikorska B, Wagrowska-Danilewicz M, Danilewicz M. Prognostic significance of apoptosis in laryngeal cancer. A quantitative immunomorphological study. Acta Histochem 2000; 102(4):413–425.
38. Xie X, De Angelis P, Clausen OP, et al. Prognostic significance of proliferative and apoptotic markers in oral tongue squamous cell carcinomas. Oral Oncol 1999; 35(5):502–509.
39. Pulkkinen JO, Klemi P, Martikainen P, et al. Apoptosis in situ, p53, bcl-2 and AgNOR counts as prognostic factors in laryngeal carcinoma. Anticancer Res 1999; 19(1B):703–707.
40. Hirvikoski P, Kumpulainen E, Virtaniemi J, et al. Enhanced apoptosis correlates with poor survival in patients with laryngeal cancer but not with cell proliferation, bcl-2 or p53 expression. Eur J Cancer 1999; 35(2):231–237.
41. Lera J, Lara PC, Perez S, et al. Tumor proliferation, p53 expression, and apoptosis in laryngeal carcinoma: relation to the results of radiotherapy. Cancer 1998; 83(12): 2493–2501.
42. Smilek P, Dusek L, Vesely K, et al. Prognostic significance of mitotic and apoptotic index and the DNA cytometry in head and neck cancer. Neoplasma 2005; 52(3): 199–207.
43. Hwang JH, Lim SC, Kim YC, et al. Apoptosis and bcl-2 expression as predictors of survival in radiation-treated non-small-cell lung cancer. Int J Radiat Oncol Biol Phys 2001; 50(1):13–18.
44. Macluskey M, Baillie R, Chandrachud LM, et al. High levels of apoptosis are associated with improved survival in non-small cell lung cancer. Anticancer Res 2000; 20(3B):2123–2128.
45. Matturri L, Colombo B, Lavezzi AM. Evidence for apoptosis in non-small cell lung carcinoma. Relationship with cell kinetics and prognosis. Anal Quant Cytol Histol 1999; 21(3):240–244.
46. Puglisi F, Minisini AM, Aprile G, et al. Balance between cell division and cell death as predictor of survival in patients with non-small-cell lung cancer. Oncology 2002; 63(1):76–83.
47. Tanaka F, Kawano Y, Li M, et al. Prognostic significance of apoptotic index in completely resected non-small-cell lung cancer. J Clin Oncol 1999; 17(9): 2728–2736.
48. Dworakowska D, Jassem E, Jassem J, et al. Clinical significance of apoptotic index in non-small cell lung cancer: correlation with p53, mdm2, pRb and p21WAF1/CIP1 protein expression. J Cancer Res Clin Oncol 2005; 131(9):617–623.
49. Langendijk H, Thunnissen E, Arends JW, et al. Cell proliferation and apoptosis in stage III inoperable non-small cell lung carcinoma treated by radiotherapy. Radiother Oncol 2000; 56(2):197–207.
50. Hanaoka T, Nakayama J, Haniuda M, et al. Immunohistochemical demonstration of apoptosis-regulated proteins, Bcl-2 and Bax, in resected non-small-cell lung cancers. Int J Clin Oncol 2002; 7(3):152–158.
51. Chyle V, Pollack A, Czerniak B, et al. Apoptosis and downstaging after preoperative radiotherapy for muscle-invasive bladder cancer. Int J Radiat Oncol Biol Phys 1996; 35(2):281–287.
52. Lavezzi AM, Biondo B, Cazzullo A, et al. The role of different biomarkers (DNA, PCNA, apoptosis and karyotype) in prognostic evaluation of superficial transitional cell bladder carcinoma. Anticancer Res 2001; 21(2B):1279–1284.

53. Abe T, Sakaguchi Y, Ohno S, et al. Apoptosis and p53 overexpression in human rectal cancer; relationship with response to hyperthermo-chemo-radiotherapy. Anticancer Res 2001; 21(3C):2115–2120.
54. Adell GC, Zhang H, Evertsson S, et al. Apoptosis in rectal carcinoma: prognosis and recurrence after preoperative radiotherapy. Cancer 2001; 91(10):1870–1875.
55. Bendardaf R, Ristamaki R, Kujari H, et al. Apoptotic index and bcl-2 expression as prognostic factors in colorectal carcinoma. Oncology 2003; 64(4):435–442.
56. Garrity MM, Burgart LJ, Mahoney MR, et al. Prognostic value of proliferation, apoptosis, defective DNA mismatch repair, and p53 overexpression in patients with resected Dukes' B2 or C colon cancer: a North Central Cancer Treatment Group Study. J Clin Oncol 2004; 22(9):1572–1582.
57. Marijnen CA, Nagtegaal ID, Mulder-Stapel AA, et al. High intrinsic apoptosis, but not radiation-induced apoptosis, predicts better survival in rectal carcinoma patients. Int J Radiat Oncol Biol Phys 2003; 57(2):434–443.
58. Michael-Robinson JM, Reid LE, Purdie DM, et al. Proliferation, apoptosis, and survival in high-level microsatellite instability sporadic colorectal cancer. Clin Cancer Res 2001; 7(8):2347–2356.
59. Noguchi T, Kikuchi R, Ono K, et al. Prognostic significance of p27/kip1 and apoptosis in patients with colorectal carcinoma. Oncol Rep 2003; 10(4):827–831.
60. Paradiso A, Simone G, Lena MD, et al. Expression of apoptosis-related markers and clinical outcome in patients with advanced colorectal cancer. Br J Cancer 2001; 84(5):651–658.
61. Rodel C, Grabenbauer GG, Papadopoulos T, et al. Apoptosis as a cellular predictor for histopathologic response to neoadjuvant radiochemotherapy in patients with rectal cancer. Int J Radiat Oncol Biol Phys 2002; 52(2):294–303.
62. Sakakura C, Koide K, Ichikawa D, et al. Analysis of histological therapeutic effect, apoptosis rate and p53 status after combined treatment with radiation, hyperthermia and 5-fluorouracil suppositories for advanced rectal cancers. Br J Cancer 1998; 77(1):159–166.
63. Schwandner O, Schiedeck TH, Bruch HP, et al. Apoptosis in rectal cancer: prognostic significance in comparison with clinical histopathologic, and immunohistochemical variables. Dis Colon Rectum 2000; 43(9):1227–1236.
64. Sinicrope FA, Hart J, Hsu HA, et al. Apoptotic and mitotic indices predict survival rates in lymph node-negative colon carcinomas. Clin Cancer Res 1999; 5(7): 1793–1804.
65. Sugamura K, Makino M, Kaibara N. Apoptosis as a prognostic factor in colorectal carcinoma. Surg Today 1998; 28(2):145–150.
66. Smith FM, Reynolds JV, Kay EW, et al. COX-2 overexpression in pretreatment biopsies predicts response of rectal cancers to neoadjuvant radiochemotherapy. Int J Radiat Oncol Biol Phys 2006; 64(2):466–472.
67. Birner P, Piribauer M, Fischer I, et al. Prognostic relevance of p53 protein expression in glioblastoma. Oncol Rep 2002; 9(4):703–707.
68. Heesters MA, Koudstaal J, Go KG, et al. Analysis of proliferation and apoptosis in brain gliomas: prognostic and clinical value. J Neurooncol 1999; 44(3):255–266.
69. Kuriyama H, Lamborn KR, O'Fallon JR, et al. Prognostic significance of an apoptotic index and apoptosis/proliferation ratio for patients with high-grade astrocytomas. Neuro Oncol 2002; 4(3):179–186.

70. Ribeiro Mde C, Coutinho LM, Hilbig A. The role of apoptosis, cell proliferation index, bcl-2, and p53 in glioblastoma prognosis. Arq Neuropsiquiatr 2004; 62(2A): 262–270.
71. Vaquero J, Zurita M, Coca S, et al. Imbalance between apostain expression and proliferative index can predict survival in primary glioblastoma. Acta Neurochir (Wien) 2002; 144(2):151–155; discussion 155–156.
72. Kato T, Kameoka S, Kimura T, et al. p53, mitosis, apoptosis and necrosis as prognostic indicators of long-term survival in breast cancer. Anticancer Res 2002; 22(2B):1105–1112.
73. de Jong JS, van Diest PJ, Baak JP. Number of apoptotic cells as a prognostic marker in invasive breast cancer. Br J Cancer 2000; 82(2):368–373.
74. Hilska M, Collan YU, O Laine VJ, et al. The significance of tumor markers for proliferation and apoptosis in predicting survival in colorectal cancer. Dis Colon Rectum 2005; 48(12):2197–2208.
75. Reed JC. Bcl-2 family proteins. Oncogene 1998; 17(25):3225–3236.
76. Crawford RA, Caldwell C, Iles RK, et al. Prognostic significance of the bcl-2 apoptotic family of proteins in primary and recurrent cervical cancer. Br J Cancer 1998; 78(2):210–214.
77. Dimitrakakis C, Kymionis G, Diakomanolis E, et al. The possible role of p53 and bcl-2 expression in cervical carcinomas and their premalignant lesions. Gynecol Oncol 2000; 77(1):129–136.
78. Graflund M, Sorbe B, Hussein A, et al. The prognostic value of histopathologic grading parameters and microvessel density in patients with early squamous cell carcinoma of the uterine cervix. Int J Gynecol Cancer 2002; 12(1):32–41.
79. Harima Y, Harima K, Shikata N, et al. Bax and Bcl-2 expressions predict response to radiotherapy in human cervical cancer. J Cancer Res Clin Oncol 1998; 124 (9):503–510.
80. Pillai MR, Jayaprakash PG, Nair MK. bcl-2 immunoreactivity but not p53 accumulation associated with tumour response to radiotherapy in cervical carcinoma. J Cancer Res Clin Oncol 1999; 125(1):55–60.
81. Rajkumar T, Rajan S, Baruah RK, et al. Prognostic significance of Bcl-2 and p53 protein expression in stage IIB and IIIB squamous cell carcinoma of the cervix. Eur J Gynaecol Oncol 1998; 19(6):556–560.
82. Tjalma W, De Cuyper E, Weyler J, et al. Expression of bcl-2 in invasive and in situ carcinoma of the uterine cervix. Am J Obstet Gynecol 1998; 178(1 pt 1):113–117.
83. Uehara T, Kuwashima Y, Izumo T, et al. Expression of the proto-oncogene bcl-2 in uterine cervical squamous cell carcinoma: its relationship to clinical outcome. Eur J Gynaecol Oncol 1995; 16(6):453–460.
84. Wootipoom V, Lekhyananda N, Phungrassami T, et al. Prognostic significance of Bax, Bcl-2, and p53 expressions in cervical squamous cell carcinoma treated by radiotherapy. Gynecol Oncol 2004; 94(3):636–642.
85. Adhya AK, Srinivasan R, Patel FD. Radiation therapy induced changes in apoptosis and its major regulatory proteins, Bcl-2, Bcl-XL, and Bax, in locally advanced invasive squamous cell carcinoma of the cervix. Int J Gynecol Pathol 2006; 25(3): 281–287.
86. Harima Y, Nagata K, Harima K, et al. Bax and Bcl-2 protein expression following radiation therapy versus radiation plus thermoradiotherapy in stage IIIB cervical carcinoma. Cancer 2000; 88(1):132–138.

87. Fouret P, Temam S, Charlotte F, et al. Tumour stage, node stage, p53 gene status, and bcl-2 protein expression as predictors of tumour response to platin-fluorouracil chemotherapy in patients with squamous-cell carcinoma of the head and neck. Br J Cancer 2002; 87(12):1390–1395.
88. Gasparini G, Bevilacqua P, Bonoldi E, et al. Predictive and prognostic markers in a series of patients with head and neck squamous cell invasive carcinoma treated with concurrent chemoradiation therapy. Clin Cancer Res 1995; 1(11):1375–1383.
89. Homma A, Furuta Y, Oridate N, et al. Prognostic significance of clinical parameters and biological markers in patients with squamous cell carcinoma of the head and neck treated with concurrent chemoradiotherapy. Clin Cancer Res 1999; 5(4): 801–806.
90. Lo Muzio L, Falaschini S, Farina A, et al. Bcl-2 as prognostic factor in head and neck squamous cell carcinoma. Oncol Res 2005; 15(5):249–255.
91. Nagy B, Tiszlavicz L, Eller J, et al. Ki-67, cyclin D1, p53 and bcl-2 expression in advanced head and neck cancer. In Vivo 2003; 17(1):93–96.
92. Wilson GD, Saunders MI, Dische S, et al. bcl-2 expression in head and neck cancer: an enigmatic prognostic marker. Int J Radiat Oncol Biol Phys 2001; 49(2): 435–441.
93. Gallo O, Chiarelli I, Boddi V, et al. Cumulative prognostic value of p53 mutations and bcl-2 protein expression in head-and-neck cancer treated by radiotherapy. Int J Cancer 1999; 84(6):573–579.
94. Jackel MC, Sellmann L, Youssef S, et al. [Prognostic significance of expression of p53, bcl-2 and bax in squamous epithelial carcinoma of the larynx—a multivariate analysis]. HNO 2001; 49(3):204–211.
95. Casado S, Forteza J, Dominguez S, et al. Predictive value of P53, BCL-2, and BAX in advanced head and neck carcinoma. Am J Clin Oncol 2002; 25(6):588–590.
96. Dijkema IM, Struikmans H, Dullens HF, et al. Influence of p53 and bcl-2 on proliferative activity and treatment outcome in head and neck cancer patients. Oral Oncol 2000; 36(1):54–60.
97. Friedman M, Lim JW, Manders E, et al. Prognostic significance of Bcl-2 and p53 expression in advanced laryngeal squamous cell carcinoma. Head Neck 2001; 23(4):280–285.
98. Klatka J. Prognostic value of the expression of p53 and bcl-2 in patients with laryngeal carcinoma. Eur Arch Otorhinolaryngol 2001; 258(10):537–541.
99. Kunz C, Bosch FX, Klein-Kuhne W, et al. [Immunohistochemical determination of cell cycle regulatory proteins. No prognostic significance in advanced squamous epithelial carcinomas of the head-neck area]. HNO 2003; 51(10):800–805.
100. Kuropkat C, Venkatesan TK, Caldarelli DD, et al. Abnormalities of molecular regulators of proliferation and apoptosis in carcinoma of the oral cavity and oropharynx. Auris Nasus Larynx 2002; 29(2):165–174.
101. Spafford MF, Koeppe J, Pan Z, et al. Correlation of tumor markers p53, bcl-2, CD34, CD44H, CD44v6, and Ki-67 with survival and metastasis in laryngeal squamous cell carcinoma. Arch Otolaryngol Head Neck Surg 1996; 122(6): 627–632.
102. Veneroni S, Silvestrini R, Costa A, et al. Biological indicators of survival in patients treated by surgery for squamous cell carcinoma of the oral cavity and oropharynx. Oral Oncol 1997; 33(6):408–413.

103. Uchida T, Minei S, Gao JP, et al. Clinical significance of p53, MDM2 and bcl-2 expression in transitional cell carcinoma of the bladder. Oncol Rep 2002; 9(2): 253–259.
104. Cooke PW, James ND, Ganesan R, et al. Bcl-2 expression identifies patients with advanced bladder cancer treated by radiotherapy who benefit from neoadjuvant chemotherapy. BJU Int 2000; 85(7):829–835.
105. Kong G, Shin KY, Oh YH, et al. Bcl-2 and p53 expressions in invasive bladder cancers. Acta Oncol 1998; 37(7–8):715–720.
106. Ong F, Moonen LM, Gallee MP, et al. Prognostic factors in transitional cell cancer of the bladder: an emerging role for Bcl-2 and p53. Radiother Oncol 2001; 61(2): 169–175.
107. Pollack A, Wu CS, Czerniak B, et al. Abnormal bcl-2 and pRb expression are independent correlates of radiation response in muscle-invasive bladder cancer. Clin Cancer Res 1997; 3(10):1823–1829.
108. Wolf HK, Stober C, Hohenfellner R, et al. Prognostic value of p53, p21/WAF1, Bcl-2, Bax, Bak and Ki-67 immunoreactivity in pT1 G3 urothelial bladder carcinomas. Tumour Biol 2001; 22(5):328–336.
109. Hussain SA, Ganesan R, Hiller L, et al. Proapoptotic genes BAX and CD40L are predictors of survival in transitional cell carcinoma of the bladder. Br J Cancer 2003; 88(4):586–592.
110. Lipponen PK, Aaltomaa S, Eskelinen M. Expression of the apoptosis suppressing bcl-2 protein in transitional cell bladder tumours. Histopathology 1996; 28(2): 135–140.
111. Matsumoto H, Wada T, Fukunaga K, et al. Bax to Bcl-2 ratio and Ki-67 index are useful predictors of neoadjuvant chemoradiation therapy in bladder cancer. Jpn J Clin Oncol 2004; 34(3):124–130.
112. Nakanishi K, Tominaga S, Hiroi S, et al. Expression of survivin does not predict survival in patients with transitional cell carcinoma of the upper urinary tract. Virchows Arch 2002; 441(6):559–563.
113. Nakopoulou L, Zervas A, Lazaris AC, et al. Predictive value of topoisomerase II alpha immunostaining in urothelial bladder carcinoma. J Clin Pathol 2001; 54(4): 309–313.
114. Plastiras D, Moutzouris G, Barbatis C, et al. Can p53 nuclear over-expression, Bcl-2 accumulation and PCNA status be of prognostic significance in high-risk superficial and invasive bladder tumours? Eur J Surg Oncol 1999; 25(1):61–65.
115. Rodel C, Grabenbauer GG, Rodel F, et al. Apoptosis, p53, bcl-2, and Ki-67 in invasive bladder carcinoma: possible predictors for response to radiochemotherapy and successful bladder preservation. Int J Radiat Oncol Biol Phys 2000; 46(5): 1213–1221.
116. Stavropoulos NE, Filiadis I, Ioachim E, et al. Prognostic significance of p53, bcl-2 and Ki-67 in high risk superficial bladder cancer. Anticancer Res 2002; 22(6B): 3759–3764.
117. Theodoropoulos VE, Lazaris AC, Kastriotis I, et al. Evaluation of hypoxia-inducible factor 1alpha overexpression as a predictor of tumour recurrence and progression in superficial urothelial bladder carcinoma. BJU Int 2005; 95(3):425–431.
118. Wu TT, Chen JH, Lee YH, et al. The role of bcl-2, p53, and ki-67 index in predicting tumor recurrence for low grade superficial transitional cell bladder carcinoma. J Urol 2000; 163(3):758–760.

119. Bhatavdekar JM, Patel DD, Ghosh N, et al. Coexpression of Bcl-2, c-Myc, and p53 oncoproteins as prognostic discriminants in patients with colorectal carcinoma. Dis Colon Rectum 1997; 40(7):785–790.
120. Baretton GB, Diebold J, Christoforis G, et al. Apoptosis and immunohistochemical bcl-2 expression in colorectal adenomas and carcinomas. Aspects of carcinogenesis and prognostic significance. Cancer 1996; 77(2):255–264.
121. Buglioni S, D'Agnano I, Cosimelli M, et al. Evaluation of multiple bio-pathological factors in colorectal adenocarcinomas: independent prognostic role of p53 and bcl-2. Int J Cancer 1999; 84(6):545–552.
122. Elkablawy MA, Maxwell P, Williamson K, et al. Apoptosis and cell-cycle regulatory proteins in colorectal carcinoma: relationship to tumour stage and patient survival. J Pathol 2001; 194(4):436–443.
123. Kaklamanis L, Savage A, Whitehouse R, et al. Bcl-2 protein expression: association with p53 and prognosis in colorectal cancer. Br J Cancer 1998; 77(11):1864–1869.
124. Krajewska M, Kim H, Kim C, et al. Analysis of apoptosis protein expression in early-stage colorectal cancer suggests opportunities for new prognostic biomarkers. Clin Cancer Res 2005; 11(15):5451–5461.
125. Leahy DT, Mulcahy HE, O'Donoghue DP, et al. bcl-2 protein expression is associated with better prognosis in colorectal cancer. Histopathology 1999; 35(4): 360–367.
126. Manne U, Weiss HL, Grizzle WE. Bcl-2 expression is associated with improved prognosis in patients with distal colorectal adenocarcinomas. Int J Cancer 2000; 89(5):423–430.
127. Mottolese M, Buglioli S, Piperno G, et al. Bio-pathological factors of prognostic value in colorectal adenocarcinomas. J Exp Clin Cancer Res 2003; 22(4 suppl): 163–166.
128. Ofner D, Riehemann K, Maier H, et al. Immunohistochemically detectable bcl-2 expression in colorectal carcinoma: correlation with tumour stage and patient survival. Br J Cancer 1995; 72(4):981–985.
129. Sun XF, Bartik Z, Zhang H. Bcl-2 expression is a prognostic factor in the subgroups of patients with colorectal cancer. Int J Oncol 2003; 23(5):1439–1443.
130. Zavrides H, Zizi-Sermpetzoglou A, Elemenoglou I, et al. Immunohistochemical expression of bcl-2 in Dukes' stage B and C colorectal carcinoma patients: correlation with p53 and ki-67 in evaluating prognostic significance. Pol J Pathol 2005; 56(4):179–185.
131. Bosari S, Moneghini L, Graziani D, et al. bcl-2 oncoprotein in colorectal hyperplastic polyps, adenomas, and adenocarcinomas. Hum Pathol 1995; 26(5):534–540.
132. Giatromanolaki A, Sivridis E, Stathopoulos GP, et al. Bax protein expression in colorectal cancer: association with p53, bcl-2 and patterns of relapse. Anticancer Res 2001; 21(1A):253–259.
133. Lustosa SA, Logullo A, Artigiani R, et al. Analysis of the correlation between p53 and bcl-2 expression with staging and prognosis of the colorectal adenocarcinoma. Acta Cir Bras 2005; 20(5):353–357.
134. Rodel F, Hoffmann J, Grabenbauer GG, et al. High survivin expression is associated with reduced apoptosis in rectal cancer and may predict disease-free survival after preoperative radiochemotherapy and surgical resection. Strahlenther Onkol 2002; 178(8):426–435.

135. Rosati G, Chiacchio R, Reggiardo G, et al. Thymidylate synthase expression, p53, bcl-2, Ki-67 and p27 in colorectal cancer: relationships with tumor recurrence and survival. Tumour Biol 2004; 25(5–6):258–263.
136. Sarela AI, Scott N, Ramsdale J, et al. Immunohistochemical detection of the anti-apoptosis protein, survivin, predicts survival after curative resection of stage II colorectal carcinomas. Ann Surg Oncol 2001; 8(4):305–310.
137. Schneider HJ, Sampson SA, Cunningham D, et al. Bcl-2 expression and response to chemotherapy in colorectal adenocarcinomas. Br J Cancer 1997; 75(3):427–431.
138. Schwandner O, Schiedeck TH, Bruch HP, et al. p53 and Bcl-2 as significant predictors of recurrence and survival in rectal cancer. Eur J Cancer 2000; 36(3): 348–356.
139. Watson NF, Madjd Z, Scrimegour D, et al. Evidence that the p53 negative / Bcl-2 positive phenotype is an independent indicator of good prognosis in colorectal cancer: a tissue microarray study of 460 patients. World J Surg Oncol 2005; 3:47.
140. Zhao DP, Ding XW, Peng JP, et al. Prognostic significance of bcl-2 and p53 expression in colorectal carcinoma. J Zhejiang Univ Sci B 2005; 6(12):1163–1169.
141. Deininger MH, Weller M, Streffer J, et al. Antiapoptotic Bcl-2 family protein expression increases with progression of oligodendroglioma. Cancer 1999; 86(9): 1832–1839.
142. Heesters MA, Koudstaal J, Go KG, et al. Proliferation and apoptosis in long-term surviving low grade gliomas in relation to radiotherapy. J Neurooncol 2002; 58(2): 157–165.
143. Kraus JA, Wenghoefer M, Glesmann N, et al. TP53 gene mutations, nuclear p53 accumulation, expression of Waf/p21, Bcl-2, and CD95 (APO-1/Fas) proteins are not prognostic factors in de novo glioblastoma multiforme. J Neurooncol 2001; 52(3):263–272.
144. Rieger L, Weller M, Bornemann A, et al. BCL-2 family protein expression in human malignant glioma: a clinical-pathological correlative study. J Neurol Sci 1998; 155(1):68–75.
145. Strik H, Deininger M, Streffer J, et al. BCL-2 family protein expression in initial and recurrent glioblastomas: modulation by radiochemotherapy. J Neurol Neurosurg Psychiatry 1999; 67(6):763–768.
146. Huang A, Gandour-Edwards R, Rosenthal SA, et al. p53 and bcl-2 immunohistochemical alterations in prostate cancer treated with radiation therapy. Urology 1998; 51(2):346–351.
147. Mackey TJ, Borkowski A, Amin P, et al. bcl-2/bax ratio as a predictive marker for therapeutic response to radiotherapy in patients with prostate cancer. Urology 1998; 52(6):1085–1090.
148. Pollack A, Cowen D, Troncoso P, et al. Molecular markers of outcome after radiotherapy in patients with prostate carcinoma: Ki-67, bcl-2, bax, and bcl-x. Cancer 2003; 97(7):1630–1638.
149. Scherr DS, Vaughan ED Jr., Wei J, et al. BCL-2 and p53 expression in clinically localized prostate cancer predicts response to external beam radiotherapy. J Urol 1999; 162(1):12–16; discussion 16–17.
150. Bylund A, Stattin P, Widmark A, et al. Predictive value of bcl-2 immunoreactivity in prostate cancer patients treated with radiotherapy. Radiother Oncol 1998; 49(2): 143–148.

151. Khor LY, Desilvio M, Li R, et al. Bcl-2 and bax expression and prostate cancer outcome in men treated with radiotherapy in Radiation Therapy Oncology Group protocol 86–10. Int J Radiat Oncol Biol Phys 2006; 66(1):25–30.
152. Brooks KR, To K, Joshi MB, et al. Measurement of chemoresistance markers in patients with stage III non-small cell lung cancer: a novel approach for patient selection. Ann Thorac Surg 2003; 76(1):187–193; discussion 193.
153. Ludovini V, Gregorc V, Pistola L, et al. Vascular endothelial growth factor, p53, Rb, Bcl-2 expression and response to chemotherapy in advanced non-small cell lung cancer. Lung Cancer 2004; 46(1):77–85.
154. van de Vaart PJ, Belderbos J, de Jong D, et al. DNA-adduct levels as a predictor of outcome for NSCLC patients receiving daily cisplatin and radiotherapy. Int J Cancer 2000; 89(2):160–166.
155. Yaren A, Oztop I, Kargi A, et al. Bax, bcl-2 and c-kit expression in non-small-cell lung cancer and their effects on prognosis. Int J Clin Pract 2006; 60(6):675–682.
156. Gregorc V, Ludovini V, Pistola L, et al. Relevance of p53, bcl-2 and Rb expression on resistance to cisplatin-based chemotherapy in advanced non-small cell lung cancer. Lung Cancer 2003; 39(1):41–48.
157. Harada T, Ogura S, Yamazaki K, et al. Predictive value of expression of P53, Bcl-2 and lung resistance-related protein for response to chemotherapy in non-small cell lung cancers. Cancer Sci 2003; 94(4):394–399.
158. Berrieman HK, Cawkwell L, O'Kane SL, et al. Hsp27 may allow prediction of the response to single-agent vinorelbine chemotherapy in non-small cell lung cancer. Oncol Rep 2006; 15(1):283–286.
159. Krug LM, Miller VA, Filippa DA, et al. Bcl-2 and bax expression in advanced non-small cell lung cancer: lack of correlation with chemotherapy response or survival in patients treated with docetaxel plus vinorelbine. Lung Cancer 2003; 39(2): 139–143.
160. Kawaguchi T, Yamamoto S, Naka N, et al. Immunohistochemical analysis of Bcl-2 protein in early squamous cell carcinoma of the bronchus treated with photodynamic therapy. Br J Cancer 2000; 82(2):418–423.
161. Moldvay J, Scheid P, Wild P, et al. Predictive survival markers in patients with surgically resected non-small cell lung carcinoma. Clin Cancer Res 2000; 6(3): 1125–1134.
162. Ohsaki Y, Toyoshima E, Fujiuchi S, et al. bcl-2 and p53 protein expression in non-small cell lung cancers: correlation with survival time. Clin Cancer Res 1996; 2(5):915–920.
163. Chen HH, Su WC, Guo HR, et al. p53 and c-erbB-2 but not bcl-2 are predictive of metastasis-free survival in breast cancer patients receiving post-mastectomy adjuvant radiotherapy in Taiwan. Jpn J Clin Oncol 2002; 32(9):332–339.
164. Koukourakis MI, Giatromanolaki A, Galazios G, et al. Molecular analysis of local relapse in high-risk breast cancer patients: can radiotherapy fractionation and time factors make a difference? Br J Cancer 2003; 88(5):711–717.
165. Silvestrini R, Veneroni S, Benini E, et al. Expression of p53, glutathione S-transferase-pi, and Bcl-2 proteins and benefit from adjuvant radiotherapy in breast cancer. J Natl Cancer Inst 1997; 89(9):639–645.
166. Bottini A, Berruti A, Bersiga A, et al. p53 but not bcl-2 immunostaining is predictive of poor clinical complete response to primary chemotherapy in breast cancer patients. Clin Cancer Res 2000; 6(7):2751–2758.

167. Kymionis GD, Dimitrakakis CE, Konstadoulakis MM, et al. Can expression of apoptosis genes, bcl-2 and bax, predict survival and responsiveness to chemotherapy in node-negative breast cancer patients? J Surg Res 2001; 99(2):161–168.
168. Sjostrom J, Blomqvist C, von Boguslawski K, et al. The predictive value of bcl-2, bax, bcl-xL, bag-1, fas, and fasL for chemotherapy response in advanced breast cancer. Clin Cancer Res 2002; 8(3):811–816.
169. Ogston KN, Miller ID, Schofield AC, et al. Can patients' likelihood of benefiting from primary chemotherapy for breast cancer be predicted before commencement of treatment? Breast Cancer Res Treat 2004; 86(2):181–189.
170. Gurova KV, Gudkov AV. Paradoxical role of apoptosis in tumor progression. J Cell Biochem 2003; 88(1):128–137.
171. Fontanini G, Vignati S, Bigini D, et al. Bcl-2 protein: a prognostic factor inversely correlated to p53 in non-small-cell lung cancer. Br J Cancer 1995; 71(5):1003–1007.
172. Hu YX, Watanabe H, Ohtsubo K, et al. Bcl-2 expression related to altered p53 protein and its impact on the progression of human pancreatic carcinoma. Br J Cancer 1999; 80(7):1075–1079.
173. Mitsudomi T, Hamajima N, Ogawa M, et al. Prognostic significance of p53 alterations in patients with non-small cell lung cancer: a meta-analysis. Clin Cancer Res 2000; 6(10):4055–4063.
174. Steels E, Paesmans M, Berghmans T, et al. Role of p53 as a prognostic factor for survival in lung cancer: a systematic review of the literature with a meta-analysis. Eur Respir J 2001; 18(4):705–719.
175. Malats N, Bustos A, Nascimento CM, et al. P53 as a prognostic marker for bladder cancer: a meta-analysis and review. Lancet Oncol 2005; 6(9):678–686.
176. Pharoah PD, Day NE, Caldas C. Somatic mutations in the p53 gene and prognosis in breast cancer: a meta-analysis. Br J Cancer 1999; 80(12):1968–1973.
177. Thames HD, Petersen C, Petersen S, et al. Immunohistochemically detected p53 mutations in epithelial tumors and results of treatment with chemotherapy and radiotherapy. A treatment-specific overview of the clinical data. Strahlenther Onkol 2002; 178(8):411–421.
178. Altieri DC, Marchisio PC, Marchisio C. Survivin apoptosis: an interloper between cell death and cell proliferation in cancer. Lab Invest 1999; 79(11):1327–1333.
179. Li F, Altieri DC. The cancer antiapoptosis mouse survivin gene: characterization of locus and transcriptional requirements of basal and cell cycle-dependent expression. Cancer Res 1999; 59(13):3143–3151.
180. Sah NK, Khan Z, Khan GJ, et al. Structural, functional and therapeutic biology of survivin. Cancer Lett 2006; 244(2):164–171.
181. Rodel F, Hoffmann J, Distel L, et al. Survivin as a radioresistance factor, and prognostic and therapeutic target for radiotherapy in rectal cancer. Cancer Res 2005; 65(11):4881–4887.
182. Sah NK, Munshi A, Hobbs M, et al. Effect of downregulation of survivin expression on radiosensitivity of human epidermoid carcinoma cells. Int J Radiat Oncol Biol Phys 2006; 66(3):852–859.
183. Li F, Yang J, Ramnath N, et al. Nuclear or cytoplasmic expression of survivin: what is the significance? Int J Cancer 2005; 114(4):509–512.
184. Mahotka C, Liebmann J, Wenzel M, et al. Differential subcellular localization of functionally divergent survivin splice variants. Cell Death Differ 2002; 9(12): 1334–1342.

185. O'Driscoll L, Linehan R, M Kennedy S, et al. Lack of prognostic significance of survivin, survivin-deltaEx3, survivin-2B, galectin-3, bag-1, bax-alpha and MRP-1 mRNAs in breast cancer. Cancer Lett 2003; 201(2):225–236.
186. Sohn DM, Kim SY, Baek MJ, et al. Expression of survivin and clinical correlation in patients with breast cancer. Biomed Pharmacother 2006; 60(6):289–292.
187. Ryan BM, Konecny GE, Kahlert S, et al. Survivin expression in breast cancer predicts clinical outcome and is associated with HER2, VEGF, urokinase plasminogen activator and PAI-1. Ann Oncol 2006; 17(4):597–604.
188. Span PN, Sweep FC, Wiegerinck ET, et al. Survivin is an independent prognostic marker for risk stratification of breast cancer patients. Clin Chem 2004; 50(11): 1986–1993.
189. Kennedy SM, O'Driscoll L, Purcell R, et al. Prognostic importance of survivin in breast cancer. Br J Cancer 2003; 88(7):1077–1083.
190. Span PN, Tjan-Heijnen VC, Manders P, et al. High survivin predicts a poor response to endocrine therapy, but a good response to chemotherapy in advanced breast cancer. Breast Cancer Res Treat 2006; 98(2):223–230.
191. Miller M, Smith D, Windsor A, et al. Survivin gene expression and prognosis in recurrent colorectal cancer. Gut 2001; 48(1):137–138.
192. Sarela AI, Macadam RC, Farmery SM, et al. Expression of the antiapoptosis gene, survivin, predicts death from recurrent colorectal carcinoma. Gut 2000; 46(5): 645–650.
193. Kawasaki H, Altieri DC, Lu CD, et al. Inhibition of apoptosis by survivin predicts shorter survival rates in colorectal cancer. Cancer Res 1998; 58(22):5071–5074.
194. Knutsen A, Adell G, Sun XF. Survivin expression is an independent prognostic factor in rectal cancer patients with and without preoperative radiotherapy. Int J Radiat Oncol Biol Phys 2004; 60(1):149–155.
195. Mega S, Miyamoto M, Li L, et al. Immunohistochemical analysis of nuclear survivin expression in esophageal squamous cell carcinoma. Dis Esophagus 2006; 19(5):355–359.
196. Rosato A, Pivetta M, Parenti A, et al. Survivin in esophageal cancer: an accurate prognostic marker for squamous cell carcinoma but not adenocarcinoma. Int J Cancer 2006; 119(7):1717–1722.
197. Grabowski P, Kuhnel T, Muhr-Wilkenshoff F, et al. Prognostic value of nuclear survivin expression in oesophageal squamous cell carcinoma. Br J Cancer 2003; 88(1):115–119.
198. Kato J, Kuwabara Y, Mitani M, et al. Expression of survivin in esophageal cancer: correlation with the prognosis and response to chemotherapy. Int J Cancer 2001; 95(2):92–95.
199. Warnecke-Eberz U, Hokita S, Xi H, et al. Overexpression of survivin mRNA is associated with a favorable prognosis following neoadjuvant radiochemotherapy in esophageal cancer. Oncol Rep 2005; 13(6):1241–1246.
200. Ikeguchi M, Kaibara N. survivin messenger RNA expression is a good prognostic biomarker for oesophageal carcinoma. Br J Cancer 2002; 87(8):883–887.
201. Dabrowski A, Filip A, Zgodzinski W, et al. Assessment of prognostic significance of cytoplasmic survivin expression in advanced oesophageal cancer. Folia Histochem Cytobiol 2004; 42(3):169–172.
202. Atikcan S, Unsal E, Demirag F, et al. Correlation between survivin expression and prognosis in non-small cell lung cancer. Respir Med 2006;100(12):2220–2226.

203. Wang DF, Zeng CG, Lin YB, et al. [Expression and clinical significance of apoptosis-related oncogenes in stage I-II non-small cell lung cancer]. Ai Zheng 2006; 25(3):359–362.
204. Shinohara ET, Gonzalez A, Massion PP, et al. Nuclear survivin predicts recurrence and poor survival in patients with resected non small cell lung carcinoma. Cancer 2005; 103(8):1685–1692.
205. Lu B, Gonzalez A, Massion PP, et al. Nuclear survivin as a biomarker for non-small-cell lung cancer. Br J Cancer 2004; 91(3):537–540.
206. Karczmarek-Borowska B, Filip A, Wojcierowski J, et al. Survivin antiapoptotic gene expression as a prognostic factor in non-small cell lung cancer: in situ hybridization study. Folia Histochem Cytobiol 2005; 43(4):237–242.
207. Vischioni B, van der Valk P, Span SW, et al. Nuclear localization of survivin is a positive prognostic factor for survival in advanced non-small-cell lung cancer. Ann Oncol 2004; 15(11):1654–1660.
208. Ling X, Yang J, Tan D, et al. Differential expression of survivin-2B and survivin-DeltaEx3 is inversely associated with disease relapse and patient survival in non-small-cell lung cancer (NSCLC). Lung Cancer 2005; 49(3):353–361.
209. Bache M, Holzapfel D, Kappler M, et al. Survivin protein expression and hypoxia in advanced cervical carcinoma of patients treated by radiotherapy. Gynecol Oncol 2006.
210. Espinosa M, Cantu D, Herrera N, et al. Inhibitors of apoptosis proteins in human cervical cancer. BMC Cancer 2006; 6:45.
211. Preusser M, Gelpi E, Matej R, et al. No prognostic impact of survivin expression in glioblastoma. Acta Neuropathol 2005; 109(5):534–538.
212. Schneider BL, Kulesz-Martin M. Destructive cycles: the role of genomic instability and adaptation in carcinogenesis. Carcinogenesis 2004; 25(11):2033–2044.
213. Hail N Jr., Carter BZ, Konopleva M, et al. Apoptosis effector mechanisms: a requiem performed in different keys. Apoptosis 2006; 11(6):889–904.
214. McShane LM, Altman DG, Sauerbrei W, et al. Reporting recommendations for tumor marker prognostic studies (REMARK). J Natl Cancer Inst 2005; 97(16): 1180–1184.

3

Response of Solid Tumors to Cancer Therapy: How Relevant Is Apoptosis?

J. Martin Brown
Division of Radiation and Cancer Biology, Stanford University Medical Center, Stanford, California, U.S.A.

INACTIVATION OF APOPTOSIS DURING CANCER DEVELOPMENT DOES NOT LEAD TO LOSS OF APOPTOSIS PRODUCED BY CANCER THERAPY

Apoptosis is a powerful cellular defense against cancer development. This conclusion has derived largely from studies of the tumor suppressor gene p53. Not only do a large majority of human cancers have either mutations in p53 or defects in the pathway, but p53 null mice are highly prone to developing cancers (1) and this is correlated, at least in some cases, with loss of the apoptotic function of p53 (1). So important is this inactivation of apoptosis to cancer development that evasion of apoptosis is considered to be one of the six fundamental hallmarks of cancer (2). The argument is then made that if apoptosis is the important mode of cell death following cancer treatment, then inactivation of apoptosis during tumorigenesis would be expected to render tumors resistant to treatment with anticancer agents. Though this is a powerful argument made by numerous observers, it is certainly not absolutely or universally true as both experimental and clinical cancers show a range of levels of apoptosis following treatment with anticancer agents. Further, the basic tenet that normal cells are sensitive to apoptosis is also not true; indeed, the cells of many normal tissues do not undergo apoptosis following DNA damage. In fact, there is considerable evidence that the large variation in apoptosis between different tumor types is a reflection of the sensitivity of induction of apoptosis in the normal cells from which the tumors arose (3). Thus, inactivation of the apoptotic pathway during tumorigenesis does not necessarily mean that the pathway is inactivated following DNA damage. Nonetheless, this does not invalidate the assumption that the variation in apoptosis induction to anticancer therapy leads to the variation in sensitivity of individual tumors to cancer therapy. We will examine the evidence for this assumption in the following sections.

APOPTOSIS AND TREATMENT SENSITIVITY: THE CURRENT PARADIGM

The current paradigm for how cells die following treatment with anticancer drugs is that radiation and anticancer drugs kill cells by apoptosis, and cells resistant to apoptosis (e.g., by increased Bcl-2 levels or with mutant p53) will be resistant to therapy. This is well articulated by Robert Weinberg in his excellent book "One Renegade Cell."

> For years, it was assumed that radiation therapy and many chemotherapeutic drugs killed malignant cells directly, by wreaking widespread havoc in their DNA. We now know that the treatments often harm DNA to a relatively minor extent. Nevertheless, the affected cells perceive that the inflicted damage cannot be repaired easily, and they actively kill themselves. This discovery implies that cancer cells able to evade apoptosis will be far less responsive to treatment (4).

It is worthwhile to examine how this concept arose. The availability of cells from p53 knockout mice, which were produced in the early 1990s, played a significant role. First, the finding that p53-deficient thymocytes were resistant to radiation-induced apoptosis (5,6) led Lowe and colleagues to examine the role of p53 in the response of a model tumor cell to a variety of anticancer agents both in vitro (7) and in vivo (8). Their model cancer cells were mouse embryo fibroblasts (MEFs) transformed by coexpression of the adenovirus early region 1A (E1A) and activated *ras*. Cotransfection of these two oncogenes into mouse MEFs dramatically sensitized them in a p53-dependent manner to apoptosis induced by radiation and other DNA-damaging agents (7). Tumors formed from these cell lines in mice were found to be sensitive to radiation and to adriamycin in a p53-dependent manner (8). Another tumor model that strengthened the hypothesis that sensitivity to apoptosis determines tumor response to anticancer therapy and has allowed much of the genetics of the elucidated is the *Eμ-myc*-driven B-cell lymphoma (9–12). In addition, the publication of Weinstein and colleagues (13), who showed that the response of the NCI-60 cell line panel of tumor cells to 86 clinically used anticancer agents was highly correlated with wild-type p53 status also

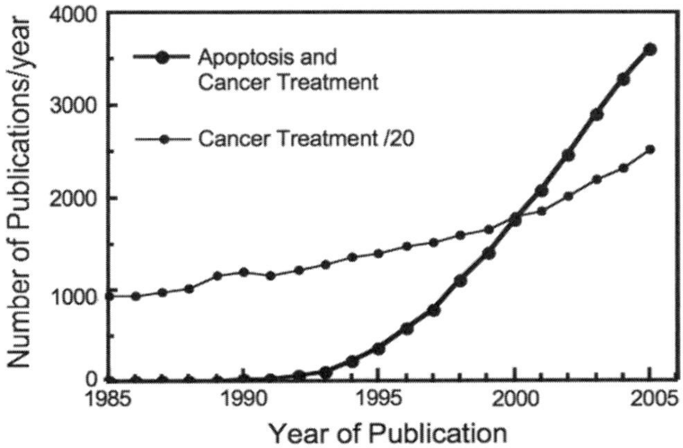

Figure 1 Interest in apoptosis and cancer therapy has risen rapidly in the past decade and continues to rise. This is shown by the number of publications per year on apoptosis and cancer therapy. The number of total publications per year on cancer therapy (divided by 20) is also shown as a control.

strengthened the connection between p53 and chemotherapy response. But perhaps more than specific data, the attractiveness of being able to attribute cancer therapy response for the first time to a molecular pathway that could be manipulated brought increasing numbers of investigators to the field. This accelerated tremendously in the mid- to late-90s and continues to this day (Fig. 1).

However, it is now generally recognized that the cell can die following cancer treatment of causes other than apoptosis. Though this has been known for many years, the fact that these pathways are not as well characterized in molecular terms has led them to be largely overlooked. In the following sections we examine the data on the different forms of cell killing and develop some overall conclusions.

CELLS DIE BY DIFFERENT PATHWAYS FOLLOWING DNA-DAMAGING AGENTS

As noted elsewhere in this book, the clinical data "do not provide definitive evidence for prime role for the apoptotic pathway in the response of tumors to cancer therapy" (14). However, the lack of such definitive clinical data does not mean that the apoptotic model for cell killing in response to cancer therapy is wrong because the many confounding variables with clinical material could obscure a real relationship. For this reason it is appropriate to examine preclinical models in which the variables can be isolated and tested individually.

Groundbreaking studies by Lowe and colleagues with matched pairs of primary lymphomas from *Eμ-myc* transgenic mice with the tumor cells transduced with a vector overexpressing the antiapoptotic Bcl-2 protein have

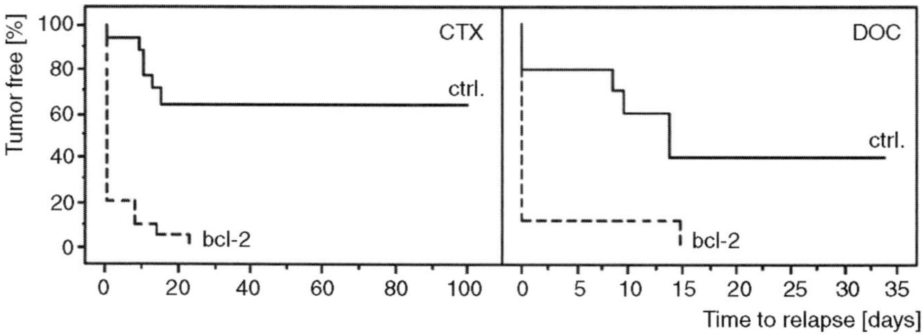

Figure 2 Apoptosis provides the basis for treatment sensitivity of a mouse lymphoma. Kaplan-Meier curves illustrating relapse after single treatments with either cyclophosphamide or doxorubicin of mice with lymphomas driven by the *Eμ-myc* transgene and bearing either MSCV-ctrl or MSCV-Bcl-2 vectors. The differences between the two curves are highly significant. *Source*: Ref. 15.

demonstrated the importance of apoptosis in tumor sensitivity (15). Lymphoma cells overexpressing Bcl-2 were resistant to apoptosis both in vitro and in vivo and the tumor response to cyclophosphamide or doxorubicin was highly attenuated in the Bcl-2 overexpressing tumors (Fig. 2). In this model, apoptosis occurs rapidly after treatment and was assessed four hours after drug application in the studies shown. This B-cell lymphoma model is an excellent example of a tumor for which apoptosis determines treatment sensitivity.

However, there are many examples of studies with cells both in vitro and with tumors in situ where changing apoptosis does not change overall cell kill or tumor response. Two such examples from our own work are shown in Figure 3. In the upper figure, we show that HCT116 tumor cells in which p21 has been deleted are much more sensitive to the development of apoptosis following treatment with three different DNA-damaging agents than their isogenic wild-type counterparts. On the other hand, when assessed by colony formation or clonogenic survival, an assay that integrates all forms of cell death and permanent growth arrest, there is no significant difference between the two cell lines in their sensitivities to the agents. When assessed as tumors in vivo, the HCT116 p21−/− cells show a high degree of apoptosis following a single dose of radiation and this can be abrogated by overexpression of the antiapoptotic protein Bcl-2 in these cells. However, despite eliminating apoptosis, the response of the tumors is larger in the cells overexpressing Bcl-2. This lack of correlation between the levels of apoptosis and overall cell kill or tumor response has been noted in many other studies (16–23). Despite this the focus of modifying apoptosis to modify the sensitivity of cancers to therapy remains strong with little or no mention of alternative pathways of cell killing (24).

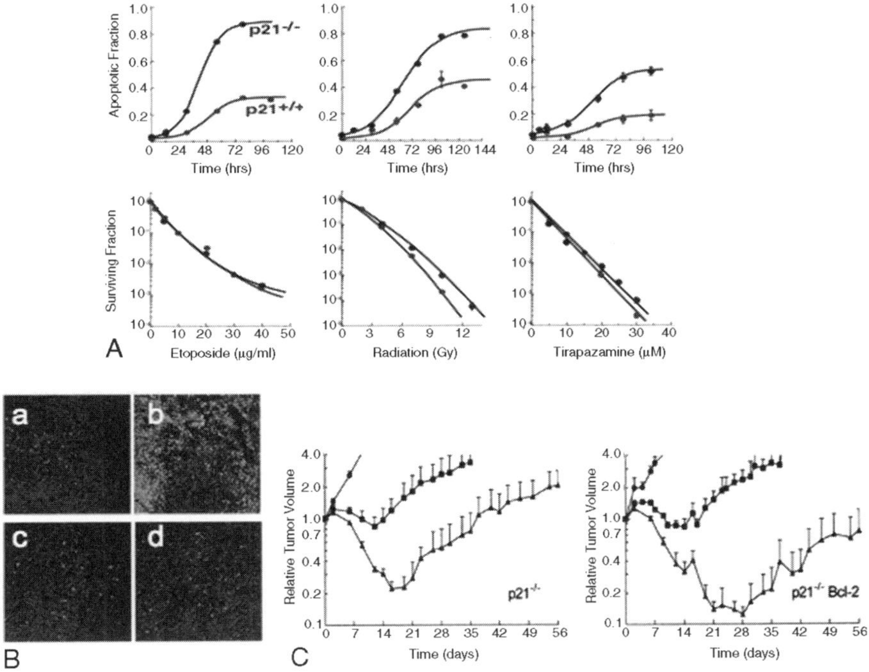

Figure 3 The level of apoptosis does not predict overall sensitivity of cells in vitro or of tumors in vivo. (**A**) The response of HCT116 cell lines isogenic for the cyclin-dependent kinase inhibitor $p21^{waf1}$ is shown following three different genotoxic treatments—etoposide [(5 μg/mL, 1 h), radiation (10 Gy), and tirapazamine (20 μM, 1 h)] under hypoxia. The upper frames show that the loss of p21 sensitizes cells to death by apoptosis for each of these treatments in comparison to the wild-type cell line. However, the overall sensitivity as assessed by clonogenic assay, as shown in lower frames, is unchanged (51). (**B**) Expression of Bcl-2 inhibits apoptosis in p21–/– tumors after irradiation. Tumors were removed four days after irradiation and apoptotic cells identified in frozen sections using TUNEL staining (fluorescein). (**a**,**b**) HCT116 p21–/– tumors, (**c**,**d**) tumors from HCT116 p21–/– cells transfected with the Bcl-2. The tumors are either unirradiated (**a**,**c**) or received a dose of 15 Gy. (**C**) Loss of apoptosis by expression of Bcl-2 in HCT116 p21–/– tumors does not result in less antitumor effect of radiation. (*Left*) p21–/– tumors, (*right*) tumors from p21–/– cells transfected with Bcl-2. Solid squares and solid triangles show the tumor response to 7.5 and 15 Gy, respectively. *Source*: Ref. 50 (*The color version of this figure is provided on the DVD*).

The situation is now perhaps changing. There is emerging recognition that cells can die following cancer therapy by multiple pathways (20,25–29), and this is well documented by many of the chapters in this book. One of these major forms of death is variously referred to as "mitotic catastrophe," "mitotically linked death," or simply "mitotic death." These terms can be misleading. "Mitotic catastrophe" carries an implication that the cell death can be seen as severely aberrant mitosis with centrosome fragmentation and multipolar spindles probably caused by a defective G2/M arrest following DNA damage (30). However, "mitotic catastrophe" has also been used to include any death that occurs after mitosis resulting from aberrations in mitosis. As such death can occur by apoptosis, this leads to the confusion. In this chapter, we will use the broader term "mitotic death" to refer to cell death caused by events occurring in mitosis that later become incompatible with life of the cell. Such death could occur following mitosis by a variety of processes including apoptosis or necrosis. It is assumed that mitotic catastrophe is a subdivision of mitotic death, though it could be that if the mechanisms are entirely different, they should be considered different forms of death. A good example of classical mitotic death has come from studies performed by several investigators who demonstrated that there is a remarkably strong correlation between lethal chromosome aberrations observed at the first mitosis after irradiation and subsequent cell death measured by clonogenic survival (31–33). This applies to cells at different oxygen tensions (34), radiations of different ionizing densities (35) and dose rates (36,37), and mutants with different DNA repair capacities (38,39). All of these data show a remarkable correlation of cell killing with chromosome aberrations measured at the first mitosis. Specifically, it is the creation of dicentrics, which produces both anaphase bridges and micronuclei (which produces loss of genetic material) that leads to cell death. This was nicely illustrated in early work of Revell and colleagues and most recently by Dewey and colleagues, who showed that cells containing a micronucleus at the first or subsequent divisions after irradiation were invariably unable to form colonies (i.e., did not survive), whereas those without micronuclei did survive to form colonies (40,41).

We have also shown using a range of tumor cells in vitro that the presence of a dicentric chromosome aberration in a specific chromosome leads to loss of those cells from the population, whereas surrounding cells in the same population given the same radiation dose with a nonlethal reciprocal translocation in the same chromosome has no deleterious effect on the cell relative to its neighbors (42). This subtle determinant of whether a cell lives or dies in a population depending on whether it has a reciprocal or nonreciprocal translocation in a given chromosome is incompatible with apoptosis, or indeed, any programmed cell death since the chance of whether a translocation in a given chromosome is reciprocal or nonreciprocal is 50:50 and is essentially a stochastic (random) event.

What determines the form of cell death induced by a particular anticancer agent? The answer is not simple: It depends on the context, including cell type, the genotype of the cell, the type of DNA damage to which the cell is exposed, and the dose of the agent used (43). Clearly, the situation is more complex than the originally proposed concept that cells either repair their damage or undergo apoptosis following treatment with DNA-damaging agents. A further complexity arises because the actual way in which a cell manifests its death, may not be the primary reason for the death of the cell. This is a little understood characteristic of apoptosis and is perhaps one of the main reasons why there is often such a disconnect between the levels of apoptosis and overall response to cancer therapy. This is discussed in the following section.

APOPTOSIS CAN BE BOTH A MODE OF CELL KILLING AND A TYPE OF "FUNERAL"

Surprisingly, the time period over which apoptosis occurs in different cell types following treatment with DNA-damaging agents has received only minimal attention (44). We believe, however, that this is a crucial determinant of whether genes that affect apoptosis will also affect overall cell kill. Essentially, if cells undergo apoptosis following treatment with a DNA damaging agent, they do so either rapidly (within 2–6 hours after treatment) and prior to undergoing mitosis, or they first go through mitosis and then (usually 24–72 hours later) undergo apoptosis. A classic example of these two different time periods for different cells undergoing apoptosis following the same treatment is shown in Figure 4. Both of these cells are of lymphoid origin and undergo apoptosis following irradiation. However, the L5178Y-S cells typically undergo apoptosis only after they have gone through mitosis. It thus seems reasonable to conclude that these cells are undergoing mitotic death even though their death is indistinguishable from that of cells undergoing a rapid apoptosis.

Cells that undergo an early apoptosis following treatment with DNA-damaging agents include thymocytes, spermatogonia, hair follicle cells, stem cells of the small intestine and bone marrow, and tissues in developing embryos (5,45–47). Tumors arising from these tissues, including T-cell lymphomas and some hematological tumors, are often sensitive to the induction of apoptosis and experience a marked overall response following treatment with DNA-damaging agents. Apoptosis in these tumors is invariably p53 dependent and if inhibited, for example, by inactivating p53 or by overexpressing the antiapoptotic protein Bcl-2, not only impedes apoptosis but also changes the sensitivity of the organ or tissue to the DNA-damaging agent (48). This is illustrated by the elegant work of Lowe and colleagues shown in Figure 2.

However, early apoptosis does not usually occur in tumors of epithelial or mesenchymal origin following administration of DNA-damaging agents. This is true both for cells in vitro and for tumors of spontaneous origin in mice (3,49). In cell lines in vitro derived from such tumors or in the tumors themselves, DNA-damaging agents may induce apoptosis, but this invariably occurs much later and subsequent to mitosis. Such secondary or postmitotic apoptosis is not the rate-limiting step in determining overall cell killing, and manipulation of the genes affecting apoptosis can dramatically change the overall levels of apoptosis without changing overall survival in vitro or in vivo (50). These data, therefore, suggest that apoptosis subsequent to mitosis is not triggered by the primary DNA-damaging event, but by a signal subsequent to mitosis that

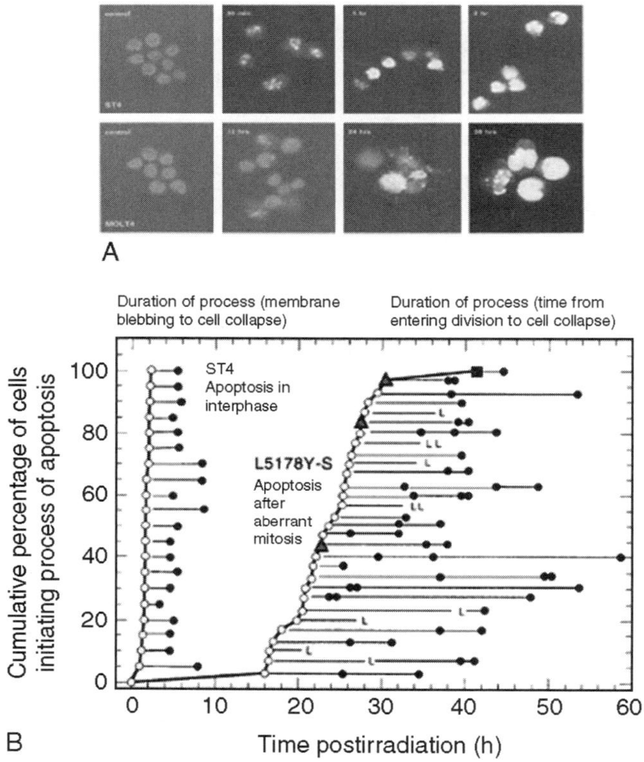

Figure 4 Time-lapse microscopy studies show two types of apoptosis after DNA damage. (**A**) Visualization of apoptosis using Hoechst 33342 and propidium iodide (PI) staining in two cell lines each given a single dose of 4 Gy. The STA4 mouse lymphoma cell line (*upper panels*) undergoes rapid apoptosis with most of the cells in late apoptosis (with permeable cell membrane) by two hours after irradiation. On the other hand, the MOLT-4 human lymphoid cell line undergoes delayed apoptosis, with no cells showing apoptosis at 12 hours and 50% showing apoptosis at 24 hours. (**B**) Time course of cumulative percentage of cells undergoing apoptosis for populations of ST4 and L5178Y-S cells illustrating early and late apoptosis. A similar pattern to that of the L5178Y-S cells was obtained for MOLT-4 cells. Initiation of apoptosis for ST4 cells is defined as the time after irradiation when membrane blebbing commenced and two to four hours later, the cells collapsed. For the L5178Y-S cells, initiation of apoptosis is defined as the time when the cells entered an abortive mitosis or in the absence of mitosis when membrane blebbing commenced. The cells collapsed at 3 to 23 hours after entering an abortive mitosis. *Source*: Ref. 52 (*The color version of this figure is provided on the DVD*).

the cell can no longer survive. Thus, the cell is destined to die because of an event triggered by mitosis, and whether it dies by apoptosis or by some other process will not alter the overall level of cell killing. A simple diagram to illustrate how changing apoptosis may or may not change overall cell killing is shown in Figure 5.

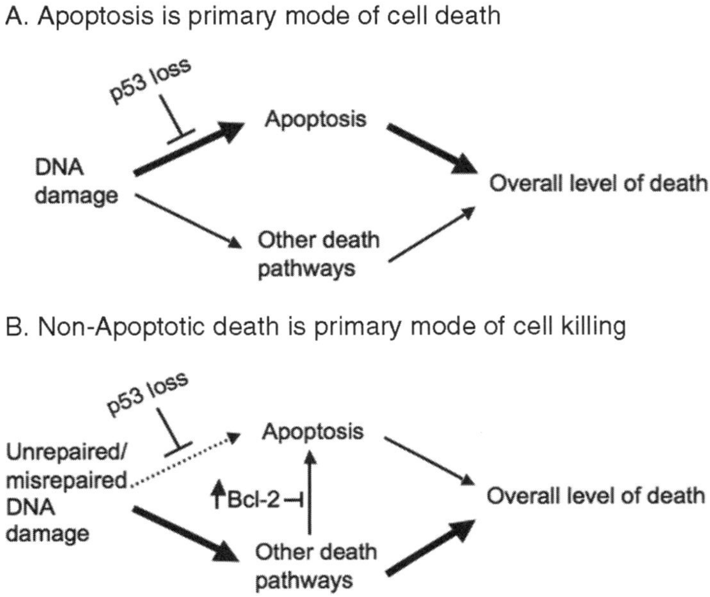

Figure 5 Changing the sensitivity to apoptosis may or may not change the overall level of cell kill depending on the primary reason for cell death. In some lymphomas, apoptosis is the primary reason for cell killing by DNA-damaging anticancer agents, whereas for most epithelial tumors, it is not. In the former case (**A**), DNA damage itself appears to trigger the apoptotic pathway, which is the main form of death, and the cells are very sensitive to genotoxic agents. Loss of the apoptotic pathway (e.g., by inactivation of p53), reduces the overall level of cell death. On the other hand, most cells of epithelial origin (**B**) do not have apoptosis as their primary reason for cell death, and changing the sensitivity to apoptosis does not change the overall level of cell death. Also shown in B is the fact that the mode of cell death can be apoptosis but as a consequence of other death pathways such as mitotic catastrophe (53). This also illustrates how inactivating p53 or increasing the levels of Bcl-2 can inhibit apoptosis without changing overall cell killing. However, changes that affect the initial level of DNA damage or its repair would be expected to change apoptosis and the overall level of cell death in parallel. *Source*: Ref. 27.

SUMMARY

The dominant paradigm over the past decade has been that tumor cells treated with anticancer agents die from apoptosis, and that cells resistant to apoptosis are resistant to cancer treatment. We examine in this review the history and data that underlie this tenet and conclude that it is likely to be true only for a subset of tumors of hematological origin. For most solid tumors, or the cells of these tumors, there is usually little correlation between the response of the tumors in vivo, or the overall killing of the cells in vitro, and the level of apoptosis. This can be reconciled by the growing acceptance that there are several ways for a cell to die following treatment with anticancer agents, and apoptosis appears to be relatively unimportant for the majority of tumors. Complicating the situation, however, is the fact that there are two types of apoptosis; primary apoptosis, which occurs within 4 to 6 hours of treatment, and secondary apoptosis, which usually occurs following the first division and some 24 to 72 hours after treatment. As secondary apoptosis is not the cause of death of the cell (but only the consequence of another lethal lesion), genetic or other modifications that change the level of apoptosis do not change the overall level of cell death, but rather change the way in which the cell dies. A lack of appreciation of these two types of apoptosis has led to tremendous confusion in the literature as to agents that might affect the response of solid tumors to cancer treatment.

CONCLUSION

Apoptosis has dominated the landscape of cell killing by anticancer drugs for the past decade and indeed, has become synonymous with "cell death." It is now becoming more accepted that apoptosis is but one mode of cell death. However, the ease of measurement of apoptosis and the multiple ways in which its frequency can be modified have led to numerous studies to identify genes or anticancer agents that might sensitize tumors to therapy. Apoptosis is probably the most important mode of death following cancer therapy for a subset of human tumors of hematopoietic origin that undergo rapid apoptosis following therapy. However, for the majority of solid tumors, which are of epithelial origin, apoptosis is not the primary reason for cell death, even though some or all of the tumor cells may undergo death by apoptosis (have an apoptotic "funeral").

The fact that genes controlling apoptosis may not be as important in determining tumor sensitivity to anticancer treatment as has been thought until recently does not mean that treatment sensitivity is not affected by the cell's genetic makeup. The fact that most anticancer drugs are DNA-damaging agents is evidence that the critical pathway for tumor cell killing is DNA damage. The characteristic hallmark of cancer is its genomic instability, which is often produced by defects in the repair of DNA lesions. This suggests that these tumors with defects in DNA repair will be particularly sensitive to one or more types of DNA-damaging agents. The challenge for the future will be to identify which DNA-damage pathway is compromised in any individual tumor and how this can best be exploited by presently available or future DNA-damaging agents.

REFERENCES

1. Attardi LD. The role of p53-mediated apoptosis as a crucial anti-tumor response to genomic instability: lessons from mouse models. Mutat Res 2005; 569(1–2):145–157.
2. Hanahan D, Weinberg RA. The hallmarks of cancer. Cell 2000; 100(1):57–70.
3. Kemp CJ, Sun S, Gurley KE. p53 induction and apoptosis in response to radio- and chemotherapy in vivo is tumor-type-dependent. Cancer Res 2001; 61(1):327–332.

4. Weinberg RA. One Renegade Cell: How Cancer Begins. Perseus Publishing, 1999.
5. Lowe SW, Schmitt EM, Smith SW, et al. p53 is required for radiation-induced apoptosis in mouse thymocytes. Nature 1993; 362:847–849.
6. Clarke AR, Purdie CA, Harrison DJ, et al. Thymocyte apoptosis induced by p53-dependent and independent pathways. Nature 1993; 362(6423):849–852.
7. Lowe SW, Ruley HE, Jacks T, et al. p53-dependentapoptosis modulates the cytotoxicity of anticancer agents. Cell 1993; 74(6):957–967.
8. Lowe SW, Bodis S, McClatchey A, et al. p53 status and the efficacy of cancer therapy in vivo. Science 1994; 266(5186):807–810.
9. Schmitt CA, Lowe SW. Apoptosis and therapy. J Pathol 1999; 187(1):127–137.
10. Schmitt CA, McCurrach ME, De Stanchina E, et al. INK4a/ARF mutations accelerate lymphomagenesis and promote chemoresistance by disabling p53. Genes Dev 1999; 13(20):2670–2677.
11. Johnstone RW, Ruefli AA, Lowe SW. Apoptosis: a link between cancer genetics and chemotherapy. Cell 2002; 108(2):153–164.
12. Lowe SW, Lin AW. Apoptosis in cancer. Carcinogenesis 2000; 21(3):485–495.
13. Weinstein JN, Myers TG, O'Connor PM, et al. An information-intensive approach to the molecular pharmacology of cancer. Science 1997; 275(5298):343–349.
14. Wilson GD. What do the clinical data tell us about the role of apoptosis in sensitivity to cancer therapy? In: Roninson IB, Brown JM, Bredesen DE, eds. Beyond Apoptosis: Cellular Outcomes of Cancer Therapy. New York: Informa Health Care, 2007.
15. Schmitt CA, Rosenthal CT, Lowe SW. Genetic analysis of chemoresistance in primary murine lymphomas. Nat Med 2000; 6(9):1029–1035.
16. Aldridge DR, Arends MJ, Radford IR. Increasing the susceptibility of the rat 208F fibroblast cell line to radiation-induced apoptosis does not alter its clonogenic survival dose-response. Br J Cancer 1995; 71(3):571–577.
17. Han JW, Dionne CA, Kedersha NL, et al. p53 status affects the rate of the onset but not the overall extent of doxorubicin-induced cell death in rat-1 fibroblasts constitutively expressing c-Myc. Cancer Res 1997; 57(1):176–182.
18. Kyprianou N, King ED, Bradbury D, et al. bcl-2 over-expression delays radiation-induced apoptosis without affecting the clonogenic survival of human prostate cancer cells. Int J Cancer 1997; 70(3):341–348.
19. Lock RB, Stribinskiene L. Dual modes of death induced by etoposide in human epithelial tumor cells allow Bcl-2 to inhibit apoptosis without affecting clonogenic survival. Cancer Res 1996; 56(17):4006–4012.
20. Ruth AC, Roninson IB. Effects of the multidrug transporter P-glycoprotein on cellular responses to ionizing radiation. Cancer Res 2000; 60(10):2576–2578.
21. Tannock IF, Lee C. Evidence against apoptosis as a major mechanism for reproductive cell death following treatment of cell lines with anti-cancer drugs. Br J Cancer 2001; 84(1):100–105.
22. Clarke PA, Pestell KE, Di Stefano F, et al. Characterisation of molecular events following cisplatin treatment of two curable ovarian cancer models: contrasting role for p53 induction and apoptosis in vivo. Br J Cancer 2004; 91(8):1614–1623.
23. Dewey WC, Ling CC, Meyn RE. Radiation-induced apoptosis: relevance to radiotherapy. Int J Radiat Oncol Biol Phys 1995; 33(4):781–796.
24. Stenner-Liewen F, Reed JC. Meeting report: apoptosis and cancer: basic mechanisms and therapeutic opportunities in the postgenomic era. Cancer Res 2003; 63:263–268.

25. Leist M, Jaattela M. Four deaths and a funeral: from caspases to alternative mechanisms. Nat Rev Mol Cell Biol 2001; 2(8):589–598.
26. Okada H, Mak TW. Pathways of apoptotic and non-apoptotic death in tumour cells. Nat Rev Cancer 2004; 4(8):592–603.
27. Brown JM, Attardi LD. The role of apoptosis in cancer development and treatment response. Nat Rev Cancer 2005; 5(3):231–237.
28. Broker LE, Kruyt FA, Giaccone G. Cell death independent of caspases: a review. Clin Cancer Res 2005; 11(9):3155–3162.
29. Morse DL, Gray H, Payne CM, et al. Docetaxel induces cell death through mitotic catastrophe in human breast cancer cells. Mol Cancer Ther 2005; 4(10):1495–1504.
30. Loffler H, Lukas J, Bartek J, et al. Structure meets function–centrosomes, genome maintenance and the DNA damage response. Exp Cell Res 2006; 312(14): 2633–2640.
31. Dewey WC, Miller HH, Leeper DB. Chromosomal aberrations and mortality of X-irradiated mammalian cells: Emphasis on repair. Proc Natl Acad Sci U S A 1971; 68:667–671.
32. Carrano AV. Chromosome aberrations and radiation-induced cell death. II. Predicted and observed cell survival. Mutat Res 1973; 17:355–366.
33. Bedford JS, Mitchell JB, Griggs HG, et al. Radiation-induced cellular reproductive death and chromosome aberrations. Radiat Res 1978; 76(3):573–586.
34. Deschner EE, Gray LH. Influence of oxygen on the sensitivity of Tradescantia pollen tube chromosomes to X-rays. Radiat Res 1959; 11:115–147.
35. Schneider DO, Whitmore GF. Comparative effects of neutrons and x-rays on mammalian cells. Radiat Res 1963; 18:286–306.
36. Lloyd DC, Edwards AA. Chromosome aberrations in human lymphocytes: effect of radiation quality, dose, and dose-rate. In: Ishihara T, Sasaki MS, eds. Radiation-Induced Chromosome Damage in Man. New York: Alan R. Liss, 1983:23–49.
37. Bedford JS, Cornforth MN. Relationship between the recovery from sublethal X-ray damage and the rejoining of chromosome breaks in normal human fibroblasts. Radiat Res 1987; 111(3):406–423.
38. Cornforth MN, Bedford JS. On the nature of a defect in cells from individuals with ataxia-telangiectasia. Science 1985; 227(4694):1589–1591.
39. Stackhouse MA, Bedford JS. An ionizing radiation-sensitive mutant of CHO cells: irs-20. III. Chromosome aberrations, DNA breaks and mitotic delay. Int J Radiat Biol 1994; 65(5):571–582.
40. Revell SH. The breakage-and-reunion theory and the exchange theory for chromosomal aberrations induced by ionizing radiations: a short history. Adv Radiat Biol 1974; 367–416.
41. Forrester HB, Albright N, Ling CC, et al. Computerized video time-lapse analysis of apoptosis of REC:Myc cells X- irradiated in different phases of the cell cycle. Radiat Res 2000; 154(6):625–639.
42. Kovacs MS, Yudoh K, Evans JW, et al. Stable translocations detected by fluorescence in situ hybridization: a rapid surrogate end point to evaluate the efficacy of a potentiator of tumor response to radiotherapy. Cancer Res 1997; 57(4):672–677.
43. Abend M. Reasons to reconsider the significance of apoptosis for cancer therapy. Int J Radiat Biol 2003; 79(12):927–941.
44. Held KD. Radiation-induced apoptosis and its relationship to loss of clonogenic survival. Apoptosis 1997; 2(3):265–282.

45. Potten CS, Grant HK. The relationship between ionizing radiation-induced apoptosis and stem cells in the small and large intestine. Br J Cancer 1998; 78(8):993–1003.
46. Komarova EA, Chernov MV, Franks R, et al. Transgenic mice with p53-responsive lacZ: p53 activity varies dramatically during normal development and determines radiation and drug sensitivity in vivo. Embo J 1997; 16(6):1391–1400.
47. Song S, Lambert PF. Different responses of epidermal and hair follicular cells to radiation correlate with distinct patterns of p53 and p21 induction. Am J Pathol 1999; 155(4):1121–1127.
48. Gudkov AV, Komarova EA. The role of p53 in determining sensitivity to radiotherapy. Nat Rev Cancer 2003; 3(2):117–129.
49. Bearss DJ, Subler MA, Hundley JE, et al. Genetic determinants of response to chemotherapy in transgenic mouse mammary and salivary tumors. Oncogene 2000; 19(8):1114–1122.
50. Wouters BG, Denko NC, Giaccia AJ, et al. A p53 and apoptotic independent role for p21waf1 in tumour response to radiation therapy. Oncogene 1999; 18(47): 6540–6545.
51. Brown JM, Wouters BG. Apoptosis, p53, and tumor cell sensitivity to anticancer agents. Cancer Res 1999; 59(7):1391–1399.
52. Endlich B, Radford IR, Forrester HB, et al. Computerized video time-lapse microscopy studies of ionizing radiation-induced rapid-interphase and mitosis-related apoptosis in lymphoid cells. Radiat Res 2000; 153(1):36–48.
53. Castedo M, Perfettini JL, Roumier T, et al. Cell death by mitotic catastrophe: a molecular definition. Oncogene 2004; 23(16):2825–2837.

4

Historical Studies of Various Forms of Cell Death

Richard A. Lockshin
Department of Biological Sciences, St. John's University, Jamaica, New York, U.S.A.

Zahra Zakeri
Department of Biology, Queens College and Graduate Center of City University of New York, Flushing, New York, U.S.A.

INTRODUCTION

The history of our ideas is that, when a concept appears, it first gives a sense of clarity to a rather diffuse set of observations. For some time following the acceptance of the idea, new observations that support the hypothesis accumulate, while for several very human reasons, ambiguities are denied, tortuously forced into the hypothesis, or ruled exceptional or irrelevant. Finally, the exceptions accumulate, their impact can no longer be denied, and the hypothesis is revised to become a more tolerant, encompassing tenet—perhaps less elegant, but truer to biology. Evolution has provided cells with redundancy, fail-safe, backup, and far more options than we can imagine. If we are lucky, we can unravel the cell's secrets one step at a time.

The story of programmed cell death, apoptosis, and nonapoptotic deaths follows this trajectory. Today we have moved past the earlier certainty—not to say dogmatism—that all cell deaths are alike. In doing so, we recognize how wonderfully fertile the hypothesis of apoptosis has been, and we are ready to focus on the clinically important questions of what establishes the threshold of death for a given cell, and what, in vivo, triggers a cell to cross that threshold.

HISTORICAL PERSPECTIVE

The death of cells was noted by early cell biologists almost as soon as cells were clearly identified and, toward the end of the 19th century, there was even a bit of a *fin de siècle* fascination with cell death. However, it was the developmental biologists who took a real interest in the subject, noting that in the metamorphosis of amphibians and insects, larval tissues had to be destroyed. Embryologists, pursuing an interest in embryonic development driven by the quest to understand evolutionary relationships, established that many vestigial organs were identifiable in embryos, only to die and disappear in later development (1–3). Likewise, the differentiation of sexes was noted to include cell death, and later, the differentiation of mammalian erythrocytes and keratinocytes led to their inevitable death. All this was admirably summarized by Glücksmann in 1951, though with only a pseudo-functional overtone. Glücksmann described the (teleonomic) purpose of the death—to sculpt an organ, to remove a vestigial tissue—but suggested no mechanism by which death occurred (4,5).

The first efforts to determine mechanisms came from several fronts. John Saunders at the University of Albany identified regions of cell death in chick embryos, notably in the formation of digits and in the axial separation of the limb from the trunk. He asked how this was brought about, leading to a series of experiments in the 1960s, in which he established (a) that cells that appeared to be in good condition could be excised and placed in a Petri dish, but that they would die on schedule; and (b) that, if instead of excising them, he transplanted them to the back of another embryo, the cells would heal and survive. He therefore concluded that they could have the information to die before they passed the point of no return (6,7).

In the late 1950s, Christian de Duve and J. Berthet, in Louvain, were studying the enzymes of particulate organelles as examined by centrifugal subfractionation of rat hepatocytes; the organelles were then considered to be only mitochondria. One evening they stored a sample to study the next day and were surprised to find a marked increase in acid phosphatase activity. This observation led to the discovery of lysosomes, organelles replete with acid hydrolases. Lysosomes were originally named because, unfortunately, an early attempt to understand their function included exposure of the rats to carbon tetrachloride, known as a hepatotoxin. When the de Duve group observed the solubilization of the acid phosphatase after the CCl_4 treatment (since CCl_4 destroys membranes), they hypothesized that the rupture of lysosomes could kill cells and that, indeed, this might be a common mechanism of cell death (8–11). We, of course, have since learned that lysosomal rupture is only in the rarest of circumstances a likely scenario for cell death. Nevertheless, the role of the lysosome is, as is discussed below, an important factor in physiological cell death.

Meanwhile, one of us (RAL) had joined the laboratory of the great insect physiologist Carroll M. Williams at Harvard and was casting around for a topic for a doctoral thesis. One seemed particularly promising, in part because it could be conducted on stored pupae, which would be available year-round and therefore, not limited by season. It involved a delayed metamorphic change: larval abdominal muscles that were retained in the pupa to generate the force necessary for the emerging moth to inflate its wings. Immediately after eclosion (the escape of the moth from its pupal shell) the muscles degenerated. There were many possibilities with this project: to test the Louvain idea about lysosomes as killers of cells; to examine how lysosomes might be activated; to examine the mechanisms by which the process was controlled. We undertook these experiments and came up with several very interesting observations. The death of the cells was ordained at the beginning of metamorphosis, three weeks prior to eclosion, by the hormones that initiated the metamorphosis; it was triggered at eclosion by neurally controlled activity; and it did not become irreversible until there was clear activation of lysosomal activity. This period in history (early 1960s) was one in which early computers were an exciting field of research, and computer terminology had a certain cachet. Williams was renowned for his colorful expressions—his description of a moth as "a flying machine devoted to sex" comes to mind—and members of the laboratory tended, however feebly, to emulate his style. Thus it was apparent from the study that the death of the intersegmental muscles of the moth was a biologically planned, carefully executed, fully physiological process, developmental in the same sense that any growth process was. It was self-evident that the death mechanism was ultimately imbedded in the genes. We searched for a means of describing it. Multistep pathways such as fermentation pathways or Krebs cycle did not catch the deliberateness or potential inexorability of the sequence leading to death, while the structure and goal of a computer program appeared to be at least formally analogous. Thus the phrase "programmed cell death" became the shorthand by which we summarized our understanding (12–16). Later, following the leads of Jamshed Tata (17,18) and Allan Munck (19,20), we demonstrated that the death of the muscles required the synthesis of new proteins (21,22). This latter point, though confirmed for many embryonic and developmental situations such as the death of excess neurons, is not generally true for cell deaths in mature animals. Cells in mature animals generally possess complete machinery for apoptosis or other death, and the issue is one of activation.

The concept or hypothesis of programmed cell death was inherently one of function, a statement that the control and execution of cell death was biological. The hypothesis of apoptosis approached the issue of function from a different direction: morphology of death that was not explicable by conventional understanding of osmotic lysis, but was common to many frequently physiological forms of death. The shrinkage and fragmentation of cells, and the condensation and margination of the chromatin, begged an explanation. This explanation was ultimately found through the genetic analysis of cell deaths in the nematode

Caenorhabditis elegans (23–27), and it is now recognized that much of the morphology of apoptosis can be traced to the caspase cascades.

As Kerr described "shrinkage necrosis" and his idea gradually expanded into "apoptosis" (28–31), we recognized that the morphology that we saw, while not identical to, for instance, an apoptotic thymocyte, was similar in many respects to apoptosis, and in any event we were all examining physiological processes of cell death. Over the years, in common interpretation, the terms "programmed cell death" and "apoptosis" melded, often accompanied by heated debate over definitions.

These debates acknowledged but did not focus on variations of what we can now describe as apoptosis or type I cell death. In general, the issue was whether or not apoptosis was truly programmed and whether any controlled cell death could exist if the dying cell did not manifest completely and exclusively the characteristics of apoptosis. And yet it was completely evident that other types of biological cell death existed (32). Even disregarding the deaths of differentiation characteristic of keratinocytes (33) and platelets (34), the necrosis-appearing deaths of isolated cells such as osteoblasts (35–38) and osteoclasts (39,40), or the partial apoptosis and eventual death of lens fibers (41–45) and mammalian erythrocytes (46–50), many types of cell death were known that were not classically apoptotic.

LYSOSOMAL CELL DEATH

Most studies of biological cell death in the 1960s and 1970s focused on activation of lysosomes. Although today it appears probable that at least some of the lysosomal activity was attributable to phagocytes, it is notable that two prominent examples were involution of prostate following castration and postweaning involution of mammary epithelium. Today these two models are cited as characteristic apoptosis. Tail muscle in metamorphosing tadpoles breaks into large segments comprising a few sarcomeres, termed sarcolytes, which undergo some self-digestion prior to being phagocytosed (30,51–54). In metamorphosing insects, much evidence accumulated, including especially careful electron microscopy by Jacques Beaulaton and Michael Locke, which demonstrated that the bulk of the cytoplasm in the cells of metamorphosing insects was destroyed through lysosomal activity (55,56). We have shown that lysosomes play a major role in destruction of the labial glands of *Manduca sexta* (57–59). Today we understand that, especially in large sedentary cells, the bulk of the cytoplasm is destroyed through lysosomal or autophagic activity in a process that has become known as lysosomal or autophagic cell death (60–71).

AUTOPHAGY

Autophagy is, however, a function of normal physiological regulation and homeostasis. Autophagy can be encountered if a cell is undernourished or deprived of growth factors, or if it suffers damage to mitochondria or other organelles (70,72–79). Indeed, culture media are presumably not perfectly equivalent to the in situ environment of most cells, and cells in culture frequently manifest active autophagy. Research on this point has not identified a distinction between the autophagy that can be encountered in these conditions and the autophagy that terminates in the death of the cell. Furthermore, there are many indications that the two or more forms of autophagy are not different. First, in synchronous cell deaths such as those during insect metamorphosis, autophagy destroys virtually all the cytoplasm of the cell, but the cell then activates processes that appear to be apoptosis, including exteriorization of phosphatidylserine, condensation of chromatin, internucleosomal degradation of DNA, and perhaps even activation of insect caspases (80–88). Second, and completely consonant with this argument, when PC12 cells are deprived of nerve growth factor (NGF), they undergo autophagy but may be rescued by NGF until the time at which autophagy has eliminated mitochondria and the cells have no further means of generating adequate ATP (89–91). Third, again in synchronous cells, autophagy progresses in waves, eliminating specific organelles at different times. This pattern suggests that autophagy is a spontaneous and undirected response to deterioration or other changes occurring in these organelles (55,56,92–94). This consideration changes the focus of our quest. Perhaps the sequence, in what is termed "autophagic cell death," is as follows: The doomed cell encounters a deprivation of unknown character. Subsequently, the cell attempts to compensate for this lack by generating substrates through autophagy. This mechanism protects the cell for an extended but ultimately limited time. If the lack is not compensated, ultimately, the cell consumes its own seed corn, as it were, and is no longer capable of surviving. At this time, if sufficient energy remains, it undergoes apoptosis; otherwise it dies by other mechanisms (95–100). This viewpoint suggests that the most interesting question is what is missing from the doomed cell. In insect metamorphosis, for instance, larval tissues undergo autophagy in a milieu in which adult cells are rapidly growing and indeed, in a milieu substantially enriched by the digested remnants of the dying cells. One might presume that the hormones of metamorphosis have inactivated key receptors or transport mechanisms on the larval cells, or that the adult cells compete far more effectively for limiting nutrients; but there is no adequate experimental evidence to defend a specific argument. For situations of hormonoprivy such as sex hormone–dependent cells in culture, or differentiating PC12 cells deprived variously of NGF or glucocorticoids, we need to understand in more detail what impact the hormone has on the subtleties of intermediary metabolism. This may be the same question as what glucocorticoids turn off in cells of the lymphatic system. The different versions of this question fall between those normally asked—either "How does the hormone activate caspases?" or "How does the hormone stimulate growth responses?"—but ultimately, the subtlety of intermediary metabolism may well be the most relevant issue.

Cell death is manifest in many forms, and there is some propensity to assign new names to the different forms. While, as is discussed in several other chapters, many of these names contribute new understanding, it is the biology of the cell that is paramount. Many years ago, Levi-Montalcini and Aloe argued that the mode of death of a neuron deprived of NGF depended on its prior history and state of differentiation, and that, for the same state of differentiation, different causes of death produced different forms of death (101). This is not surprising. Mitotic cells, subject to DNA damage or subversion by viral invasion, must prioritize the disarming and destruction of DNA should they choose to commit suicide (102). Sedentary, bulky, postmitotic cells have a considerable amount of cytoplasm to eliminate; while cells trapped in locations inaccessible to phagocytes may not be able to present fragments for phagocytosis. Furthermore, apoptosis and other mechanisms of controlled cell death operate through the use of energy; but the event that triggered the death may limit the energy available to complete it, so that a cell may initiate a controlled death but, using up its remaining energy, finally lyse in a necrotic termination. In this situation, there may be no meaningful distinction of process, in the same sense that the death of a heart attack victim on the street or following extensive intervention at a hospital does not change the fundamental process.

FORMS OF CELL DEATH

Many terms are used to describe different modes of cell death. Among the most widely used are the following:

Apoptosis

Apoptosis is the most widely used term. Coined by Kerr, Wyllie, and Currie in 1972, apoptosis refers to a caspase-mediated cell death characterized by nuclear condensation, fragmentation, and margination of chromatin, detachment, rounding, and shrinkage of the cell, followed by cell fragmentation and phagocytosis of the fragments. The fragments are marked by the appearance, by means of an active, energy-requiring movement of phosphatidylserine to the external leaflet of the cell membrane, where it can be identified by use of suitably labeled annexin V (103). The DNA is cut between nucleosomes, leading to a ladderlike appearance when it is electrophoresed, with the cut ends identifiable by 3′ end labeling [terminal deoxynucleotidyl transferase–mediated dUTP nick end labeling (TUNEL)] technique (104). There are several means of activating the major effector caspases, caspases 3 and 7; most of these activations manage to activate either a metabolic pathway, and thus caspase 9, or a cell membrane–linked pathway, and thus caspase 8 (105–107). Most of the changes in an apoptotic cell can be directly or indirectly attributed to proteolysis by caspases.

Programmed Cell Death

Programmed cell death is a term now considered to be synonymous with apoptosis. Originally, as described above, it referred to a distinct series of events that led to cell death during development (embryogenesis and metamorphosis), but today it reflects recognition of intrinsic machinery capable of destroying a cell. The term has an inherently more functional connotation than apoptosis, as apoptosis described the morphology of certain dying cells more specifically, but at present the terms are used nearly interchangeably.

Necrosis

Necrosis was considered originally to be a completely uncontrolled death (33,105–110). Necrosis typically occurs following irreparable damage to membranes or any failure of oxidative phosphorylation or ionic pumps (oncosis). Following either of these failures, the cell imbibes water, often because glycolysis produces lactate, which, at near neutral pH, cannot readily escape from the cell, leading to osmotic uptake of water. Thus, the cell undergoes osmotic lysis or, owing to increased cytoplasmic [Ca^{2+}] or change of pH, proteins precipitate and denature (109,110). Although the original assumption was that a necrotic death was completely uncontrolled, it is now known that, for instance, apoptosis can begin but fail to be completed. For instance, the cells of nonmammalian vertebrate eggs are considered to be capable of undergoing apoptosis before the 8th to 10th cell division, when zygotic messenger RNA finally replaces the maternal mRNA that has driven development to that point (111,112). In reality, zebrafish eggs, when exposed to toxins, can at all times activate caspase 3, but the earlier eggs lyse before they can manifest the morphological characteristics of apoptosis (113). Likewise, in hepatotoxicity, depending on the concentration of the toxin and the kinetics of its distribution, hepatocytes may manifest necrosis, apoptosis, or an abortive apoptosis that terminates in necrosis (114,115).

There is even evidence that what appears to be uncontrolled necrosis may be far more controlled, and therefore subject to experimental and therapeutic manipulation, than was previously suspected. The evidence comes from several sources. First, it has long been known that cells undergoing physiological deaths in locations inaccessible to phagocytes, such as osteoblasts, undergo a death whose morphology appears to be more necrotic than apoptotic (35). Second, even in traumatic deaths such as those deriving from infarct, the anoxic cells are capable of surviving and recovering for far longer periods than was previously suspected. Their condition had been missed because reperfusion provided fluid without oxygen or before the cells were able to reestablish their pumps, allowing the cells to imbibe water and undergo osmotic lysis (116–118). Finally, and most intriguingly, deaths that appear to be necrotic in neurons or other cells may be inhibited by the use of tricyclic (heterocyclic) molecules (119–122). Although their mechanism of action is not understood, the ability of these small molecules to block an apparently uncontrolled fate indicates that there is some level of biological control even in this situation. This type of death has been termed *necroptosis*, to recognize that it may represent a form of physiologically controlled death that functions without caspases (119–122).

Paraptosis

Paraptosis likewise has often been ignored. Paraptosis, like necroptosis, is a form of death, apparently under physiological control, that does not rely on the activation of caspases (123,124). Paraptosis is characterized by massive vacuolization of the dying cell. In earlier literature, this was termed "vacuolar cell death," and it was assumed that the vacuoles represented a pathological manifestation of a very sick cell. Today it is clearer that the vacuoles arise from the endoplasmic reticulum, and the mechanism remains under study (125).

Anoikis

Anoikis refers to the death of a cell that loses contact with other cells and thus loses whatever chemical or physical support that the other cells provide. Cells described as dying by anoikis frequently display morphologies, and therefore presumptive metabolic pathways, similar to but not identical with classical apoptosis (126–131). It is not clear that anoikis is a death process functionally distinct from other forms of cell death. It may be that the difference resides only in the presumptive origin of the death. By this interpretation, any meaningful distinctions would more likely derive from a loss of morphological constraints, as opposed to a cell in contact with substratum or other cells. In any event, early apoptotic cells frequently round, lose contact with other cells and substratum, and bleb violently.

Autophagic Cell Death

Autophagic cell death is a type of death in which the biochemistry and morphology are very distinct from apoptosis. The cytoplasm is destroyed with the appearance of all phases of lysosomal and autophagic vacuoles, though calpains and proteasomal proteases may also be active. The nucleus condenses very late, if at all, and similarly, DNA ladders, DNA fragmentation as detected by TUNEL, and exteriorization of phosphatidylserine occur very late, if at all. This type of death is encountered in large, sedentary, differentiated, and frequently postmitotic cells, including muscle, mammary epithelium, and neurons, as well as other types of cells (50,132). As is described below, although the term "autophagic cell death" is commonly used, it is not clear that the autophagy marks a death process as opposed to a means of coping with metabolic limitation of undefined origin, with the death ensuing only if the metabolic limitation fails to be resolved.

An illustrative example is seen in insect metamorphosis. Larval tissues that die during metamorphosis have caspase genes, but they undergo massive autophagy before dying. This process is generally called autophagic cell death. However, toward the end of the autophagy, the cells finally activate caspases and show signs of apoptosis (58,59,133). Does this mean that the entire period of autophagy is, from the point of view of the cell, not a death process, but a starvation-induced preservation mechanism, with apoptosis ensuing only after so much of the cytoplasm has been consumed that survival is no longer possible? If this is the case, why are these tissues starving when others are thriving? Have receptors for nutrients been lost? Are means of transporting or processing critical nutrients defective? How has this situation come about? This, too, is a question of what determines the threshold at which cells die.

CONTROL OF ACTIVATION OF DEATH PROGRAM IS CRITICAL

For biological systems, commitment to an irreversible step must be very tightly controlled, in the same sense that all sources of fire, spark, and friction must be completely controlled in a factory that manufactures explosives or volatile, flammable chemicals. Thus, processes such as commitment to male or female development, engagement of the fetus in the birth canal, commitment to tissue type such as muscle (activation of myoD), or other processes that ultimately become positive feedback are vigorously defended but, once engaged, become inexorable. It is not surprising that initiation of apoptosis is controlled by many interlocking, feedback mechanisms and closely regulated. Activation of suicide mechanisms is not a decision to be taken lightly. Thus, it is not surprising that many controls have been identified, and it is similarly unlikely that any single control is dominant.

WHAT DETERMINES THE THRESHOLD AT WHICH CELL DEATH IS ACTIVATED?

The hardest but perhaps most interesting question becomes the identification of the threshold at which the cell commits to die. For apoptosis, experimentally, we can establish the importance of the amounts and locations of the several pro- and antiapoptotic members of the Bcl-2 family and the competition of members of the tumor necrosis family, and we can identify the more-or-less serious consequences of mutation of the genes for these proteins; but, on a day-to-day basis of physiological regulation, we do not know how these are adjusted or whether adjustment of relative levels changes the fate of the cell, as opposed to constant levels interacting with different metabolic conditions. Is a cell under metabolic stress more susceptible to receptor-mediated apoptosis?

Another example of the importance of thresholds is the death of interdigital cells in the paws of amniote embryos. The amount of interdigital cell death in amniote embryos varies by species and can be lowered or raised by mutations such as hammertoe and biological regulators such as retinoic acid, respectively. The pattern of death, however, does not change, arguing that the patterning mechanism is intact but that either the threshold at which cells collapse has changed or the cells have moved closer to or further from the threshold (134,135). Cells in the regions of death differ from those in other regions by many criteria: nearness to neighbors, nearness to differentiating blood vessels, distribution of retinoic acid, distribution of Hox genes, and distribution of growth factors. The interrelationship of these differences and their influence on the survival of the cells is speculated but not known.

APOPTOSIS AS FIRST CHOICE FOR DEATH

In many situations, it appears that apoptosis is the most rapid and efficient means of cell death, and activation of an initiator caspase is the first action of a cell that has committed to die. There is much redundancy and internal control over this step. For instance, caspases may be directly or indirectly activated through the action of lysosomal proteases and calpains, as well as through other routes; and other lytic pathways may be activated simultaneously with activation of caspases. However, should activation of caspases not occur for any reason—as in the situation of knockout of a caspase gene, knockdown of a caspase message, upregulation of antiapoptotic molecules such as Bcl-2, or by other means—many researchers observe that death still ensues, albeit following an independent, "caspase-independent" route. Such a result is often heralded as illustrating a previously unknown or occult death pathway. However, the result is neither remarkable nor unexpected, much less illustrative. Many such reports arise from a situation in which a cell is challenged by a lethal toxin such as cycloheximide, staurosporine, or another drug that deprives the cell of its ability to manufacture or maintain a molecule essential for its survival. Unless this condition is rectified, the cell cannot continue indefinitely and must eventually succumb. What has been learned is that the material or pathway that was disrupted was essential to life, nothing more. The fact that stopping the heart will kill an animal quickly but destruction of the digestive tract will also kill the animal does not mean that starvation is an evolved and tightly regulated death pathway. Faced with an irreparable toxic condition, a cell may well activate calpain, proteasomal proteases, or lysosomal proteases. Whether it does so to kill itself, to generate more resources, or simply as an uncontrolled aspect of its deteriorating physiology, is not clear. A death pathway becomes a death pathway when it can be shown to be biologically relevant and when specific blockage of the pathway prevents death and does not simply delay it briefly.

CELLS CONTAIN SELF-DESTRUCTION MECHANISMS OTHER THAN APOPTOSIS

The most striking philosophical point to be drawn from the hypotheses of programmed cell death and apoptosis is that the death as well as the life of cells is very tightly controlled. In hindsight, such control is necessary; the unmonitored destruction of cells potentially releases lytic enzymes, viruses, toxins, or other noxious agents into the body or into the environment of single-celled organisms, and it makes more biological sense to assure the ordered removal of such materials. Scientific attitudes have changed from the image of uncontrolled cell death in organisms being corrected by compensating mitoses, to the image that often the deaths are controlled and that the compensating mitoses are the more adaptive or secondary phenomena.

Perhaps the most efficient, rapid, controlled, and highly evolved pathway is the several mechanisms and pathways that lead to the activation of the sequence-specific protease named caspase 3, and ultimately, to the fragmentation and phagocytosis of the cell. This process is called apoptosis. However, as has been acknowledged the last few years, apoptosis is not the only way to die. Cells with quite massive cytoplasm, such as muscle fibers, neurons, or large secretory cells, employ autophagy or other means of eliminating cytoplasm. Cells that are completely or relatively inaccessible to phagocytes, such as osteoblasts, may die in a controlled procedure but ultimately display characteristics of necrosis. Likewise, since conversion to apoptotic morphology requires energy, some cells initiate apoptosis but fail to complete it, terminating with an odd mixture of characteristics that include characteristics of both apoptosis and necrosis. Finally, highly differentiated cells, such as red blood cells, lens fibers, and keratinocytes, complete some aspects of apoptosis, eliminating organelles, but can persist (in the case of lens fibers, for years) with functioning cell membranes and other characteristics of viability.

CONCLUSION

For biological systems, commitment to an irreversible step must be very tightly controlled and are vigorously defended, but once engaged, become inexorable. Thus, initiation of apoptosis is controlled by many interlocking feedback mechanisms and closely regulated. Many controls have been identified, but it is unlikely that any single control is dominant.

In sum, there are several clear-cut pathways to cell death, each of which has been given specific names; but these pathways merge and sometimes overlap, and ultimately, the biology of the cell determines how it will die. Rather than focusing on subdividing categories, it will be far more interesting to learn how, in vivo, healthy and viable cells read signals and respond by commiting to death.

REFERENCES

1. Clarke PGH, Clarke S. Nineteenth century research on naturally occurring cell death and related phenomena. Anat Embryol (Berl) 1996; 193:81–99.
2. Häcker G, Vaux DL. A chronology of cell death. Apoptosis 1997; 2:247–256.
3. Lockshin RA, Zakeri Z. Programmed cell death and apoptosis: origins of the theory. Nat Rev Mol Cell Biol 2001; 2:545–550.
4. Glücksmann A. Cell deaths in normal vertebrate ontogeny. Biol Rev Camb Philos Soc 1951; 26:59–86.
5. Glücksmann A. Cell death in normal development. Arch Biol (Liege) 1965; 76: 419–437.
6. Saunders JW Jr. Death in embryonic systems. Science 1966; 154:604–612.
7. Fallon JF, Saunders JW Jr. In vitro analysis of the control of cell death in a zone of prospective necrosis from the chick wing bud. Dev Biol 1968; 18:553–570.
8. Berthet J, Berthet L, Appelmans F, et al. Tissue fractionation studies. II. The nature of the linkage between acid phosphatase and mitochondria in rat-liver tissue. Biochem J 1951; 50:182–189.
9. De Duve C, Berthet J. Reproducibility of differential centrifugation experiments in tissue fractionation. Nature 1953; 172:1142.
10. Appelmans F, Wattiaux R, De Duve C. Tissue fractionation studies 5. The association of acid phosphatase with a special class of cytoplasmic granules on rat liver. Biochem J 1955; 59:448–445.
11. De Duve C. The lysosomes, a new group of cytoplasmic granules. J Physiol (Paris) 1957; 49:113–115.
12. Lockshin RA, Williams CM. Programmed cell death. II. Endocrine potentiation of the breakdown of the intersegmental muscles of silkmoths. J Insect Physiol 1964; 10:643–649.
13. Lockshin RA, Williams CM. Programmed cell death. I. Cytology of the degeneration of the intersegmental muscles of the Pernyi silkmoth. J Insect Physiol 1965; 11:123–133.
14. Lockshin RA, Williams CM. Programmed cell death. III. Neural control of the breakdown of the intersegmental muscles. J Insect Physiol 1965; 11:605–610.
15. Lockshin RA, Williams CM. Programmed cell death. IV. The influence of drugs on the breakdown of the intersegmental muscles of silkmoths. J Insect Physiol 1965; 11:803–809.
16. Lockshin RA, Williams CM. Programmed cell death. V. Cytolytic enzymes in relation to the breakdown of the intersegmental muscles of silkmoths. J Insect Physiol 1965; 11:831–844.
17. Tata JR. Inhibition of the biological action of thyroid hormones by actinomycin D and puromycin. Nature 1963; 197:1167–1168.
18. Tata JR. Requirement for RNA and protein synthesis for induced regression of the tadpole tail in organ culture. Dev Biol 1966; 13:77–94.
19. Munck A. Glucocorticoid inhibition of glucose uptake by peripheral tissues: old and new evidence, molecular mechanisms, and physiological significance. Perspect Biol Med 1971; 14:265–289.
20. Makman MH, Dvorkin B, White A. Alterations in protein and nucleic acid metabolism of thymocytes produced by adrenal steroids in vitro. J Biol Chem 1966; 241:1646–1648.

21. Lockshin RA. Programmed cell death. Activation of lysis of a mechanism involving the synthesis of protein. J Insect Physiol 1969; 15:1505–1516.
22. Lockshin RA, Wadewitz AG. Degeneration of myofibrillar proteins during programmed cell death in Manduca sexta. In: Finch CA, Johnson T, eds. Molecular Biology of Aging, UCLA Symposia in Molecular and Cell Biology. New York: Wiley-Liss, 1990:283–297.
23. Ellis HM, Horvitz HR. Genetic control of programmed cell death in the nematode C. elegans. Cell 1986; 44:817–829.
24. Ellis RE, Yuan J, Horvitz HR. Mechanisms and functions of cell death. Annu Rev Cell Biol 1991; 7:663–698.
25. Hengartner MO, Horvitz HR. Programmed cell death in *Caenorhabditis elegans*. Curr Opin Genet Dev 1994; 4:581–586.
26. Horvitz HR, Shaham S, Hengartner MO. The genetics of programmed cell death in the nematode *Caenorhabditis elegans*. Cold Spring Harb Symp Quant Biol 1994; 59:377–386.
27. Horvitz HR. Nobel lecture. Worms, life and death. Biosci Rep 2003; 23:239–303.
28. Kerr JFR. Shrinkage necrosis: a distinct mode of cellular death. J Pathol 1971; 105:13–20.
29. Kerr JFR, Wyllie AH, Currie AR. Apoptosis: a basic biological phenomenon with wide-ranging implications in tissue kinetics. Br J Cancer 1972; 26:239–257.
30. Kerr JFR, Harmon BV, Searle JW. An electron microscope study of cell deletion in the anuran tadpole tail during spontaneous metamorphosis with special reference to apoptosis of striated muscle fibers. J Cell Sci 1974; 14:571–585.
31. Kerr JFR, Harmon BV. Definition and incidence of apoptosis: an historical perspective. In: Tomei LD, Cope FO, eds. Apoptosis: The Molecular Biology of Cell Death. Cold Spring Harbor, New York: Cold Spring Harbor Press, 1991:5–30.
32. Bowen ID, Lockshin RA. Cell Death in Biology and Pathology. London: Chapman and Hall, 1981.
33. Mammone T, Gan D, Collins D, et al. Successful separation of apoptosis and necrosis pathways in HaCaT keratinocyte cells induced by UVB irradiation. Cell Biol Toxicol 2000; 16:293–302.
34. Joseph R, Li W, Han E. Neuronal death, cytoplasmic calcium and internucleosomal DNA fragmentation: evidence for DNA fragments being released from cells. Brain Res Mol Brain Res 1993; 17:70–76.
35. Manolagas SC. Birth and death of bone cells: basic regulatory mechanisms and implications for the pathogenesis and treatment of osteoporosis. Endocr Rev 2000; 21:115–137.
36. Adams CS, Shapiro IM. Mechanisms by which extracellular matrix components induce osteoblast apoptosis. Connect Tissue Res 2003; 44(suppl 1):230–239, 230–239.
37. Cerri PS, Boabaid F, Katchburian E. Combined TUNEL and TRAP methods suggest that apoptotic bone cells are inside vacuoles of alveolar bone osteoclasts in young rats. J Periodontal Res 2003; 38:223–226.
38. Mogi M, Togari A. Activation of caspases is required for osteoblastic differentiation. J Biol Chem 2003; 278(48):47477–47482 [Epub September 3, 2003].
39. Kameda T, Ishikawa H, Tsutsui T. Detection and characterization of apoptosis in osteoclasts *in vitro*. Biochem Biophys Res Commun 1995; 207:753–760.

40. Lutton JD, Moonga BS, Dempster DW. Osteoclast demise in the rat: physiological *versus* degenerative cell death. Exp Physiol 1996; 81:251–260.
41. Appleby DW, Modak SP. DNA degradation in terminally differentiating lens fiber cells from chick embryos. Proc Natl Acad Sci U S A 1977; 74:5579–5583.
42. Dahm R, Gribbon C, Quinlan RA, et al. Lens cell organelle loss during differentiation versus stress-induced apoptotic changes. Biochem Soc Trans 1997; 25:S584.
43. Ishizaki Y, Jacobson MD, Raff MC. A role for caspases in lens fiber differentiation. J Cell Biol 1998; 140:153–158.
44. Counis MF, Chaudun E, Arruti C, et al. Analysis of nuclear degradation during lens cell differentiation. Cell Death Differ 1998; 5:251–261.
45. Dahm R. Lens fibre cell differentiation - A link with apoptosis? Ophthalmic Res 1999; 31:163–183.
46. Daugas E, Cande C, Kroemer G. Erythrocytes: death of a mummy. Cell Death Differ 2001; 8:1131–1133.
47. Bratosin D, Estaquier J, Petit F, et al. Programmed cell death in mature erythrocytes: a model for investigating death effector pathways operating in the absence of mitochondria. Cell Death Differ 2001; 8:1143–1156.
48. Lang KS, Duranton C, Poehlmann H, et al. Cation channels trigger apoptotic death of erythrocytes. Cell Death Differ 2003; 10:249–256.
49. Lang F, Lang KS, Wieder T, et al. Cation channels, cell volume and the death of an erythrocyte. Pflugers Arch 2003; 121:919–927.
50. Lockshin RA, Zakeri Z. Caspase-independent cell death? Oncogene 2004; 23: 2766–2773.
51. Fox H. Degeneration of the nerve cord in the tail of Rana temporaria during metamorphic climax: study by electron microscopy. J Embryol Exp Morphol 1973; 30:377–396.
52. Fox H. Degeneration of the tail notochord of Rana temporaria at metamorphic climax. Examination by electron microscopy. Z Zellforsch Mikrosk Anat 1973; 138:371–386.
53. Nishikawa A, Hayashi H. Spatial, temporal and hormonal regulation of programmed muscle cell death during metamorphosis of the frog Xenopus laevis. Differentiation 1995; 59:207–214.
54. Nishikawa A, Murata E, Akita M, et al. Roles of macrophages in programmed cell death and remodeling of tail and body muscle of Xenopus laevis during metamorphosis. Histochem Cell Biol 1998; 109:11–17.
55. Locke M, Collins JV. Protein uptake into multivesicular bodies and storage granules in the fat body of an insect. J Cell Biol 1968; 36:453–483.
56. Locke M, McMahon JT. The origin and fate of microbodies in the fat body of an insect. J Cell Biol 1971; 48:61–78.
57. Halaby R, Zakeri Z, Lockshin RA. Metabolic events during programmed cell death in insect labial glands. Biochem Cell Biol 1994; 72:597–601.
58. Jochová J, Quaglino D, Zakeri Z, et al. Protein synthesis, DNA degradation, and morphological changes during programmed cell death in labial glands of *Manduca sexta*. Dev Genet 1997; 21:249–257.
59. Jochová J, Zakeri Z, Lockshin RA. Rearrangement of the tubulin and actin cytoskeleton during programmed cell death in *Drosophila* salivary glands. Cell Death Differ 1997; 4:140–149.

60. Kimchi A. DAP kinase and DAP-3: novel positive mediators of apoptosis. Ann Rheum Dis 1999; 58(suppl 1):I14–I19.
61. Foghsgaard L, Wissing D, Mauch D, et al. Cathepsin B acts as a dominant execution protease in tumor cell apoptosis induced by tumor necrosis factor. J Cell Biol 2001; 153:999–1010.
62. Leist M, Jäättelä M. Four deaths and a funeral: from caspases to alternative mechanisms. Nat Rev Mol Cell Biol 2001; 2:589–598.
63. Leist M, Jäättelä M. Triggering of apoptosis by cathepsins. Cell Death Differ 2001; 8:324–326.
64. Inbal B, Bialik S, Sabanay I, et al. DAP kinase and DRP-1 mediate membrane blebbing and the formation of autophagic vesicles during programmed cell death. J Cell Biol 2002; 157:455–468.
65. Boya P, Andreau K, Poncet D, et al. Lysosomal membrane permeabilization induces cell death in a mitochondrion-dependent fashion. J Exp Med 2003; 197:1323–1334.
66. Boya P, Gonzalez-Polo RA, Poncet D, et al. Mitochondrial membrane permeabilization is a critical step of lysosome-initiated apoptosis induced by hydroxychloroquine. Oncogene 2003; 22:3927–3936.
67. Jäättelä M, Tschopp J. Caspase-independent cell death in T lymphocytes. Nat Immunol 2003; 4:416–423.
68. Jäättelä M. Multiple cell death pathways as regulators of tumour initiation and progression. Oncogene 2004; 23:2746–2756.
69. Gonzalez-Polo RA, Boya P, Pauleau AL, et al. The apoptosis/autophagy paradox: autophagic vacuolization before apoptotic death. J Cell Sci 2005; 118:3091–3102.
70. Kroemer G, Jäättelä M. Lysosomes and autophagy in cell death control. Nat Rev Cancer 2005; 5:886–897.
71. Kroemer G, Martin SJ. Caspase-independent cell death. Nat Med 2005; 11:725–730.
72. Klionsky DJ, Emr SD. Autophagy as a regulated pathway of cellular degradation. Science 2000; 290:1717–1721.
73. Lee CY, Baehrecke EH. Steroid regulation of autophagic programmed cell death during development. Development 2001; 128:1443–1455.
74. Larsen KE, Sulzer D. Autophagy in neurons: a review. Histol Histopathol 2002; 17:897–908.
75. Kadowaki M, Kanazawa T. Amino acids as regulators of proteolysis. J Nutr 2003; 133:2052S–2056S.
76. Cuervo AM. Autophagy: in sickness and in health. Trends Cell Biol 2004; 14:70–77.
77. Kuma A, Hatano M, Matsui M, et al. The role of autophagy during the early neonatal starvation period. Nature 2004; 432:1032–1036.
78. Mizushima N, Yamamoto A, Matsui M, et al. In vivo analysis of autophagy in response to nutrient starvation using transgenic mice expressing a fluorescent autophagosome marker. Mol Biol Cell 2004; 15:1101–1111.
79. Lum JJ, DeBerardinis RJ, Thompson CB. Autophagy in metazoans: cell survival in the land of plenty. Nat Rev Mol Cell Biol 2005; 6:439–448.
80. Cakouros D, Daish T, Martin D, et al. Ecdysone-induced expression of the caspase DRONC during hormone-dependent programmed cell death in Drosophila is regulated by Broad-Complex. J Cell Biol 2002; 157:985–995.

81. Martin DN, Baehrecke EH. Caspases function in autophagic programmed cell death in Drosophila. Development 2004; 131:275–284.
82. Kumar S. ICE-like proteases in apoptosis. Trends Biochem Sci 1995; 20:198–202.
83. Dorstyn L, Kinoshita M, Kumar S. Caspases in cell death. Results Probl Cell Differ 1998; 24:1–24.
84. Harvey NL, Kumar S. The role of caspases in apoptosis. Adv Biochem Eng Biotechnol 1998; 62:107–128.
85. Dorstyn L, Colussi PA, Quinn LM, et al. DRONC, an ecdysone-inducible Drosophila caspase. Proc Natl Acad Sci U S A 1999; 96:4307–4312.
86. Dorstyn L, Read SH, Quinn LM, et al. DECAY, a novel drosophila caspase related to mammalian caspase-3 and caspase-7. J Biol Chem 1999; 274:30778–30783.
87. Dorstyn L, Read SH, Quinn LM, et al. DECAY, a novel Drosophila caspase related to mammalian caspase-3 and caspase-7. J Biol Chem 1999; 274:30778–30783.
88. Harvey NL, Daish T, Mills K, et al. Characterization of the Drosophila caspase, DAMM. J Biol Chem 2001; 276:25342–25350.
89. Xue L, Fletcher GC, Tolkovsky AM. Autophagy is activated by apoptotic signalling in sympathetic neurons: an alternative mechanism of death execution. Mol Cell Neurosci 1999; 14:180–198.
90. Xue L, Fletcher GC, Tolkovsky AM. Mitochondria are selectively eliminated from eukaryotic cells after blockade of caspases during apoptosis. Curr Biol 2001; 11:361–365.
91. Tolkovsky AM, Xue L, Fletcher GC, et al. Mitochondrial disappearance from cells: a clue to the role of autophagy in programmed cell death and disease? Biochimie 2002; 84:233–240.
92. Lockshin RA, Beaulaton J. Programmed cell death. Cytochemical evidence for lysosomes during the normal breakdown of the intersegmental muscles. J Ultrastruct Res 1974; 46:43–62.
93. Lockshin RA, Beaulaton J. Programmed cell death. Cytochemical appearance of lysosomes when the death of the intersegmental muscles is prevented. J Ultrastruct Res 1974; 46:63–78.
94. Beaulaton J, Lockshin RA. Ultrastructural study of the normal degeneration of the intersegmental muscles of Antheraea polyphemus and Manduca sexta (Insecta, Lepidoptera) with particular reference to cellular autophagy. J Morphol 1977; 154: 39–58.
95. Lum JJ, Bauer DE, Kong M, et al. Growth factor regulation of autophagy and cell survival in the absence of apoptosis. Cell 2005; 120:237–248.
96. Lum JJ, DeBerardinis RJ, Thompson CB. Autophagy in metazoans: cell survival in the land of plenty. Nat Rev Mol Cell Biol 2005; 6:439–448.
97. Thompson CB, Bauer DE, Lum JJ, et al. How do cancer cells acquire the fuel needed to support cell growth? Cold Spring Harb Symp Quant Biol 2005; 70: 357–362.
98. Rodriguez-Enriquez S, Kim I, Currin RT, et al. Tracker dyes to probe mitochondrial autophagy (mitophagy) in rat hepatocytes. Autophagy 2006; 2:39–46.
99. Kim I, Rodriguez-Enriquez S, Lemasters JJ. Selective degradation of mitochondria by mitophagy. Arch Biochem Biophys 2007; 462(2):245–253 [Epub April 12, 2007].
100. Liang J, Shao SH, Xu ZX, et al. The energy sensing LKB1-AMPK pathway regulates p27(kip1) phosphorylation mediating the decision to enter autophagy or apoptosis. Nat Cell Biol 2007; 9:218–224.

101. Levi-Montalcini R, Aloe L. Mechanism(s) of action of nerve growth factor in intact and lethally injured sympathetic nerve cells in neonatal rodents. In: Bowen ID, Lockshin RA, eds. Cell Death in Biology and Pathology. London and New York: Chapman and Hall, 1981:295–327.
102. Lockshin RA, Zakeri Z. Caspase-independent cell deaths. Curr Opin Cell Biol 2002; 14:727–733.
103. Koopman G, Reutelingsperger CPM, Kuijten GAM, et al. Annexin V for flow cytometric detection of phosphatidylserine expression on B cells undergoing apoptosis. Blood 1994; 84:1415–1420.
104. Gavrieli Y, Sherman Y, Ben-Sasson SA. Identification of programmed cell death in situ via specific labeling of nuclear DNA fragmentation. J Cell Biol 1992; 119: 493–501.
105. Rano TA, Timkey T, Peterson EP, et al. A combinatorial approach for determining protease specificities: application to interleukin-1beta converting enzyme (ICE). Chem Biol 1997; 4:149–155.
106. Thornberry NA, Rano TA, Peterson EP, et al. A combinatorial approach defines specificities of members of the caspase family and granzyme B. Functional relationships established for key mediators of apoptosis. J Biol Chem 1997; 272: 17907–17911.
107. Nicholson DW. Caspase structure, proteolytic substrates, and function during apoptotic cell death. Cell Death Differ 1999; 6:1028–1042.
108. Lemasters JJ, Qian T, He L, et al. Role of mitochondrial inner membrane permeabilization in necrotic cell death, apoptosis, and autophagy. Antioxid Redox Signal 2002; 4:769–781.
109. Trump BF, Berezesky IK. The role of altered [Ca2+]i regulation in apoptosis, oncosis, and necrosis. Biochim Biophys Acta 1996; 1313:173–178.
110. Trump BF, Berezesky IK, Chang SH, et al. The pathways of cell death: oncosis, apoptosis, and necrosis. Toxicol Pathol 1997; 25:82–88.
111. Hensey C, Gautier J. A developmental timer that regulates apoptosis at the onset of gastrulation. Mech Dev 1997; 69:183–195.
112. Hensey C, Gautier J. Developmental regulation of induced and programmed cell death in *Xenopus* embryos. In: Zakeri Z, Lockshin RA, Benitez-Bribiesca L, eds. Mechanisms of Cell Death. New York: New York Academy of Sciences, 1999: 105–119.
113. Negrón JF, Lockshin RA. Activation of apoptosis and caspase-3 in zebrafish early gastrulae. Dev Dyn 2004; 231:161–170.
114. Ledda-Columbano GM, Coni P, Curto M, et al. Induction of two different modes of cell death, apoptosis and necrosis, in rat liver after a single dose of thioacetamide. Am J Pathol 1991; 139:1099–1109.
115. Columbano A. Cell death: current difficulties in discriminating apoptosis from necrosis in the context of pathological processes in vivo. J Cell Biochem 1995; 58:181–190.
116. Bialik S, Geenen DL, Sasson IE, et al. Myocyte apoptosis during acute myocardial infarction in the mouse localizes to hypoxic regions but occurs independently of p53. J Clin Invest 1997; 100:1363–1372.
117. Miao W, Luo Z, Kitsis RN, et al. Intracoronary, adenovirus-mediated Akt gene transfer in heart limits infarct size following ischemia-reperfusion injury in vivo. J Mol Cell Cardiol 2000; 32:2397–2402.

118. Mani K, Peng C, Rosner G, et al. The role of apoptosis in myocardial infarction and heart failure. In: Lockshin RA, Zakeri Z, eds. When Cells Die II. New York: Wiley-Liss, 2004:483–520.
119. Teng X, Degterev A, Jagtap P, et al. Structure-activity relationship study of novel necroptosis inhibitors. Bioorg Med Chem Lett 2005; 15:5039–5044.
120. Degterev A, Huang Z, Boyce M, et al. Chemical inhibitor of nonapoptotic cell death with therapeutic potential for ischemic brain injury. Nat Chem Biol 2005; 1: 112–119.
121. Wang K, Li J, Degterev A, et al. Structure-activity relationship analysis of a novel necroptosis inhibitor, Necrostatin-5. Bioorg Med Chem Lett 2007; 17:1455–1465.
122. Jagtap PG, Degterev A, Choi S, et al. Structure-activity relationship study of tri-cyclic necroptosis inhibitors. J Med Chem 2007; 50:1886–1895.
123. Chen Y, Douglass T, Jeffes EW, et al. Living T9 glioma cells expressing membrane macrophage colony-stimulating factor produce immediate tumor destruction by polymorphonuclear leukocytes and macrophages via a "paraptosis"-induced pathway that promotes systemic immunity against intracranial T9 gliomas. Blood 2002; 100:1373–1380.
124. Schneider D, Gerhardt E, Bock J, et al. Intracellular acidification by inhibition of the Na+/H+-exchanger leads to caspase-independent death of cerebellar granule neurons resembling paraptosis. Cell Death Differ 2004; 11:760–770.
125. Sperandio S, Poksay K, De Belle I, et al. Paraptosis: mediation by MAP kinases and inhibition by AIP-1/Alix. Cell Death Differ 2004; 11:1066–1075.
126. Frisch SM, Francis H. Disruption of epithelial cell-matrix interactions induces apoptosis. J Cell Biol 1994; 124:619–626.
127. Frisch SM, Ruoslahti E. Integrins and anoikis. Curr Opin Cell Biol 1997; 9: 701–706.
128. Grossmann J. Molecular mechanisms of "detachment-induced apoptosis—Anoikis". Apoptosis 2002; 7:247–260.
129. Valentijn AJ, Zouq N, Gilmore AP. Anoikis. Biochem Soc Trans 2004; 32:421–425.
130. Mannello F, Luchetti F, Falcieri E, et al. Multiple roles of matrix metalloproteinases during apoptosis. Apoptosis 2005; 10:19–24.
131. Gilmore AP. Anoikis. Cell Death Differ 2005; 12(suppl 2):1473–1477.
132. Lockshin RA, Zakeri Z. Apoptosis, autophagy, and more. Int J Biochem Cell Biol 2004; 36:2405–2419.
133. Zakeri Z, Quaglino D, Latham T, et al. Programmed cell death in the tobacco hornworm, Manduca sexta: alteration in protein synthesis. Microsc Res Tech 1996; 34:192–201.
134. Zakeri Z, Quaglino D, Ahuja HS. Apoptotic cell death in the mouse limb and its suppression in the hammertoe mutant. Dev Biol 1994; 165:294–297.
135. Zhang Q, Ahuja HS, Zakeri ZF, et al. Cyclin-dependent kinase 5 is associated with apoptotic cell death during development and tissue remodeling. Dev Biol 1997; 183:222–233.

5

Toward a Mechanistic Taxonomy for Programmed Cell Death Pathways

Dale E. Bredesen

Buck Institute for Age Research, Novato, and University of California, San Francisco, California, U.S.A.

INTRODUCTION

How many biochemical pathways triggering cell death forms are available to normal, preneoplastic, and neoplastic cells? What morphological and biochemical alterations result from each of these pathways? Which ones predominate in each type of neoplasm? Which may be triggered therapeutically, and what is the best approach to triggering these cell death pathways? What are the inhibitors and activators of each of these pathways?

Answers to these questions may offer us new avenues that will complement current chemotherapeutic, radiotherapeutic, and immunotherapeutic approaches to the treatment of various forms of cancer. However, our knowledge of the pathways underlying nonapoptotic forms of programmed cell death (PCD) is far from complete. Although over 100,000 publications have appeared on apoptosis, it has only recently become clear that alternative cell death pathways may also be utilized by at least some cell types. Here we consider what is currently known about these various alternative pathways, which ones are clearly distinct and which may involve overlap, and what is known mechanistically about these forms of cell death.

PCD plays a critical role in development and dysregulation of cell death programs featured in developmental, neoplastic, degenerative, infectious, traumatic, ischemic, metabolic, and other disease states. It has been 100 years since the first description of developmental cell death (1), and over 50 years since the demonstration that such physiological cell death is inhibited by soluble (trophic) factors (2).

In 1964, Richard Lockshin introduced the term "programmed cell death" to describe an apparently endogenous pathway or set of pathways utilized by cells to commit suicide during insect development (3). In 1966, it was shown that this process requires protein synthesis (4), suggesting an active cellular suicide process. Then in 1972, Kerr et al. coined the term "apoptosis" to describe a morphologically relatively uniform set of cell deaths that occurs in many different paradigms, from development to insult response to cell turnover (5).

As noted above, apoptosis has been studied extensively, with well over 100,000 papers published on the subject (www.pubmed.gov). Although PCD has often been equated with apoptosis, it has become increasingly clear that nonapoptotic forms of PCD also exist (6–17), for example, certain developmental cell deaths, such as "autophagic" cell death (3,6,13–15) and "cytoplasmic" cell death (6,7,10–12,15), do not resemble apoptosis morphologically. Furthermore, some diseases such as Huntington's disease and Lou Gehrig's disease (amyotrophic lateral sclerosis) display neuronal cell death that is nonapoptotic (8,17) (although some classical apoptosis occurs in these diseases, as well). Ischemia-induced cell death may also display a nonapoptotic morphology, referred to as "oncosis" (9).

How many different mammalian cell death programs can be distinguished, and what is their relationship? A number of classifications have been proposed based on morphology, but for the purposes of mechanistic insight and therapeutic intervention, it would be preferable to identify all such programs, which are available to each cell type, and to construct a mechanistic taxonomy of these cell death programs, with special attention to their specific inhibitors and activators. The data required for such a construct are currently far from complete, and thus the classification presently possible will undoubtedly be revised repeatedly over time. Nonetheless, it is informative to consider, based on currently available data, how many programs of cell death can be classified mechanistically (Table 1).

Cell death has been divided into two broad categories, programmed and nonprogrammed, or passive. The latter has often been referred to as necrotic cell death. It is important to note, however, that a semantic issue has arisen with the demonstration that some forms of nonapoptotic cell death previously labeled necrotic, and thus assumed to be passive, have turned out to be programmatic; some have referred to these as "necrosis-like" (18), whereas others use the term "programmed necrosis" (19,20). Based on the traditional view that some term should be reserved for passive, nonprogrammatic cell death, and that necrosis is the term historically applied to this form of cell death, the term "programmed necrosis" may be something of an oxymoron. However, based on a different feature of necrosis, the breach of the plasma membrane with resulting initiation of an inflammatory response by leaked cellular contents, "programmed necrosis" would indeed be an appropriate term. This phenomenon notwithstanding, reserving the term necrosis for nonprogrammatic PCD suggests that such programmatic cell deaths with necrotic morphology and other characteristics should be referred to as "necrosis-like." As biochemical data become available for each form of PCD, it should become clear which paradigms induce necrosis-like PCD and which lead to passive, nonprogrammatic (classically necrotic) cell death.

Classical developmental studies revealed three different types of cell death, based on morphological criteria: type I—nuclear, also referred to as apoptotic; type II—autophagic; and type III—cytoplasmic (6). These PCD types occur reproducibly within specific nuclei and at specific times of nervous system development. However, these apparently physiological cell death pathways may also be activated by various insults, such as ischemia or anoxia.

Neoplastic development and metastasis, as well as embryonic development, neurodegeneration, osteoporosis, and many other physiological and pathophysiological processes—all involve cell death and its control. Intrinsic cellular suicide programs described by Lockshin in 1964, were equated for many years with a specific program, apoptosis, a term introduced by John Kerr and his colleagues in 1972. However, over the last several years it has become clear that alternative, nonapoptotic cell death programs may be triggered in certain circumstances. This finding may have practical application in the understanding of cancer development and therapy, especially since resistance to chemotherapy and radiotherapy may be associated with mutations in apoptosis-associated genes, raising the question of whether specific triggering of nonapoptotic cell death programs may be advantageous under such conditions. Accordingly, it is of interest to attempt to distinguish and classify the various cell death programs available to the metazoan cell, not simply from a morphological standpoint but also from the standpoint of the required biochemical pathways. Such a classification may prove to have utility in the treatment of various cancers and other diseases that involve PCD modulation.

APOPTOSIS

Apoptosis is described in chapter 1, as well as elsewhere in this volume and therefore the focus here will be on the molecular linchpins that, when inactivated, may potentially shift the cell away from the pursuit of an apoptotic pathway and toward the utilization of any of the nonapoptotic pathways.

The activation of apoptosis is effected via two general pathways: the intrinsic pathway, triggered by mitochondrial release of cytochrome c and in turn its activation, with (d)ATP, of caspase-9; and the extrinsic pathway, triggered by the activation of cell surface death receptors (e.g., Fas), and resulting in the

Table 1 Forms of Programmed Cell Death (*The color version of this figure is provided on the DVD*)

Types →	Convicted Killers		Some of the New Suspects (*Innocent until proven guilty*)			
	Apoptosis	**Autophagic**	**Paraptosis**	**Calcium-mediated**	**AIF/PARP-dependent**	**Oncosis**
Characteristics ↓	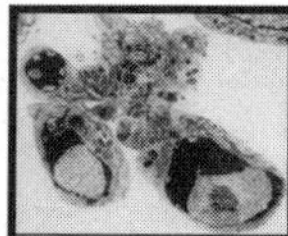	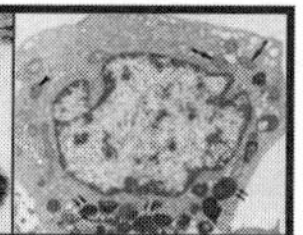	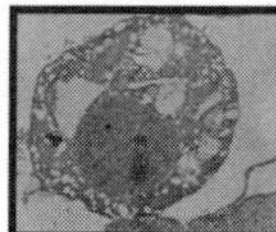	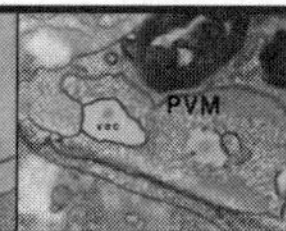	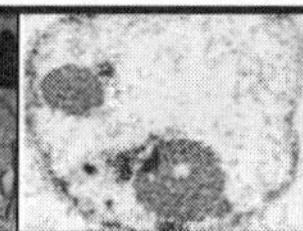	
Morphology	Chromatin condensation, nuclear fragmentation, apoptotic bodies	Autophagic vacuoles	ER swelling, mitochondrial swelling	Membrane whorls	Mild chromatin condensation	Cellular swelling
Triggers	Death receptors, trophic factor withdrawal, DNA damage, viral infections, etc.	Serum, amino acid starvation, protein aggregates	Trophotoxicity	Calcium entry, deg mutants	DNA damage, glutamate, NO	Ischemia, excitotoxicity
Mediators	Caspases, BH family, etc.	JNK, MKK7, Atg orthologs	ERK2, Nur77	Calpains, cathepsins	PARP, AIF	JNK
Inhibitors	Caspase inhibitors, BH family, etc.	JNK inhibitors, Atg reduction	U0126 DN Nur77	Calreticulin, some calpain inhibitors?	PARP inhibitors	JNK inhibitors
Examples	Type I PCD, nuclear PCD	Type II PCD	Type III PCD, cytoplasmic PCD	*C. elegans* deg mutants	Some excitotoxic PCD	Ischemic PCD

Abbreviations: AIF, apoptosis-inducing factor; PARP, poly-(ADP-ribose) polymerase; ER, endoplasmic reticulum; deg, BH, Bcl-2 homology; JNK, c-jun N-terminal kinase; MKK, mitogen-activated protein kinase kinase; Atg, autophagy; ERK2, extracellular signal-regulated kinase 2; deg, degeneration; DN, dominant negative; PCD, programmed cell death.

activation of caspase-8 or -10 (21). A third general pathway (essentially a second intrinsic pathway) is activated by endoplasmic reticulum (ER) stress due to unfolded or misfolded proteins, such as occur in the poorly vascularized ischemic regions of tumors (22–27). In addition, other organelles such as the nucleus and Golgi apparatus also display damage sensors that link to apoptotic pathways (28).

Many different pathways, both physiological and pathological, converge on the intrinsic pathway of apoptosis. Whether by transcriptional regulation (e.g., via p53) or posttranscriptional regulation, this results in a shift in the balance of a set of related proteins of the Bcl-2 family. These are major determinants of the apostat—the propensity of the cell to undergo apoptosis. Triggers of the intrinsic pathway shift the apostat toward the pro-apoptotic members: (*i*) the multidomain members (BH1–3: proteins that display Bcl-2 homology domains 1–3) such as Bax and Bak are capable of permeabilizing mitochondrial outer membranes (29,30); (*ii*) the BH3 only activators such as Bim and tBid [a product of Bid resulting from cleavage by caspases, calpains, or other proteases, and providing communication between the extrinsic and intrinsic systems (31)] and Bim activate Bax and Bak, and may participate in their pore formation; and (*iii*) the BH3 only derepressors such as Puma, Noxa, and Bad sequester the anti-apoptotic block created by Bcl-2, Bcl-X_L, and related anti-apoptotic proteins that display BH1–4 domains, freeing the BH1–3 and activators to permeabilize the mitochondrial outer membrane, thus allowing the release of multiple intermembrane proteins, such as cytochrome c, Smac/DIABLO, Omi/HtrA2, endonuclease G, and apoptosis-inducing factor (AIF). An anti-apoptotic modulator peptide, humanin, may also affect this balance by binding and inhibiting Bax, Bid, and BimEL (32–34). Conversely, the Bax effect may be augmented by non-Bcl-2 family member modulators such as p53. Thus, p53 exhibits both transcriptional and nontranscriptional pro-apoptotic effects, and p53 inhibitors such as pifithrin alpha show promise as potential therapeutic agents for ischemic insults (35,36). Another pro-PCD mechanism is exhibited by Nur77/TR3, which binds Bcl-2 exposing its pro-apoptotic BH3 domain resulting in a pro-PCD effect (37). How many other transcriptional factors will prove to complement their activities by providing nontranscriptional modulation of PCD by this or other mechanisms remains an open question.

As noted above, the anti-apoptotic members of the Bcl-2 family proteins, such as Bcl-2 and Bcl-X_L, include BH1–4 domains. These anti-apoptotic proteins interact with the pro-apoptotic members and prevent the mitochondrial outer membrane permeability (MOMP) that would otherwise be created by the pro-apoptotic members; thus the anti-apoptotic Bcl-2 family proteins prevent the mitochondrial release of cytochrome c and other pro-apoptotic mitochondrial proteins (38).

The net effect of the interplay of these Bcl-2 family and functionally related proteins is to set, along with other critical determinants, the apostat. Another form of apostat modulation is the inhibition of caspases by inhibitor of apoptosis (IAP) proteins such as XIAP (see below), and block of this inhibition by Smac/DIABLO and the serine protease Omi/HtrA2.

Following release from the mitochondria, cytochrome c interacts with the cytosolic protein Apaf-1 via the WD-40 repeats of Apaf-1, exposing a (d)ATP-binding site on Apaf-1, and binding at that site leads to a conformational change resulting in heptamerization of Apaf-1. The resultant exposure of the Apaf-1 caspase activation and recruitment domain (CARD) recruits caspase-9 into this apoptosomal complex, inducing a proximity of caspase-9 molecules that leads to their activation (39). Apical caspase-9 activation leads to the activation of a cascade of downstream (nonapical) caspases, including the effector caspases such as caspase-3 and caspase-7. However, active caspases-3, -7, and -9 are inhibitable by the IAP proteins, such as XIAP (40), which may function as both direct inhibitors of caspase activity (in the case of caspase-9, by inhibiting dimerization) and as ubiquitin E3 ligases that mediate caspase degradation (41). XIAP and other IAP proteins (e.g., CIAP-1 and CIAP-2) display three BIR (baculovirus IAP repeat) domains, BIR-1, -2, and -3, which, in addition to their caspase interaction, may also interact with other proteins (42). The IAP-mediated caspase block may, however, be reversed by additional mitochondrial proteins that, themselves, interact with IAPs—Smac/DIABLO (43,44) and Omi/HtrA2 (45,46).

In contrast to the intrinsic pathway, which activates caspase-9, the extrinsic pathway activates caspase-8 or caspase-10. In the best characterized example, Fas, which is trimerized even prior to ligand binding (47,48), is bound by trimeric Fas ligand, resulting in the recruitment of FADD (Fas-associated death domain protein) though its death domain, and then caspase-8 through FADD's death effector domain (DED) (49). As with caspase-9, the induced proximity of the apical caspase leads to activation with subsequent activation of effector caspases such as caspase-3 and caspase-7. In addition, FLIP(L) (FLICE-like inhibitory protein, long form), which may function as an inhibitor of extrinsic pathway activation, may also act as a caspase-8 activator by functioning as a higher affinity dimeric partner of caspase-8 than caspase-8 itself, leading to activation by heterodimerization in preference to homodimerization (50).

The intrinsic and extrinsic pathways of apoptosis (as well as an alternative intrinsic pathway that is triggered by ER stress due to the accumulation of misfolded proteins) thus, both converge on the activation of effector caspases, cysteine aspartyl-specific proteases that cleave with high specificity at a small subset of aspartic acid residues in proteins. Their substrates, which total somewhere between a few hundred and one thousand, contribute to the apoptotic phenotype via proteolytic cascade activation, structural alterations, repair inactivation, internucleosomal DNA cleavage, phagocytic uptake signaling, mitochondrial permeabilization, and other effects. Caspases are synthesized as zymogens, but differ markedly in their activation: the apical caspases (caspase-8, -9, and -10) reside in the cytosol as monomers until dimerization is effected by adaptor molecules such as heptameric Apaf-1 (as part of the apoptosome) or trimeric FADD, whereas the effector caspases are proteolytically activated by the apical caspases.

Contrary to earlier belief, cleavage of apical caspases is neither sufficient nor required for activation (51). Their zymogenicity (the ratio of activity of the active form to that of the zymogen) is relatively low—from 10 (caspase-9) to 100 (caspase-8) (51)—and thus the (monomeric) zymogens themselves are relatively active. These caspases display large prodomains that are utilized in the protein-protein interactions that mediate activation—CARD in caspase-9 and DED in caspase-8 and -10. The substrates of the apical caspases typically display I/L/V-E-X-D in the P4-P1 positions, with a preference for small or aromatic residues in the P1' position (51).

Effector caspases such as caspase-3 and -7 are activated, in turn, by the apical caspases. Unlike the apical caspases, the effector caspases are cytosolic dimers, display high zymogenicity (greater than 10,000 for caspase-3) and short prodomains, and are activated by cleavage rather than induced proximity. Cleavage produces a heterotetramer with two large subunits (17–20 kDa) and two small subunits (10–12 kDa). Because of a difference in the S_4 pocket structure of these caspases (in comparison to the apical caspases), their substrate preference is D-E-X-D, with a marked preference for Asp over Glu (and Glu over all other residues) in the P4 position (51).

Not all caspases fit neatly within these two groups: caspase-2 has the long prodomain characteristic of an initiator caspase but has a substrate preference similar to effector caspases (with the exception that, unlike other caspases, it has a P5 preference [for small hydrophobic residues)]; caspase-6 has the short prodomain of an effector caspase but has a substrate preference similar to apical caspases; the inflammatory caspases (-1, -4, -5) are involved in the processing of interleukin-1β and interleukin-18. The inflammatory caspases are thought not to play a role in PCD, however, inhibition in some paradigms such as cerebral ischemia has been associated with a reduction in infarct size (52).

Programmed Cell Death Induced by Loss of Trophic Support

One of the key features of cancerous cells is their ability to metastasize to sites that do not provide the same trophic support that exists at the site of the cells' origin. In contrast, nonneoplastic cells (and some tumor cells, as well) depend for their survival on stimulation that is mediated by various receptors and sensors, and in these cells PCD may be induced in response to the withdrawal of trophic factors, hormonal support, electrical activity, extracellular matrix support, or other trophic stimuli (53). It has generally been assumed that cells dying as a result of the withdrawal of required stimuli do so because of the loss of a positive survival signal, typically mediated by receptor tyrosine kinases and their downstream targets (54). Clearly such positive survival signals are extremely important; however data obtained over the past 15 years argue for a complementary signal that is pro-apoptotic, activated or propagated by trophic stimulus withdrawal, and mediated by pro-PCD receptors dubbed "dependence receptors," such as DCC (deleted in colorectal cancer) and Unc5H2 (uncoordinated gene 5 homologue 2)

(55–65). The intracytoplasmic domains of these receptors have been shown to interact with caspases, including apical caspases such as caspase-9, and may therefore serve as sites of induced proximity and activation of these caspases. Caspase activation leads in turn to receptor cleavage, producing pro-apoptotic fragments (55,62). However, mutation of the caspase cleavage sites of dependence receptors suppresses PCD mediated by such receptors (62,66).

Thus cellular dependence on specific survival signals is mediated partly by specific "dependence receptors," which trigger apoptosis in the absence of the required stimulus (when unoccupied by a trophic ligand, or when bound by a competing, antitrophic ligand), but inhibit apoptosis following binding to their specific ligands (53,63,64). Expression of these dependence receptors thus creates cellular states of dependence on their respective ligands. These states of dependence are not absolute, since they can be blocked downstream in some cases by the expression of anti-apoptotic genes such as Bcl-2 or p35 (53,60,67); however, they result in a shift of the apostat (21,68) toward an increased probability of initiating apoptosis. Considered in the aggregate, therefore, these receptors may serve as a molecular integration system for trophic signals, analogous to the electrical integration system comprised of the dendritic arbors within the nervous system.

Dependence receptors may play a role in development, oncogenesis, and neurodegeneration, among other processes. The expression of such pro-apoptotic receptors may potentially prevent metastasis of cells outside the region of trophic ligand availability (62). Furthermore, these receptors create a novel class of tumor suppressor genes—conditional tumor suppressor genes (69,70): in the presence of limiting concentrations of the trophic ligands, these receptors mediate a tumor suppressive effect; however, if the ligand is present at high concentrations, then cell death is inhibited and tumor formation is actually supported (69). Interestingly, the notion of conditional tumor suppressor genes extends beyond dependence receptors: Arakawa and colleagues have shown that the classical tumor suppressor p53 actually functions as a conditional tumor suppressor, based on the upregulation of Unc5B by p53, which induces apoptosis unless blocked by netrin-1 binding to Unc5B (71).

Autophagic Programmed Cell Death

Autophagy (Greek, "self eating"), which includes macroautophagy, microautophagy, and chaperone-mediated autophagy, is a regulated lysosomal pathway that complements the proteasomal pathway by degrading long-lived proteins, protein aggregates, and organelles. Targets for degradation, such as damaged mitochondria or aggregates of misfolded proteins, first become encircled by an autophagosome, which then fuses with a lysosome, resulting in the degradation of the contents of the autophagosome. The biochemical pathway utilized in this process has been characterized best in yeast, in which a number of autophagy (Atg) genes, most of which have orthologs in higher eukaryotes, have been identified.

Since energy and amino acids for protein synthesis result from the degradation of molecules and organelles by autophagy, it is a protective cellular pathway. Furthermore, although active constitutively at low level, autophagy can be upregulated markedly by nutrient starvation via the TOR (target of rapamycin) pathway. Nutrient withdrawal inactivates TOR, activating a complex of autophagy proteins (72). Although the roles of the autophagic process in protein and organellar degradation, and in cellular protection during nutrient starvation are well accepted, the role of autophagy in PCD is much less clear (72–74). This is in part because the term "autophagic cell death" has been applied both to situations in which cell death is associated with (but not necessarily requiring) autophagy and to situations in which a form of PCD requires the autophagic machinery. The majority of examples of autophagic cell death represent the former rather than the latter. However, increasing evidence suggests that the autophagic process is indeed required for at least some of what have been referred to as autophagic cell deaths. For example, mouse embryo fibroblasts (MEFs) that are null for both Bax and Bak, when treated with either of two standard apoptosis inducers, staurosporine and etoposide, undergo autophagic PCD dependent on autophagy genes Atg5 and Beclin-1, and inhibited by the autophagy/class III PI3 kinase inhibitor, 3-methyladenine (73). This autophagic PCD may, however, play out more slowly than apoptosis (75), or may potentially be triggered by an apoptosis block. Indeed, caspase inhibition by zVAD.fmk in L929 cells results in autophagy-dependent PCD (76), mediated by RIP (receptor-interacting protein), MKK7 (mitogen-activated protein kinase kinase 7), JNK (c-jun N-terminal kinase), c-jun, and degradation of catalase (77). On the one hand, this may serve as an admonishment that anti-apoptotic therapies carry the potential risk of inducing autophagic PCD; on the other hand, it may argue that therapeutics directed at multiple cell death pathways will be required for optimal efficacy in diseases involving PCD. However, the initial conclusion that caspase-8 inhibition is sufficient to induce autophagic PCD (presumably by decreasing caspase-8-mediated turnover of RIP, triggering the RIP-dependent pathway of autophagic PCD) has been challenged recently (78); Madden et al. (see chapter 7 of this volume) found that induction of autophagic PCD in L929 cells required not only caspase-8 inhibition but also the inhibition of an as-yet-unidentified calpain-like cysteine protease.

The requirement for inhibition of both caspase-8 and a calpain-like protease by zVAD.fmk to induce autophagic cell death suggests cross talk between mediators of apoptosis and autophagy. Further support for this notion has been provided by the finding that the anti-apoptotic protein Bcl-2 interacts with the autophagy mediator Beclin-1, and that this interaction blocks Beclin-1-mediated autophagy (79). In addition, Atg5 is cleaved by calpain, leading to a pro-apoptotic N-terminal fragment that associates with mitochondria (80).

Thus, there is growing evidence of an extensive interaction between apoptosis mediators and autophagy mediators. However, this does not necessarily imply that autophagic PCD is mechanistically similar to apoptotic PCD—indeed, the executioners operative in autophagic PCD (i.e., the caspase analogues) are not yet defined. One possibility is simply that runaway autophagy results in a lethal loss of functional organelles. Another is that autophagy upregulation leads to a disproportionate loss of protective proteins, such as catalase (81). If these are the predominant mechanisms underlying autophagic PCD, then it will be a much more indirect program of cell death than apoptosis. Furthermore, if such indirect mechanisms are involved, then cell death associated with autophagy may prove to be a diverse set of cell death programs involving both apoptotic and necrotic (and potentially other) mediators. However, it is still possible that direct executioners for autophagic PCD will be identified.

Many questions about autophagic PCD remain unanswered: if autophagy is indeed a cellular protective program that, like the UPR, at some point activates PCD, what is the signal for the "switch" to PCD initiation? How important is the role of autophagic PCD in ischemic neuronal cell death? Does autophagic PCD occur in vivo in the absence of apoptosis inhibition? Are there executioners analogous to caspases in autophagic PCD?

Other Cell Death Programs

In comparison to apoptosis, relatively little is known about autophagic PCD and even less is known about other nonapoptotic forms of PCD. Furthermore, most of what is known is based on morphological descriptions. Mechanistic requirements within type I include two general groups: caspase-dependent apoptosis (extrinsic and intrinsic, as noted above) and caspase-independent apoptosis. Types II (autophagic PCD, described immediately above) and III do not require caspase activation.

Type III PCD, which was subdivided by Clarke (6) into type A and B, is a "necrosis-like" form of PCD that includes swelling of ER and mitochondria, and an absence of typical apoptotic features such as apoptotic bodies and nuclear fragmentation. One process that leads to apparent type III PCD is the hyperactivation of the tyrosine kinase receptor insulin-like growth factor I receptor (IGFR). The nonapoptotic form of PCD induced in this way has been dubbed paraptosis (16). This form of PCD was found to require transcription and translation and to be indistinguishable morphologically from type III PCD, with swelling of the ER and mitochondria and an absence of apoptotic features. It is noteworthy that this morphology is observed in many cell deaths labeled necrotic and therefore the mechanistic implications of this particular morphological pattern remain an open question. Caspases are not activated during paraptosis and neither Bcl-2 nor caspase inhibitors block this form of PCD. However, inhibitors of extracellular signal-regulated kinase 2 (ERK2) (but not ERK1) have been shown to inhibit paraptosis (82), as has AIP-1/Alix. One of the ERK-discriminating targets—i.e., a substrate of ERK2 but not ERK1—is Nur77, and dominant negative Nur77 also blocks paraptosis. In addition, antisense oligonucleotides directed against JNK1 were partially inhibitory.

Induction of PCD by hyperactivation of a trophic factor receptor—essentially "trophotoxicity"—as observed in this paraptosis paradigm, is compatible with earlier observations that some trophic factors may increase neuronal cell death, for example, induced by excitotoxicity (83). Such an effect might potentially be protective against neoplasia, in that it may eliminate cells that would otherwise undergo autocrine loop-stimulated oncogenesis. The resulting program would reasonably be nonapoptotic, since trophic factor signaling inactivates apoptotic signaling.

Other forms of PCD have been described that do not fit the criteria for either type I, type II, or type III PCD. One example is a nonapoptotic, caspase-independent form of cell death that has been described by Driscoll et al. (84,85) in *Caenorhabditis elegans* that express mutant channel proteins such as mec-4 (d). A uniform, necrosis-like cell death is induced, characterized morphologically by membranous whorls lacking in other cell death types. This form of PCD is triggered by calcium entry, mediated by specific calpains and cathepsins, and inhibited by calreticulin. While it is possible that this alternative form of PCD will ultimately turn out to be mediated by one of the previously described pathways (e.g., given the cathepsin requirement, lysosomes may be involved, suggesting a relationship to autophagic PCD, perhaps as a final common pathway), the morphological characteristics suggest that it is indeed a fourth form of PCD, distinct from type I, II, or III.

A fifth apparent form of PCD has been described by the Dawsons and their colleagues (chapter 9), who demonstrated that a nonapoptotic form of cell death, parthanatos, is triggered by the activation of poly-(ADP-ribose) polymerase (PARP) and the consequent translocation of AIF from the mitochondria to the nucleus (81). AIF is a flavoprotein that is involved with DNA fragmentation along with endonuclease G and DNA fragmentation factor. This form of PCD is induced by agents that induce DNA damage, such as hydrogen peroxide, *N*-methyl-D-aspartate, and *N*-methyl-*N'*-nitro-*N*-nitrosoguanidine. PARP-dependent PCD displays a morphology and biochemistry (to the extent currently known) that is distinct from types I, II, and III PCD, and from the calcium-induced PCD described by Driscoll et al.

A possible sixth form of cell death, activated by ischemia, has been dubbed oncosis . An extensive literature on the morphological criteria for oncosis exists, but the biochemical pathway(s) of oncosis have not yet been described in detail. Oncosis is thought to be mediated by the failure of plasma membrane ionic pumps. One potential mediator of oncosis is a calpain-family protease [possibly a mitochondrial calpain (86)], which suggests that oncosis may turn out to be related to, or synonymous with, the calcium-activated necrosis-like cell death described by Driscoll et al.

"Aponecrosis" is a term that has been applied to a combination of apoptosis and necrosis (87). Many cytotoxins induce PCD at low concentrations, but apparently necrotic cell death at higher concentrations, presumably due to overwhelming the cellular homeostatic processes prior to completing the apoptotic cell death program. In fact, this is a common pattern seen with cellular toxins, from hydrogen peroxide and other oxidants to mitochondrial toxins such as antimycin A (87). However, it should be noted that the necrotic morphology associated with aponecrosis has not been proven to be nonprogrammatic, so it is still unknown whether aponecrosis represents a combination of apoptosis and a nonapoptotic form of PCD, as opposed to a combination of apoptosis and nonprogrammatic cell death.

Since so little is known mechanistically about these various forms of caspase-independent PCD, it is difficult to know at present how many will prove to employ similar biochemical pathways. In addition to those forms of PCD noted above, a few other forms have been described, for example, a caspase-independent form of PCD that features nuclear shrinkage (which occurs with apoptosis but not paraptosis or oncosis, for example) is dependent on phospholipase A2 (PLA2) (88). Shinzawa and Tsujimoto described a caspase-independent form of PCD induced by hypoxia and uninhibited by Bcl-2 or Apaf-1 reduction. Interestingly, they found that PLA2 inhibitors prevented both the nuclear atrophy and the PCD, and further that hypoxia led to an elevation in PLA2 activity as well as PLA2 nuclear translocation.

Another form of nonapoptotic PCD has been referred to as necroptosis (89). Tumor necrosis factor receptor I mediates both an apoptotic pathway (via FADD) and a nonapoptotic pathway, and inhibitors for this latter pathway have recently been developed (89).

Yet another potential form of PCD has been dubbed autoschizis. When some cancer cells (e.g., bladder carcinoma cells) are exposed to oxidants (in this paradigm, ascorbate and menadione), a form of cell death ensues that involves nuclear excision from the cytoplasm (90). The biochemical mediators of autoschizis are unknown.

Mitochondrial permeability may be involved in both apoptotic and nonapoptotic PCD. Ling et al. described a nonapoptotic form of PCD mediated by the mitochondrial permeability transition pore (91). In their studies, honokiol treatment of HL60, MCF-7, and HEK293 cell lines was shown to induce a nonapoptotic form of PCD unaccompanied by phosphatidylserine flipping, caspase activation, or internucleosomal DNA fragmentation, and uninhibited by zVAD.fmk, Bcl-2, or Bcl-X_L. This cell death was associated with the mitochondrial permeability transition pore regulated by cyclophilin D (CypD).

Apoptotic vs. Nonapoptotic Cell Death Programs: Comparison of Mechanisms

What of the biochemical pathways common to these different forms of PCD? In the intrinsic pathway of apoptosis, holocytochrome c and other PCD mediators are released from the intermembranous space of mitochondria secondary to outer membrane permeability that is induced by pro-apoptotic members of the Bcl-2 family such as Bax and Bak, in concert with BH3 proteins Bim or tBid.

However, mitochondrial proteins may also gain egress in association with the mitochondrial membrane permeability transition (MPT) (92). Whether by consequent swelling and rupture of the mitochondrial outer membrane or by another mechanism, activation of the MPT by calcium, oxidants, or other activators offers a Bcl-2-independent [or at least partially independent: Bax may interact with components of the MPT such as the adenine nucleotide translocator (93) and the voltage-dependent anion channel (94)], at least partially CypD-dependent, route for the release of mitochondrially-derived pro-apoptotic factors.

Beyond these two general categories of mitochondrial pro-apoptotic factor release, novel pathways may also exist, for example, work from Polster et al. showed that the release of AIF (95) from mitochondria requires a combination of mitochondrial membrane permeabilization (e.g., by tBid or by the MPT) and active calpain. This work suggests that an endogenous mitochondrial calpain may be involved in AIF release (96).

These results suggest that combinatorial paths to PCD may be dissected, for example, Bcl-2 inhibitable (i.e., BH1–3 mediated) versus independent (presumably related to MPT); caspase-dependent versus caspase-independent; calpain-dependent versus calpain-independent; AIF-dependent versus AIF-independent; PARP-dependent versus PARP-independent; and so on for other critical factors such as cathepsins, JNK, and the autophagy-mediating gene products (Atg6, etc.). Using this type of dissection, classical apoptosis would fall predominantly into three groups: caspase-dependent and Bcl-2 inhibitable (intrinsic pathway, and extrinsic pathway with amplification via the intrinsic pathway), caspase-dependent and Bcl-2 resistant (some extrinsic pathway paradigms without amplification, and some MPT activators that lead to caspase-dependent PCD), and caspase-independent, Bcl-2 inhibitable (e.g., some paradigms of ER stress (97), and intracellular pathogen–induced PCD (98).

In contrast, toxins that inactivate caspases directly or indirectly, such as diethylmaleate and buthionine sulfoximine, should induce PCD that is Bcl-2 inhibitable and caspase-independent (99). On the other hand, an increase in cytosolic calcium, such as occurs with the mec-4(d) mutants of *C. elegans* (85) could induce MPT, which would explain the Bcl-2 (ced-9) independence, and activate calpains, which would potentially inactivate caspases (100), compatible with the caspase independence of this form of PCD. As noted above, the cathepsin dependence suggests lysosomal involvement, and thus a potential relationship with type II PCD. Adding DNA damage to the calcium entry (e.g., with excitotoxicity or nitric oxide) should trigger a similar scenario with the addition of PARP activation with the combination of calcium-activated MPT and calpain activation explaining the AIF activation (96,101)

These alternative pathways share the common feature of caspase inhibition, whether direct (e.g., by zVAD.fmk or diethylmaleate) or indirect (e.g., via receptor tyrosine kinase or calpain activation). It will be of interest to determine how many inhibitors of these various pathways are expressed in neoplastic cells.

ACKNOWLEDGMENTS

This work was supported in part by NIH grants AG12282, NS45093, and NS33376, a grant from the Joseph Drown Foundation, and a grant from American Bioscience, Inc. I thank Molly Susag, Loretta Sheridan, and Rowena Abulencia for manuscript preparation.

REFERENCES

1. Studnicka FK. Die Parietalorgane. In: Oppel A, ed. Lehrbuch der vergleichende mikroskopischen Anatomie der Wirbeltiere. Jena: SG Fischer Verlag, 1905.
2. Levi-Montalcini R. The nerve growth factor: its mode of action on sensory and sympathetic nerve cells. Harvey Lect 1966; 60:217–259.
3. Lockshin RA, Williams CM. Programmed cell death. II. Endocrine potentiation of the breakdown of the intersegmental muscles of silkmoths. J Insect Physiol 1964; 10:643–649.
4. Tata JR. Requirement for RNA and protein synthesis for induced regression of the tadpole tail in organ culture. Dev Biol 1966; 13(1):77–94.
5. Kerr JF, Wyllie AH, Currie AR. Apoptosis: a basic biological phenomenon with wide-ranging implications in tissue kinetics. Br J Cancer 1972; 26(4):239–257.
6. Clarke PG. Developmental cell death: morphological diversity and multiple mechanisms. Anat Embryol 1990; 181(3):195–213.
7. Cunningham TJ. Naturally occurring neuron death and its regulation by developing neural pathways. Int Rev Cytol 1982; 74:163–186.
8. Dal Canto MC, Gurney ME. Development of central nervous system pathology in a murine transgenic model of human amyotrophic lateral sclerosis. Am J Pathol 1994; 145(6):1271–1279.
9. Majno G, Joris I. Apoptosis, oncosis, and necrosis. An overview of cell death. Am J Pathol 1995; 146(1):3–15.
10. Oppenheim RW. Naturally occurring cell death during neural development. Trends Neurosci 1985; 17:487–493.
11. Oppenheim RW. Cell death during development of the nervous system. Annu Rev Neurosci 1991; 14:453–501.
12. Pilar G, Landmesser L. Ultrastructural differences during embryonic cell death in normal and peripherally deprived ciliary ganglia. J Cell Biol 1976; 68(2):339–356.
13. Schwartz LM. The role of cell death genes during development. Bioessays 1991; 13(8):389–395.
14. Schweichel JU. [Electron microscopic studies on the degradation of the apical ridge during the development of limbs in rat embryos]. Z Anat Entwicklungsgesch 1972; 136(2):192–203.
15. Schweichel JU, Merker HJ. The morphology of various types of cell death in prenatal tissues. Teratology 1973; 7(3):253–266.
16. Sperandio S, De Belle I, Bredesen DE. An alternative, non-apoptotic form of programmed cell death. Proc Natl Acad Sci U S A 2000; 97(26):14376–14381.
17. Turmaine M, Raza A, Mahal A, et al. Nonapoptotic neurodegeneration in a transgenic mouse model of Huntington's disease. Proc Natl Acad Sci U S A 2000; 97(14):8093–8097.

18. Vande Velde C, Cizeau J, Dubik D, et al. BNIP3 and genetic control of necrosis-like cell death through the mitochondrial permeability transition pore. Mol Cell Biol 2000; 20(15):5454–5468.
19. Niquet J, Liu H, Wasterlain CG. Programmed neuronal necrosis and status epilepticus. Epilepsia 2005; 46(suppl 5):43–48.
20. Zong WX, Thompson CB. Necrotic death as a cell fate. Genes Dev 2006; 20(1): 1–15.
21. Salvesen GS, Dixit VM. Caspases: intracellular signaling by proteolysis. Cell 1997; 91(4):443–446.
22. Tajiri S, Oyadomari S, Yano S, et al. Ischemia-induced neuronal cell death is mediated by the endoplasmic reticulum stress pathway involving CHOP. Cell Death Differ 2004; 11(4):403–415.
23. Morishima N, Nakanishi K, Takenouchi H, et al. An endoplasmic reticulum stress-specific caspase cascade in apoptosis. Cytochrome c-independent activation of caspase-9 by caspase-12. J Biol Chem 2002; 277(37):34287–34294.
24. Rao RV, Castro-Obregon S, Frankowski H, et al. Coupling endoplasmic reticulum stress to the cell death program. An Apaf-1-independent intrinsic pathway. J Biol Chem 2002; 277(24):21836–21842.
25. Rao RV, Hermel E, Castro-Obregon S, et al. Coupling endoplasmic reticulum stress to the cell death program. Mechanism of caspase activation. J Biol Chem 2001; 276 (36):33869–33874.
26. Rao RV, Peel A, Logvinova A, et al. Coupling endoplasmic reticulum stress to the cell death program: role of the ER chaperone GRP78. FEBS Lett 2002; 514(2–3):122–128.
27. Yuan J, Yankner BA. Caspase activity sows the seeds of neuronal death. Nat Cell Biol 1999; 1(2):E44–E45.
28. Green DR, Kroemer G. Pharmacological manipulation of cell death: clinical applications in sight? J Clin Invest 2005; 115(10):2610–2617.
29. Kuwana T, Bouchier-Hayes L, Chipuk JE, et al. BH3 domains of BH3-only proteins differentially regulate Bax-mediated mitochondrial membrane permeabilization both directly and indirectly. Mol Cell 2005; 17(4):525–535.
30. Kuwana T, Mackey MR, Perkins G, et al. Bid, Bax, and lipids cooperate to form supramolecular openings in the outer mitochondrial membrane. Cell 2002; 111(3): 331–342.
31. Stoka V, Turk B, Schendel SL, et al. Lysosomal protease pathways to apoptosis. Cleavage of bid, not pro-caspases, is the most likely route. J Biol Chem 2001; 276(5):3149–3157.
32. Guo B, Zhai D, Cabezas E, et al. Humanin peptide suppresses apoptosis by interfering with Bax activation. Nature 2003; 423(6938):456–461.
33. Zhai D, Luciano F, Zhu X, et al. Humanin binds and nullifies Bid activity by blocking its activation of Bax and Bak. J Biol Chem 2005; 280(16):15815–15824.
34. Luciano F, Zhai D, Zhu X, et al. Cytoprotective peptide humanin binds and inhibits pro-apoptotic Bcl-2/Bax family protein BimEL. J Biol Chem 2005; 280(16):15825–15835.
35. Culmsee C, Zhu X, Yu QS, et al. A synthetic inhibitor of p53 protects neurons against death induced by ischemic and excitotoxic insults, and amyloid beta-peptide. J Neurochem 2001; 77(1):220–228.

36. Leker RR, Aharonowiz M, Greig NH, et al. The role of p53-induced apoptosis in cerebral ischemia: effects of the p53 inhibitor pifithrin alpha. Exp Neurol 2004; 187(2):478–486.
37. Li H, Kolluri SK, Gu J, et al. Cytochrome c release and apoptosis induced by mitochondrial targeting of nuclear orphan receptor TR3. Science 2000; 289(5482): 1159–1164.
38. Kluck RM, Bossy-Wetzel E, Green DR, et al. The release of cytochrome c from mitochondria: a primary site for Bcl-2 regulation of apoptosis. Science 1997; 275(5303):1132–1136.
39. Boatright KM, Renatus M, Scott FL, et al. A unified model for apical caspase activation. Mol Cell 2003; 11(2):529–541.
40. Deveraux QL, Takahashi R, Salvesen GS, et al. X-linked IAP is a direct inhibitor of cell-death proteases. Nature 1997; 388(6639):300–304.
41. Holley CL, Olson MR, Colon-Ramos DA, et al. Reaper eliminates IAP proteins through stimulated IAP degradation and generalized translational inhibition. Nat Cell Biol 2002; 4(6):439–444.
42. Kurakin A, Bredesen DE. An unconventional IAP-binding motif revealed by target-assisted iterative screening (TAIS) of the BIR3-cIAP1 domain. J Mol Recognit 2007; 20(1):39–50.
43. Du C, Fang M, Li Y, et al. Smac, a mitochondrial protein that promotes cytochrome c-dependent caspase activation by eliminating IAP inhibition. Cell 2000; 102(1): 33–42.
44. Verhagen AM, Ekert PG, Pakusch M, et al. Identification of DIABLO, a mammalian protein that promotes apoptosis by binding to and antagonizing IAP proteins. Cell 2000; 102(1):43–53.
45. Martins LM, Iaccarino I, Tenev T, et al. The serine protease Omi/HtrA2 regulates apoptosis by binding XIAP through a reaper-like motif. J Biol Chem 2002; 277(1):439–444.
46. Suzuki Y, Imai Y, Nakayama H, et al. A serine protease, HtrA2, is released from the mitochondria and interacts with XIAP, inducing cell death. Mol Cell 2001; 8(3): 613–621.
47. Chan FK, Chun HJ, Zheng L, et al. A domain in TNF receptors that mediates ligand-independent receptor assembly and signaling. Science 2000; 288(5475): 2351–2354.
48. Siegel RM, Frederiksen JK, Zacharias DA, et al. Fas preassociation required for apoptosis signaling and dominant inhibition by pathogenic mutations. Science 2000; 288(5475):2354–2357.
49. Muzio M, Chinnaiyan AM, Kischkel FC, et al. FLICE, a novel FADD-homologous ICE/CED-3-like protease, is recruited to the CD95 (Fas/APO-1) death-inducing signaling complex. Cell 1996; 85(6):817–827.
50. Boatright KM, Deis C, Denault JB, et al. Activation of caspases-8 and -10 by FLIP(L). Biochem J 2004; 382(pt 2):651–657.
51. Fuentes-Prior P, Salvesen GS. The protein structures that shape caspase activity, specificity, activation and inhibition. Biochem J 2004; 384(pt 2):201–232.
52. Friedlander RM, Gagliardini V, Hara H, et al. Expression of a dominant negative mutant of interleukin-1 beta converting enzyme in transgenic mice prevents neuronal cell death induced by trophic factor withdrawal and ischemic brain injury. J Exp Med 1997; 185(5):933–940.

53. Bredesen DE, Ye X, Tasinato A, et al. p75NTR and the concept of cellular dependence: seeing how the other half die. Cell Death Differ 1998; 5(5):365–371.
54. Yao R, Cooper GM. Requirement for phosphatiylinositol-3 kinase in the prevention of apoptosis by nerve growth factor. Science 1995; 267:2003–2006.
55. Ellerby LM, Hackam AS, Propp SS, et al. Kennedy's disease: caspase cleavage of the androgen receptor is a crucial event in cytotoxicity. J Neurochem 1999; 72(1): 185–195.
56. Barrett GL, Bartlett PF. The p75 nerve growth factor receptor mediates survival or death depending on the stage of sensory neuron development. Proc Natl Acad Sci U S A 1994; 91(14):6501–6505.
57. Barrett GL, Georgiou A. The low-affinity nerve growth factor receptor p75NGFR mediates death of PC12 cells after nerve growth factor withdrawal. J Neurosci Res 1996; 45:117–128.
58. Bordeaux MC, Forcet C, Granger L, et al. The RET proto-oncogene induces apoptosis: a novel mechanism for Hirschsprung disease. EMBO J 2000; 19(15): 4056–4063.
59. Bredesen DE, Rabizadeh S. p75NTR and apoptosis: Trk-dependent and Trk-independent effects. Trends Neurosci 1997; 20(7):287–290.
60. Forcet C, Ye X, Granger L, et al. The dependence receptor DCC (deleted in colorectal cancer) defines an alternative mechanism for caspase activation. Proc Natl Acad Sci U S A 2001; 98(6):3416–3421.
61. Llambi F, Causeret F, Bloch-Gallego E, et al. Netrin-1 acts as a survival factor via its receptors UNC5H and DCC. EMBO J 2001; 20(11):2715–2722.
62. Mehlen P, Rabizadeh S, Snipas SJ, et al. The DCC gene product induces apoptosis by a mechanism requiring receptor proteolysis. Nature 1998; 395(6704):801–804.
63. Rabizadeh S, Bredesen DE. Is p75NGFR involved in developmental neural cell death? Dev Neurosci 1994; 16(3–4):207–211.
64. Rabizadeh S, Oh J, Zhong LT, et al. Induction of apoptosis by the low-affinity NGF receptor. Science 1993; 261(5119):345–348.
65. Stupack DG, Puente XS, Boutsaboualoy S, et al. Apoptosis of adherent cells by recruitment of caspase-8 to unligated integrins. J Cell Biol 2001; 155(3):459–470.
66. Bredesen DE, Mehlen P, Rabizadeh S. Apoptosis and dependence receptors: a molecular basis for cellular addiction. Physiol Rev 2004; 84(2):411–430.
67. Mah SP, Zhong LT, Liu Y, et al. The protooncogene bcl-2 inhibits apoptosis in PC12 cells. J Neurochem 1993; 60(3):1183–1186.
68. Bredesen DE. Keeping neurons alive: the molecular control of apoptosis (part I). Neuroscientist 1996; 2:181–190.
69. Mazelin L, Bernet A, Bonod-Bidaud C, et al. Netrin-1 controls colorectal tumorigenesis by regulating apoptosis. Nature 2004; 431(7004):80–84.
70. Bredesen DE, Mehlen P, Rabizadeh S. Receptors that mediate cellular dependence. Cell Death Differ 2005; 12(8):1031–1043.
71. Tanikawa C, Matsuda K, Fukuda S, et al. p53RDL1 regulates p53-dependent apoptosis. Nat Cell Biol 2003; 5(3):216–223.
72. Levine B, Yuan J. Autophagy in cell death: an innocent convict? J Clin Invest 2005; 115(10):2679–2688.
73. Shimizu S, Kanaseki T, Mizushima N, et al. Role of Bcl-2 family proteins in a nonapoptotic programmed cell death dependent on autophagy genes. Nat Cell Biol 2004; 6(12):1221–1228.

74. Yue Z, Jin S, Yang C, et al. Beclin 1, an autophagy gene essential for early embryonic development, is a haploinsufficient tumor suppressor. Proc Natl Acad Sci U S A 2003; 100(25):15077–15082.
75. Lum JJ, Bauer DE, Kong M, et al. Growth factor regulation of autophagy and cell survival in the absence of apoptosis. Cell 2005; 120(2):237–248.
76. Yu L, Alva A, Su H, et al. Regulation of an ATG7-beclin 1 program of autophagic cell death by caspase-8. Science 2004; 304(5676):1500–1502.
77. Yu L, Wan F, Dutta S, et al. Autophagic programmed cell death by selective catalase degradation. Proc Natl Acad Sci U S A 2006; 103(13):4952–4957.
78. Madden DT, Egger L, Bredesen DE. A calpain-like protease inhibits autophagic cell death. Autophagy 2007; 3(5):519–522.
79. Pattingre S, Tassa A, Qu X, et al. Bcl-2 antiapoptotic proteins inhibit Beclin 1-dependent autophagy. Cell 2005; 122(6):927–939.
80. Yousefi S, Perozzo R, Schmid I, et al. Calpain-mediated cleavage of Atg5 switches autophagy to apoptosis. Nat Cell Biol 2006; 8(10):1124–1132.
81. Yu SW, Wang H, Poitras MF, et al. Mediation of poly(ADP-ribose) polymerase-1-dependent cell death by apoptosis-inducing factor. Science 2002; 297(5579): 259–263.
82. Sperandio S, Poksay K, De Belle I, et al. Paraptosis: mediation by MAP kinases and inhibition by AIP-1/Alix. Cell Death Differ 2004; 11(10):1066–1075.
83. Koh JY, Gwag BJ, Lobner D, et al. Potentiated necrosis of cultured cortical neurons by neurotrophins. Science 1995; 268(5210):573–575.
84. Syntichaki P, Xu K, Driscoll M, et al. Specific aspartyl and calpain proteases are required for neurodegeneration in *C. elegans*. Nature 2002; 419(6910):939–944.
85. Bianchi L, Gerstbrein B, Frokjaer-Jensen C, et al. The neurotoxic MEC-4(d) DEG/ENaC sodium channel conducts calcium: implications for necrosis initiation. Nat Neurosci 2004; 7(12):1337–1344.
86. Liu X, Van Vleet T, Schnellmann RG. The role of calpain in oncotic cell death. Annu Rev Pharmacol Toxicol 2004; 44:349–370.
87. Formigli L, Papucci L, Tani A, et al. Aponecrosis: morphological and biochemical exploration of a syncretic process of cell death sharing apoptosis and necrosis. J Cell Physiol 2000; 182(1):41–49.
88. Shinzawa K, Tsujimoto Y. PLA2 activity is required for nuclear shrinkage in caspase-independent cell death. J Cell Biol 2003; 163(6):1219–1230.
89. Degterev A, Huang Z, Boyce M, et al. Chemical inhibitor of nonapoptotic cell death with therapeutic potential for ischemic brain injury. Nat Chem Biol 2005; 1(2):112–119.
90. Gilloteaux J, Jamison JM, Arnold D, et al. Cancer cell necrosis by autoschizis: synergism of antitumor activity of vitamin C: vitamin K3 on human bladder carcinoma T24 cells. Scanning 1998; 20(8):564–575.
91. Li L, Han W, Gu Y, et al. Honokiol induces a necrotic cell death through the mitochondrial permeability transition pore. Cancer Res 2007; 67(10):4894–4903.
92. Novgorodov SA, Gudz TI. Permeability transition pore of the inner mitochondrial membrane can operate in two open states with different selectivities. J Bioenerg Biomembr 1996; 28(2):139–146.
93. Brenner C, Cadiou H, Vieira HL, et al. Bcl-2 and Bax regulate the channel activity of the mitochondrial adenine nucleotide translocator. Oncogene 2000; 19(3): 329–336.

94. Adachi M, Higuchi H, Miura S, et al. Bax interacts with the voltage-dependent anion channel and mediates ethanol-induced apoptosis in rat hepatocytes. Am J Physiol Gastrointest Liver Physiol 2004; 287(3):G695–G705.
95. Susin SA, Lorenzo HK, Zamzami N, et al. Molecular characterization of mitochondrial apoptosis-inducing factor. Nature 1999; 397(6718):441–446.
96. Polster BM, Basanez G, Etxebarria A, et al. Calpain I induces cleavage and release of apoptosis-inducing factor from isolated mitochondria. J Biol Chem 2005; 280(8): 6447–6454.
97. Egger L, Schneider J, Rheme C, et al. Serine proteases mediate apoptosis-like cell death and phagocytosis under caspase-inhibiting conditions. Cell Death Differ 2003; 10(10):1188–1203.
98. Perfettini JL, Reed JC, Israel N, et al. Role of Bcl-2 family members in caspase-independent apoptosis during Chlamydia infection. Infect Immun 2002; 70(1):55–61.
99. Kane DJ, Ord T, Anton R, et al. Expression of bcl-2 inhibits necrotic neural cell death. J Neurosci Res 1995; 40(2):269–275.
100. Chua BT, Guo K, Li P. Direct cleavage by the calcium-activated protease calpain can lead to inactivation of caspases. J Biol Chem 2000; 275(7):5131–5135.
101. Lankiewicz S, Marc Luetjens C, Truc Bui N, et al. Activation of calpain I converts excitotoxic neuron death into a caspase-independent cell death. J Biol Chem 2000; 275(22):17064–17071.

6

Caspase-Independent Apoptotic Cell Death

Lotti Egger
Buck Institute for Age Research, Novato, California, U.S.A.

INTRODUCTION

One of the most important conclusions of the last decade in cell death research led to the recent acknowledgment that there are many different regulated pathways leading to cell death. However, generally accepted definitions and nomenclatures for all of the different cell death pathways are yet to emerge. In the meantime we can start by defining the common endpoint of all described pathways by determining when a cell can be considered as dead. The Nomenclature Committee on Cell Death (NCCD) suggested the following molecular or morphological criteria: (*i*) the cell has lost the integrity of the plasma membrane, as defined by vital dyes in vitro; (*ii*) the cell including its nucleus has undergone complete fragmentation into discrete bodies (which are frequently referred to as "apoptotic bodies"); and/or (*iii*) its corpse (or its fragments) have been engulfed by an adjacent cell in vivo (1). There is a limited number of well-studied mechansim-dependent definitions of cell death types for which morphological characteristics have been defined, and apoptosis is the most prominent and best studied mechanism by which a cell can die.

APOPTOSIS

Cells dying by apoptosis undergo a stereotypical series of biochemical and morphological changes. Apoptotic cells shrink, round up, and detach from the culture plate (Fig. 1B). Cytochrome c is released from mitochondria into the cytoplasma and can be seen by fluorescent microscopy in a diffuse pattern, in contrast to an elongated pattern of mitochondrial localized cytochrome c of healthy cells (Fig. 1G, H). Chromatin condenses to compact and apparently simple geometric (globular, crescent-shaped) forms (2) (Fig. 1D, E) and is cleaved into nucleosomal-sized fragments (3). In apoptotic cells, phosphatidylserine (PS) molecules are localized to the outer leaflet of the plasma membrane. This phenomenon can be visualized, with fluorescence-coupled (e.g., green fluorescent protein, GFP) annexin V, which binds specifically to PS (Fig. 1K). The cell membrane of apoptotic cells remains intact, preventing the release of intracellular compounds to the surrounding tissues. Propidium iodide (PI) is a membrane impermeable DNA stain. Thus, apoptotic cells with an intact cell membrane are usually PI negative.

Apoptosis is an irreversible process, which is precisely and tightly regulated and is often associated with the activation of a highly specific class of proteases called caspases (cystein aspartic acid proteases). Caspases are responsible for the earlier mentioned morphological changes of the dying cells. At the same time extracellular phagocytic (4) signals are exposed on the cell surface, provoking the uptake of the apoptotic cell by phagocytes. In this scenario, cellular components are not released into the surrounding tissue, thus avoiding an inflammatory response. Apoptosis is central for embryonic development and for tissue homeostasis in multicellular organisms (5,6). Dysregulation of apoptosis plays an important role in many human diseases, for example, reduced cell death may provoke tumor formation and resistance to chemotherapy (7), whereas, on the other hand, excessive apoptosis may be associated with degenerative illness like Alzheimer or autoimmune diseases.

APOPTOSIS IS A DEATH PROGRAM CONSERVED THROUGHOUT EVOLUTION

The molecular mechanisms of apoptosis were first described in the nematode *Caenorhabditis elegans* (*C. elegans*) (8), where the fate of 1090 cells arising during development is regulated and determined (9,10). Initially two death genes, *ced-3* (cell-death abnormal) and *ced-4*, were identified to be required for this developmental cell death (8). CED-3 was determined to be a caspase, and CED-4 an ATPase activating the protease CED-3. *ced-9*, a third gene, which suppresses cell suicide, was discovered subsequently, by Hengartner and Horvitz (11). The absence of *ced-3* and *ced-4* suppresses cell death during the embryonic development of the worm. This leads to the survival of 131 somatic cells that are otherwise "programmed to die." The CED-9 protein inhibits the action of the

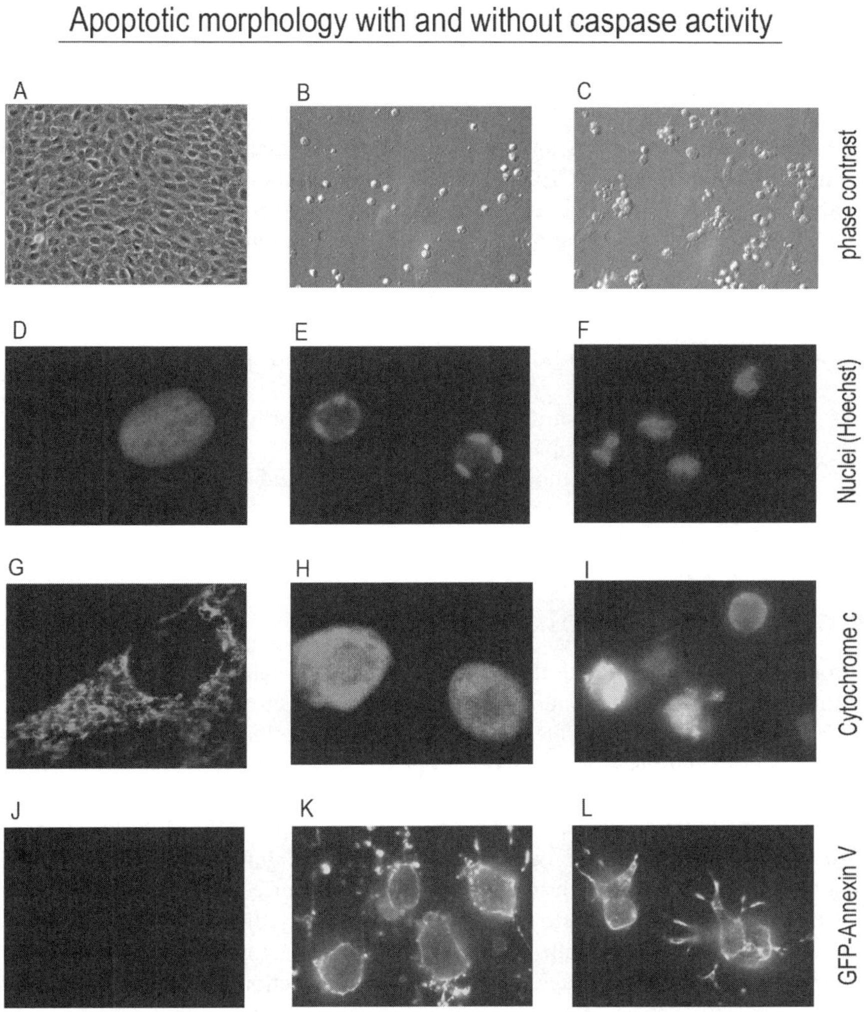

Figure 1 Morphological changes of apoptotic cells. (**A–C**): Viable cells form a monolayer, whereas apoptotic cells with and without caspase activity shrink, round up, form blebs at the cell surface, and finally detach from the culture plate. (**D–F**): In viable cells, nuclei are stained homogeneously with the DNA dye Hoechst. In apoptotic cells, chromatin gets compact and fragmented, visible as spherical masses at the nuclear periphery. (**G–I**): In viable cells, cytochrome c has a spaghetti-like pattern. In apoptotic cells with and without caspase activity, cytochrome c is released from mitochondria, detectable by immunofluorescence as a diffuse cytosolic staining. (**J–L**): Viable cells do not expose PS at the cellular surface. However, on apoptotic cells PS is exposed, and is detectable with GFP-Annexin V. (**A–C**): light microscopy, 200× magnification; (**D–L**): fluorescence microscopy, 1000× magnification. *Abbreviations*: PS, phosphatidylserine; GFP, green fluorescent protein.

CED-3, CED-4 death machinery by directly binding to CED-4, thus preventing it from activating CED-3. In 1998, Horvitz and colleagues (12) isolated a third pro-death gene from *C. elegans*, called egl-1 (egg-laying defective-1), which contains a BH3 domain (Bcl-2 homology domain 3). EGL-1 induces apoptosis of hermaphrodite-specific neurons by binding to and neutralizing CED-9, hence indirectly activating the CED-4/CED-3 death machinery. When CED-9 is overexpressed (or contains a gain-of-function mutation) it sequesters both EGL-1 and CED-4, therefore preventing CED-3-mediated programmed cell death (PCD).

APOPTOSIS IN MAMMALIAN SYSTEMS

In mammals, apoptosis can be divided into three major phases:

- *Initiation* after a death stimulus (extrinsic pathway) or after receptor engagement (intrinsic pathway)
- *Execution* with or without caspase activation and degradation of cellular compounds
- *Removal* of apoptotic cells by phagocytosis

INITIATION OF APOPTOSIS

Apoptosis can be initiated through two principal pathways. The "extrinsic pathway" is activated by binding of tumor necrosis factor (TNF)-like cytokines, such as CD95L, TNF-α, or TRAIL, to their cognate cell surface receptors (TNF-receptors, CD95/Fas, or TRAIL-receptor, respectively). The subsequent recruitment of adaptor molecules like Fas-associating protein with death domain (FADD) and initiator caspase zymogens like pro-caspase-8 leads to the formation of a death-inducing signaling complex (DISC). Initiator caspase-8 forms active dimers and subsequently cleaves downstream effector caspases, inducing the execution phase of apoptosis. The second pathway, the intrinsic pathway, is induced after stimulation with a variety of different factors like DNA damaging agents, chemotherapeutic drugs, γ-irradiation, viral infections, hypoxia, heat shock, reactive oxygen species (ROS), lipopolysaccharide (LPS), and anoikis. These stimuli lead to the mitochondrial outer membrane permeabilization (MOMP) and, by consequence, to the release of apoptogenic factors like cytochrome c, apoptosis-inducing factor (AIF) and Smac/DIABLO into the cytoplasm. Cytochrome c release induces the formation of an initiator complex called the apoptosome. The apoptosome has a heptameric structure containing cytochrome c molecules in its center, bound to the adaptor molecule apoptosis protease-activating factor 1 (Apaf-1). Apaf-1 interacts via its N-terminal caspase recruitment domain (CARD) to the CARD of pro-caspase-9. Thereby caspase-9 is activated by forming a dimer and cleaves downstream effector caspases, starting the effector phase. In order to gain full proteolytic activity, initiator caspases need to be recruited to the DISC (caspase-8) or the apoptosome (caspase-9) (13,14).

EXECUTION OF APOPTOSIS

Both initiator pathways converge on a common execution program, by which effector caspases are cleaved and activated by the upstream initiator caspases. In contrast to initiator caspases, effector caspases-3, -7, and -6 cleavage is crucial for full enzymatic activity. These enzymes are present as inactive dimers in the cytosols of healthy cells and require proteolytic cleavage for activation.

DEGRADATION AND REMOVAL OF APOPTOTIC CELLS

The final phase of apoptosis in vivo is the "silent" removal of the cells by phagocytes. The significance of this event has been underestimated for many years (4). The phagocytic clearance of cells dying by apoptosis is much more than waste disposal. Moreover, this process reflects the physiological significance of cell death in higher organisms (4). Depending on the context, the removal of apoptotic cells by phagocytes suppresses inflammation, modulates the macrophage-directed deletion of host cell or invading parasites, and critically regulates immune responses (4). The uptake of apoptotic cells actively suppresses the secretion of pro-inflammatory mediators such as TNF-α from activated macrophages. Safe clearance might therefore be doubly beneficial in inflammatory responses, preventing the secondary necrosis of apoptotic cells, with asscociated uncontrolled release of injurious contents, and reducing pro-inflammatory macrophages (4).

APOPTOSIS MEDIATED BY CASPASES

Caspases belong to a conserved family of proteases that are not only involved in metazoan PCD but also in inflammation. At least 7 of the 14 known caspases play important roles in apoptosis. Caspases involved in apoptosis are represented by members of the CED-3 group of caspases, which can be divided into two groups: initiator caspases (including caspase-2, -8, -9, -10) and effector caspases (including caspase-3, -6, -7).

A characteristic of initiator caspases is an extended N-terminal prodomain (>90 amino acids) (15), containing binding sites like death effector domain (DED) in caspase-8 and CARD domains in caspase-9, for homotypic interaction with adaptor molecules. Effector caspases contain only 20 to 30 residues in their prodomain (15). Caspases are produced as inactive zymogens. Initiator caspase activation can be achieved by a series of polyprotein complexes such as the inflammasome (16), the piddosome (17), DISC (18), and the apoptosome (19). Effector caspases-3, -7, and -6 are cleaved by upstream caspases at the junction between the large and the small subunit. The prodomain is removed by a second autocleavage process (20), leading to full catalytic activity.

CASPASE INHIBITION

Caspases exist usually in cells as inactive zymogens and are only activated by proteolytic cleavage (with the exception of initiator caspases, which are activated by dimerization rather than proteolysis). Activation of caspases is a crucial step toward the commitment of a cell to die. Thus, caspase inhibition is vital for cell survival. In order to avoid uncontrolled caspase activation, cells possess dedicated mechanisms for direct caspase inhibition called inhibitors of apoptosis proteins (IAPs). There are several IAP-like molecules known in humans, and at least four of them have been reported to inhibit caspases directly (21). IAPs interact via a homology domain (BIR domain: baculoviral IAP repeat) with caspases. Moreover, IAP family proteins have been identified by BIR domains in a variety of organisms from viruses to mammals (22). Many viruses have elaborated sophisticated mechanisms to avoid host cell death. Direct caspase inhibition is one among others. Analysis of the baculovirus genome led to the identification of caspase inhibitor p35. p35 can inhibit most mammalian caspases but not caspase-9. p35 is cleaved by caspase-8 and subsequently forms a covalent bond with it (23). The cytokine response modifier (crmA) from cowpox virus functions in a similar way as p35 on caspase-1 and caspase-8. Thus, inflammatory reactions and host cell apoptosis from the extrinsic pathway are inhibited at the same time.

SYNTHETIC CASPASE INHIBITORS

Commercially available specific caspase inhibitors often consist of a tetrapeptidic sequence like benzyloxycarbonyl-DEVD-fluoromethylketone (z-DEVD.fmk, DEVD), the inhibitor of the effector caspases (-3, -7) or z-IETD.fmk, the inhibitor of caspase-8. The tetrapeptidic sequence mimics the prefered recognition motif of a particular caspase. However, these compounds were found to be of low specificity (24). In contrast, general caspase inhibitors like z-VAD.fmk or z-D.DCB have only a sequence of three or fewer amino acids, lacking the amino acid needed for specific substrate recognition. The benzyloxycarbonyl group (z-, also known as BOC, R&D systems) at the N-terminus of the peptide sequence exhibits enhanced cellular permeability. Therefore, these inhibitors can be used in vivo as well as in vitro. Nevertheless, cell permeability was found to be of low efficiency in some cases (25,26). Linkage of the peptide sequence to aldehydes (–CHO), nitriles, and ketones form a reversible thiohemiacetal with the thiol group of the active-site cysteine. In contrast, diazomethyl ketones, acylomethyl ketones, and halomethylketones like fluoromethylketones (fmk) linked to the peptide, bind irreversibly to caspases. Both reversible and irreversible peptide-based inhibitors have been reported to be effective in animal models of stroke, myocardial ischemia/reperfusion injury, liver disease, and traumatic brain injury (27–32). Additionally z-VAD.fmk was found to be a competitive and irreversible inhibitor of caspases-1, -2, -3, -4, -5, -6, -7, -8, -9, -10 in vitro (24). However in other reports caspase inhibition with z-VAD.fmk did not block membrane blebbing and cells could not be rescued (33). Apoptosis induced by the overexpression of Bax was also not inhibited by z-VAD.fmk treatment, although caspase-3 activation and nuclear fragmentation were clearly blocked (34,35). Despite certain drawbacks, pharmacological inhibitors are a convenient method to study caspase-independent cell death (CICD) pathways.

REGULATORS OF APOPTOSIS: THE Bcl-2 FAMILY OF PROTEINS

As known for CED-9 in *C. elegans*, the Bcl-2 family of proteins in higher eukaryotes are central regulators of CED-3/caspase activation. They are therefore considered as checkpoints through which survival and death signals must pass before cell fate can be determined (36,37). In contrast to *C. elegans*, the Bcl-2 family in higher eukaryotes is vast (at least 20 members) and can be divided into the CED-9-like anti-apoptotic members Bcl-2, Bcl-X_L (38), Bcl-w (39), A1 (40), Mcl-1 (41), and the pro-apoptotic members such as Bax and Bak. Bcl-2 and Bax-like subfamilies have 3 to 4 homology domains (Bcl-2 homology domains, BH), which mediate protein-protein interactions (Fig. 2). While Bcl-2-like proteins protect cells from apoptosis, Bax-like are required for apoptosis. The latter has been shown in Bax/Bak double knockout (DKO) mice that are embryonic lethal and whose cells are resistant to a variety of apoptotic stimuli (42). In addition, higher eukaryotes contain a family of EGL-1 like proteins having only the BH3 domain in common with the other two groups—so-called BH3-only proteins such as Noxa and Puma.

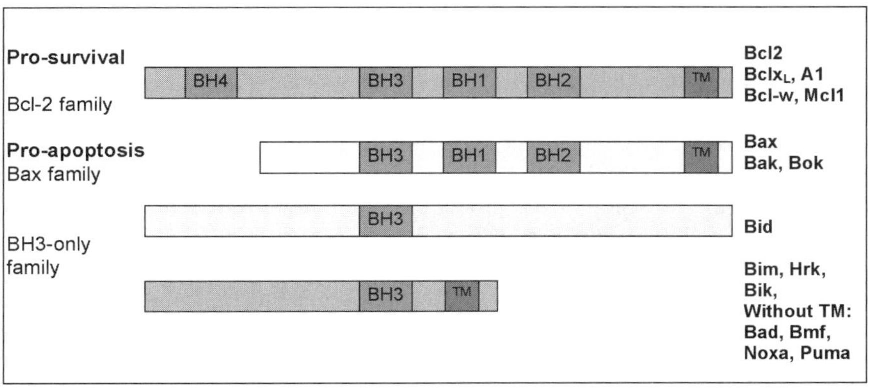

Figure 2 Bcl-2 family proteins are divided into three subfamilies: pro-survival family, pro-apoptotic Bax-family, and pro-apoptotic BH3-only family (this family is further divided into derepressors such as Puma and Noxa, and activators such as tBid and Bim). BH1–4 are Bcl-2 homology domains that are most highly conserved regions among family members. TM are transmembrane domain, caboxy-terminal hydrophobic domain that is required for the association with intracellular membranes (37). *Abbreviation*: BH3, Bcl-2 homology domain 3.

FUNCTIONS OF THE Bcl-2 FAMILY OF PROTEINS

Apoptosis is initiated when Bcl-2 and its pro-survival relatives are engaged, following a death-inducing stimulus, by pro-apoptotic BH3-only proteins via interaction of their BH3 domains with a groove on the Bcl-2-like proteins (43). The BH3 proteins Bim and Puma were found to bind and neutralize all known anti-apoptotic Bcl-2 proteins with a similar affinity, whereas Bad and Bmf preferentialy bound to Bcl-2, Bcl-X_L, and Bcl-w, Noxa to A1 and Mcl-1, and Bik and Hrk to Bcl-X_L, Bcl-w, and A1 (43). In surviving cells, Bcl-2-like factors are present in sufficient amounts to sequester pro-apoptotic factors such as Bax-like proteins or BH3-only proteins (37).

REGULATING THE REGULATORS: INACTIVATION OF BH3 PROTEINS BY PROTEASOMAL DEGRADATION

In response to apoptotic stimuli, BH3-only proteins are transcriptionally induced (like EGL-1 in *C. elegans*), post-translationally modified or otherwise activated (44). By consequence, the BH3-only proteins leave their original intracellular site and migrate to mitochondria or the endoplasmic reticulum (ER) to sequester Bcl-2-like proteins (45,46). Neutralization of anti-apoptotic Bcl-2 proteins is only one mechanism that leads to the activation of the cell death program. Recent studies have shown that proteasomal activity also regulates the cell death machinery by degrading anti-apoptotic Bcl-2 proteins. Regulation of apoptosis was achieved by specific upregulation of Mule/ARF-BP1 (Mcl-1 ubiquitin ligase E3), a BH3-only E3 ubiquitin ligase that was shown to be responsible for the polyubiquitination of anti-apoptotic Mcl-1, thereby promoting cell death. Mcl-1 degradation was blocked by proteasome inhibitors (47). Another mechanism of proteasomal regulation of cell death has been shown for the MEKK1-related protein-X (MEX). MEX undergoes self-ubiqitination as a consequence of death receptor (CD95, DR3, and DR4) induction, promoting cell death. This effect was abolished in the presence of the proteasome inhibitor MG132 (48). Moreover, our recent experiments in ER stress-induced apoptosis with Brefeldin A, Tunicamycin, or Thapsigargin have shown that cell death was blocked by the specific proteasome inhibitors Lactacystin, Epoxomycin, and MG262, whereas the general caspase inhibitors z-VAD.fmk and Q-VD failed to do so. Proteasomal inhibition led to the accumulation of the anti-apoptotic proteins Bcl-2, Bcl-X_L, and Bcl-w, as tested by Western blot analysis, suggesting that the inhibition of PCD by proteasome inhibitors acts upstream of mitochondrial cytochrome c release (49). The proteasomal degradation machinery is also used by the bacteria *Chlamydia trachomatis* for inhibition of host cell death. Cell protection is achieved by promoting the degradation of the pro-apoptotic Bcl-2 proteins Bim, Puma, and Bad via an as yet unknown mechansim that could be reversed by inhibition of the proteasome (50).

CASPASE-INDEPENDENT CELL DEATH

Since the discovery that CED-3 is required for cell death in *C. elegans*, it has been accepted that CED-3 related proteases (caspases) are essential for apoptosis. However, based on many studies in the field, it has become clear that caspase activation is not an absolute requirement for all forms of PCD, or even for all paradigms of apoptosis (51).

CICD is defined as the loss of cell viability that is induced by pro-apoptotic conditions, and which proceeds despite the inhibition or disruption of caspase function. Cell death that is caused by excessive damage that results in necrosis is excluded (51).

CICD occurs most often when caspase activity is genetically disrupted or otherwise (e.g., biochemically) inhibited. The morphological changes observed in these conditions are variable in contrast to the stereotypical changes occuring when caspases are activated. Cells undergoing CICD do not show substantial caspase-dependent chromatin condensation and DNA fragmentation (Fig. 1F). Nevertheless, the nucleus appears altered, with some chromatin condensation. Loss of mitochondrial membrane potential most often occurs at a slower rate, although cytochrome c release occurs in the absence of caspase activity (Fig. 1I) (51). Exposure of PS prior to loss of membrane integrity can also occur in the absence of caspase activity (Fig. 1L).

MOMP, which is controlled by the Bcl-2 family of proteins, seems to be central to most caspase-independent death pathways (52), as most of the pathways upstream of MOMP are caspase-independent (53–56). This means that Bcl-2 family proteins control both caspase-dependent and at least some caspase-independent signaling pathways, strengthening their role in cell death regulation.

CASPASE-INDEPENDENT SIGNALING PATHWAYS LEADING TO MOMP

Apoptosis induced via death receptor [Fig. 3(1)] activation is in most cases caspase-dependent. However, some members of the TNF superfamily (e.g., TNF itself) can also trigger caspase-independent PCD (57). Activation-induced cell death (AICD) of human and murine T cells can also occur in the presence of the pan-caspase inhibitor z-VAD.fmk (58,59). Release of cathepsins [Fig. 3(1)], especially the cysteine cathepsins B and L and the aspartyl cathepsin D, from the lysosomes were found to be responsible for CICD in response to death receptor stimulation (60,61), as well as in p53-induced apoptosis [Fig. 3(2)] (62). Cathepsins are able to trigger apoptosis-related events such as membrane blebbing, PS exposure and chromatin condensation in a caspase-independent manner (63,64). Cathepsins can also activate the caspase cascade, either by directly cleaving and activating caspases or through Bid-mediated release of cytochrome c (65,66). ER-stressing agents or an uncontrolled release of calcium lead to the accumulation of misfolded proteins in the ER lumen, and this can induce apoptosis

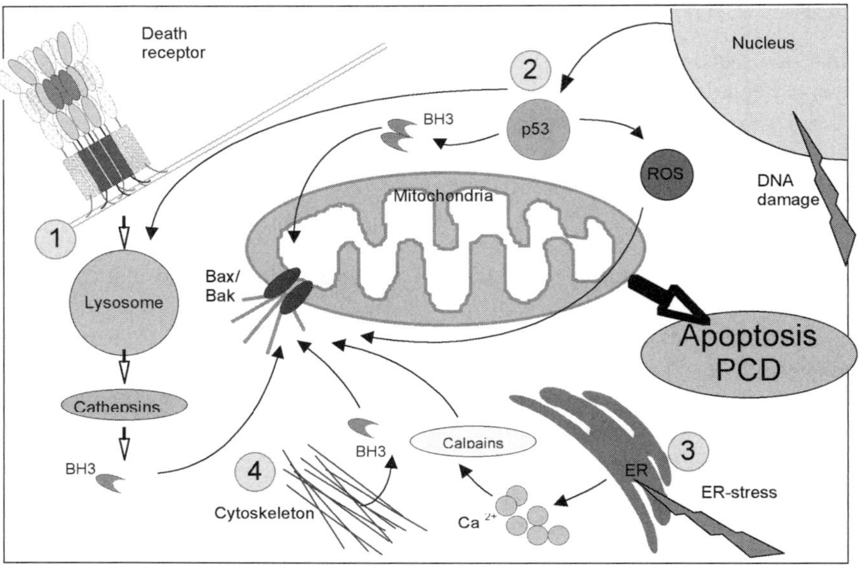

Figure 3 Caspase-independent signaling pathways leading to MOMP (52). (1) Death receptors, various oxidants, detergents, and chemotherapeutic drugs, trigger the release of cathepsins from lysosomes, which cleave the BH3 protein Bid. (2) The BH3-only proteins Bim and Bmf are released after disruption of the cytoskeleton. (3) Activation of p53 after DNA damage leads to the activation of the BH3-only proteins Puma and Noxa. (4) ER stress results in the release of calcium into the cytoplasm, which may activate Bax and Bak through calpain-mediated cleavage. The BH3 only proteins activate multidomain pro-apoptotic Bcl-2 family members (Bax and Bak), which subsequently provoke MOMP. *Abbreviations*: MOMP, mitochondrial outer membrane permeabilization; BH3, Bcl-2 homology domain 3; ER, endoplasmic reticulum.

when it becomes too excessive [Fig. 3(3)]. The latter may trigger a calpain-mediated PCD in the absence of caspase activation (67,68).

PCD TRIGGERED BY MOMP

MOMP leads to the release of various apoptogenic factors from mitochondria, which can trigger distinct death signaling pathways and hence different forms of PCD. For example, cytochrome c is often released in the absence of caspase activity (69), but it activates the classical apoptosis pathway via apoptosome formation and subsequent caspase activation. Caspase-inhibitory factors (IAPs) are removed by the action of Smac and Omi [Fig. 4(2)]. Both factors are released from mitochondria. Omi, being a serine protease, has however a second function. It is responsible for DNA fragmentation and condensation in caspase-independent apoptosis-like cell death. Another distinct death pathway initiated by MOMP is mediated by AIF [Fig. 4(3)] (55,70). Once released from the mitochondria, AIF

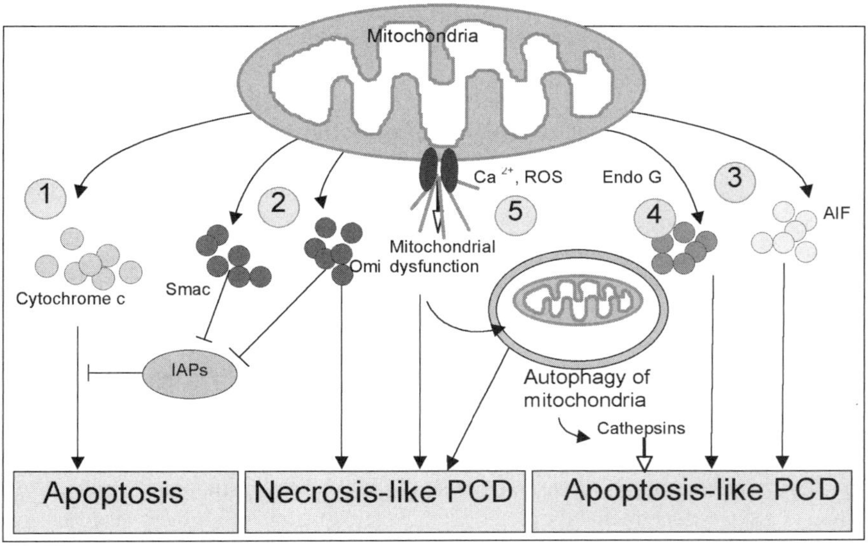

Figure 4 MOMP can trigger caspase-dependent and caspase-independent PCD (52). Mitochondrial damage leads to the release of numerous mitochondrial proteins that mediate PCD. (1) Release of cytochrome c triggers caspase activation and classic apoptosis. (2) Smac (DIABLO) and HtrA2 (Omi) assist cytochrome c–induced caspase activation by counteracting IAPs. (3) Omi can mediate caspase-independent cellular rounding and shrinkage. (4) AIF triggers a caspase-independent death pathway leading to DNA fragmentation and chromatin condensation. (5) Endo G cleaves DNA and induces chromatin condensation. (6) Calcium and ROS lead to mitochondrial dysfunction and necrosis-like PCD either directly or through autophagy of damaged mitochondria. Cathepsins may also be activated and can lead to apoptosis-like PCD. *Abbreviations*: MOMP, mitochondrial outer membrane permeabilization; PCD, programmed cell death; IAPs, inhibitors of apoptosis proteins; AIF, apoptosis-inducing factor; Endo G, Endonuclease G; ROS, reactive oxygen species.

translocates into the nucleus where it induces the caspase-independent formation of large (50 kilobase-pair) fragments of chromatin (type 1 chromatin condensation) (71). Endonuclease G (Endo G) [Fig. 4(4)] and the serine protease Htra2 (Omi) (72) can be released from mitochondria and may contribute to the caspase-independent death signaling downstream of MOMP. Omi induces PCD when expressed outside of mitochondria (72) and Endo G can cause caspase-independent DNA fragmentation in isolated nuclei (73). Calcium and ROS can lead to severe mitochondrial dysfunction and necrosis-like PCD either directly or through autophagy of damaged mitochondria (52). Besides the serine protease Omi mentioned in the previous paragraph, many studies have implicated serine proteases in CICD. Granule-mediated cytotoxicity is the main pathway for the immunological elimination of virus-infected cells and tumor cells by cytotoxic T lymphocytes (CTLs) and natural killer cells. CTLs contain a special form of granules, containing different serine proteases known as granzymes. After the CTL encouters the target cell, the granzymes (mainly A, B, and C) are released from the granules and induce apoptosis of the target cell in both caspase-dependent and caspase-independent manners (74). Granzyme B activates caspase-mediated apoptosis by cleaving caspase-3 and other caspases (75) and by cleaving Bid. By contrast, CICD is mediated by granzyme A, granzyme C, and granulysin (76–78). Another serine protease present in neutrophils, cathepsin G, cleaves and activates caspase-7 in vitro (79). A novel serine protease was also identified, which is involved in UV-induced apoptosis (80). In TNF-α induced apoptosis an AEBSF-sensitive serine protease with activity upstream of caspase-3 is responsible for PCD in human gastric cancer cells (81).

PHYSIOLOGICAL RELEVANCE OF CASPASE-INDEPENDENT PCD

Although there is no doubt that the most efficient way to execute cell suicide is by activation of caspases, CICD apparently presents back-up systems to kill cells in a regulated manner if caspase activation fails. Moreover, there are situations in which caspase-dependent apoptosis is too efficient and a slower, less energy-consuming death pathway is needed. This has been suggested by in vivo data in mice in which caspase inhibition sensitized cells to TNF treatment by increasing oxidative stress and mitochondrial damage. Mice treated with lethal doses of TNF died after 24 hours postinjection, whereas pre-treatment of those mice with z-VAD.fmk induced rapid death within four hours (82).

CONCLUSION

Cell death research of the last decade has been focussing largely on caspases as central executioners. However, considering more recent findings on caspases as important mediators of inflammatory cascades and the presence of CICD, targeting caspases for preventing cell death may no longer be the optimal choice. A major problem of developing efficient drugs for preventing CICD are the multitude of signaling cascades that are engaged in the absence of caspase activity. In the future, the identification of common events for different cell death pathways will be needed for the development of efficient blockers for CICD. One such common step could be represented by MOMP, which is regulated by proteins of the Bcl-2 family.

REFERENCES

1. Kroemer G, El-Deiry WS, Golstein P, et al. Classification of cell death: recommendations of the Nomenclature Committee on Cell Death. Cell Death Differ 2005; 12(suppl 2):1463–1467.
2. Leist M, Jaattela M. Four deaths and a funeral: from caspases to alternative mechanisms. Nat Rev Mol Cell Biol 2001; 2:589–598.
3. Kerr JF. Shrinkage necrosis: a distinct mode of cellular death. J Pathol 1971; 105: 13–20.

4. Savill J, Fadok V. Corpse clearance defines the meaning of cell death. Nature 2000; 407:784–788.
5. Jacobson MD, Weil M, Raff MC. Programmed cell death in animal development. Cell 1997; 88:347–354.
6. Horvitz HR. Genetic control of programmed cell death in the nematode *Caenorhabditis elegans*. Cancer Res 1999; 59:1701s–1706s.
7. Alnemri ES, Livingston DJ, Nicholson DW, et al. Human ICE/CED-3 protease nomenclature. Cell 1996; 87:171.
8. Ellis HM, Horvitz HR. Genetic control of programmed cell death in the nematode *C. elegans*. Cell 1986; 44:817–829.
9. Ellis RE, Yuan JY, Horvitz HR. Mechanisms and functions of cell death. Annu Rev Cell Biol 1991; 7:663–698.
10. Horvitz HR, Shaham S, Hengartner MO. The genetics of programmed cell death in the nematode *Caenorhabditis elegans*. Cold Spring Harb Symp Quant Biol 1994; 59:377–385.
11. Hengartner MO, Ellis RE, Horvitz HR. *Caenorhabditis elegans* gene CED-9 protects cells from programmed cell death. Nature 1992; 356:494–499.
12. Conradt B, Horvitz HR. The *C. elegans* protein EGL-1 is required for programmed cell death and interacts with the Bcl-2-like protein CED-9. Cell 1998; 93:519–529.
13. Boatright KM, Renatus M, Scott FL, et al. A unified model for apical caspase activation. Mol Cell 2003; 11:529–541.
14. Denault JB, Salvesen GS. Caspases: keys in the ignition of cell death. Chem Rev 2002; 102:4489–4500.
15. Shi Y. Mechanisms of caspase activation and inhibition during apoptosis. Mol Cell 2002; 9:459–470.
16. Martinon F, Tschopp J. Inflammatory caspases: linking an intracellular innate immune system to autoinflammatory diseases. Cell 2004; 117:561–574.
17. Tinel A, Tschopp J. The PIDDosome, a protein complex implicated in activation of caspase-2 in response to genotoxic stress. Science 2004; 304:843–846.
18. Krammer PH. CD95's deadly mission in the immune system. Nature 2000; 407: 789–795.
19. Jiang X, Wang X. Cytochrome *c*-mediated apoptosis. Annu Rev Biochem 2004; 73:87–106.
20. Han Z, Hendrickson EA, Bremner TA, et al. A sequential two-step mechanism for the production of the mature p17:p12 form of caspase-3 in vitro. J Biol Chem 1997; 272:13432–13436.
21. Riedl SJ, Renatus M, Schwarzenbacher R, et al. Structural basis for the inhibition of caspase-3 by XIAP. Cell 2001; 104:791–800.
22. Deveraux QL, Reed JC. IAP family proteins—suppressors of apoptosis. Genes Dev 1999; 13:239–252.
23. Xu G, Cirilli M, Huang Y, et al. Covalent inhibition revealed by the crystal structure of the caspase-8/p35 complex. Nature 2001; 410:494–497.
24. Garcia-Calvo M, Peterson EP, Leiting B, et al. Inhibition of human caspases by peptide-based and macromolecular inhibitors. J Biol Chem 1998; 273:32608–32613.
25. Nicholson DW, Ali A, Thornberry NA, et al. Identification and inhibition of the ICE/CED-3 protease necessary for mammalian apoptosis. Nature 1995; 376:37–43.
26. Enari M, Hug H, Nagata S. Involvement of an ICE-like protease in Fas-mediated apoptosis. Nature 1995; 375:78–81.

27. Yakovlev AG, Knoblach SM, Fan L, et al. Activation of CPP32-like caspases contributes to neuronal apoptosis and neurological dysfunction after traumatic brain injury. J Neurosci 1997; 17:7415–7424.
28. Rouquet N, Pages JC, Molina T, et al. ICE inhibitor YVADcmk is a potent therapeutic agent against in vivo liver apoptosis. Curr Biol 1996; 6:1192–1195.
29. Loddick SA, MacKenzie A, Rothwell NJ. An ICE inhibitor, z-VAD-DCB attenuates ischaemic brain damage in the rat. Neuroreport 1996; 7:1465–1468.
30. Hara H, Friedlander RM, Gagliardini V, et al. Inhibition of interleukin 1beta converting enzyme family proteases reduces ischemic and excitotoxic neuronal damage. Proc Natl Acad Sci U S A 1997; 94:2007–2012.
31. Rodriguez I, Matsuura K, Ody C, et al. Systemic injection of a tripeptide inhibits the intracellular activation of CPP32-like proteases in vivo and fully protects mice against Fas-mediated fulminant liver destruction and death. J Exp Med 1996; 184:2067–2072.
32. Yaoita H, Ogawa K, Maehara K, et al. Attenuation of ischemia/reperfusion injury in rats by a caspase inhibitor. Circulation 1998; 97:276–281.
33. McCarthy NJ, Whyte MK, Gilbert CS, et al. Inhibition of CED-3/ICE-related proteases does not prevent cell death induced by oncogenes, DNA damage, or the Bcl-2 homologue Bak. J Cell Biol 1997; 136:215–227.
34. Xiang J, Chao DT, Korsmeyer SJ. BAX-induced cell death may not require interleukin 1 beta-converting enzyme-like proteases. Proc Natl Acad Sci U S A 1996; 93:14559–14563.
35. Miller TM, Moulder KL, Knudson CM, et al. Bax deletion further orders the cell death pathway in cerebellar granule cells and suggests a caspase-independent pathway to cell death. J Cell Biol 1997; 139:205–217.
36. Borner C. The Bcl-2 protein family: sensors and checkpoints for life-or-death decisions. Mol Immunol 2003; 39:615–647.
37. Cory S, Adams JM. The Bcl-2 family: regulators of the cellular life-or-death switch. Nat Rev Cancer 2002; 2:647–656.
38. Boise LH, Gonzalez-Garcia M, Postema CE, et al. Bcl-x, a Bcl-2-related gene that functions as a dominant regulator of apoptotic cell death. Cell 1993; 74:597–608.
39. Gibson L, Holmgreen SP, Huang DC, et al. Bcl-w, a novel member of the Bcl-2 family, promotes cell survival. Oncogene 1996; 13:665–675.
40. Choi SS, Park IC, Yun JW, et al. A novel Bcl-2 related gene, Bfl-1, is overexpressed in stomach cancer and preferentially expressed in bone marrow. Oncogene 1995; 11:1693–1698.
41. Kozopas KM, Yang T, Buchan HL, et al. Mcl1, a gene expressed in programmed myeloid cell differentiation, has sequence similarity to Bcl-2. Proc Natl Acad Sci U S A 1993; 90:3516–3520.
42. Wei MC, Zong WX, Cheng EH, et al. Proapoptotic BAX and BAK: a requisite gateway to mitochondrial dysfunction and death. Science 2001; 292:727–730.
43. Chen L, Willis SN, Wei A, et al. Differential targeting of prosurvival Bcl-2 proteins by their BH3-only ligands allows complementary apoptotic function. Mol Cell 2005; 17:393–403.
44. Bouillet P, Strasser A. BH3-only proteins—evolutionarily conserved proapoptotic Bcl-2 family members essential for initiating programmed cell death. J Cell Sci 2002; 115:1567–1574.
45. Gross A, McDonnell JM, Korsmeyer SJ. Bcl-2 family members and the mitochondria in apoptosis. Genes Dev 1999; 13:1899–1911.

46. Martinou JC, Green DR. Breaking the mitochondrial barrier. Nat Rev Mol Cell Biol 2001; 2:63–67.
47. Zhong Q, Gao W, Du F, et al. Mule/ARF-BP1, a BH3-only E3 ubiquitin ligase, catalyzes the polyubiquitination of Mcl-1 and regulates apoptosis. Cell 2005; 121:1085–1095.
48. Nishito Y, Hasegawa M, Inohara N, et al. MEX is a testis-specific E3 ubiquitin ligase that promotes death receptor-induced apoptosis. Biochem J 2006; 396:411–417.
49. Egger L, Madden DT, Rheme C, et al. Endoplasmic reticulum stress-induced cell death mediated by the proteasome. Cell Death Differ 2007; 14(6):1172–1180.
50. Ying S, Seiffert BM, Hacker G, et al. Broad degradation of proapoptotic proteins with the conserved Bcl-2 homology domain 3 during infection with *Chlamydia trachomatis*. Infect Immun 2005; 73:1399–1403.
51. Chipuk JE, Green DR. Do inducers of apoptosis trigger caspase-independent cell death? Nat Rev Mol Cell Biol 2005; 6:268–275.
52. Jaattela M, Tschopp J. Caspase-independent cell death in T lymphocytes. Nat Immunol 2003; 4:416–423.
53. Strasser A, O'Connor L, Dixit VM. Apoptosis signaling. Annu Rev Biochem 2000; 69:217–245.
54. Kaufmann SH, Hengartner MO. Programmed cell death: alive and well in the new millennium. Trends Cell Biol 2001; 11:526–534.
55. Ferri KF, Kroemer G. Organelle-specific initiation of cell death pathways. Nat Cell Biol 2001; 3:E255–E263.
56. Mathiasen IS, Jaattela M. Triggering caspase-independent cell death to combat cancer. Trends Mol Med 2002; 8:212–220.
57. Vercammen D, Vandenabeele P, Beyaert R, et al. Tumour necrosis factor-induced necrosis versus anti-Fas-induced apoptosis in L929 cells. Cytokine 1997; 9:801–808.
58. Holler N, Zaru R, Micheau O, et al. Fas triggers an alternative, caspase-8-independent cell death pathway using the kinase RIP as effector molecule. Nat Immunol 2000; 1:489–495.
59. Hildeman DA, Mitchell T, Teague TK, et al. Reactive oxygen species regulate activation-induced T cell apoptosis. Immunity 1999; 10:735–744.
60. Guicciardi ME, Deussing J, Miyoshi H, et al. Cathepsin B contributes to TNF-alpha-mediated hepatocyte apoptosis by promoting mitochondrial release of cytochrome c. J Clin Invest 2000; 106:1127–1137.
61. Foghsgaard L, Wissing D, Mauch D, et al. Cathepsin B acts as a dominant execution protease in tumor cell apoptosis induced by tumor necrosis factor. J Cell Biol 2001; 153:999–1010.
62. Yuan XM, Li W, Dalen H, et al. Lysosomal destabilization in p53-induced apoptosis. Proc Natl Acad Sci U S A 2002; 99:6286–6291.
63. Foghsgaard L, Lademann U, Wissing D, et al. Cathepsin B mediates tumor necrosis factor-induced arachidonic acid release in tumor cells. J Biol Chem 2002; 277: 39499–39506.
64. Gobeil S, Boucher CC, Nadeau D, et al. Characterization of the necrotic cleavage of poly(ADP-ribose) polymerase (PARP-1): implication of lysosomal proteases. Cell Death Differ 2001; 8:588–594.
65. Roberg K, Kagedal K, Ollinger K. Microinjection of cathepsin D induces caspase-dependent apoptosis in fibroblasts. Am J Pathol 2002; 161:89–96.

66. Stoka V, Turk B, Schendel SL, et al. Lysosomal protease pathways to apoptosis. Cleavage of bid, not pro-caspases, is the most likely route. J Biol Chem 2001; 276:3149–3157.
67. Mathiasen IS, Sergeev IN, Bastholm L, et al. Calcium and calpain as key mediators of apoptosis-like death induced by vitamin D compounds in breast cancer cells. J Biol Chem 2002; 277:30738–30745.
68. Mathiasen H, Alpert JS. Only connect: musings on the relationship between literature and medicine. Fam Med 2001; 33:349–351.
69. Waterhouse NJ, Goldstein JC, von Ahsen O, et al. Cytochrome c maintains mitochondrial transmembrane potential and ATP generation after outer mitochondrial membrane permeabilization during the apoptotic process. J Cell Biol 2001; 153:319–328.
70. Susin SA, Daugas E, Ravagnan L, et al. Two distinct pathways leading to nuclear apoptosis. J Exp Med 2000; 192:571–580.
71. Susin SA, Lorenzo HK, Zamzam N, et al. Molecular characterization of mitochondrial apoptosis-inducing factor. Nature 1999; 397:441–446.
72. Suzuki Y, Imai Y, Nakayama H, et al. A serine protease, HtrA2, is released from the mitochondria and interacts with XIAP, inducing cell death. Mol Cell 2001; 8: 613–621.
73. van Loo G, Schotte P, van Gurp M, et al. Endonuclease G: a mitochondrial protein released in apoptosis and involved in caspase-independent DNA degradation. Cell Death Differ 2001; 8:1136–1142.
74. Lieberman J. The ABCs of granule-mediated cytotoxicity: new weapons in the arsenal. Nat Rev Immunol 2003; 3:361–370.
75. Darmon AJ, Nicholson DW, Bleackley RC. Activation of the apoptotic protease CPP32 by cytotoxic T-cell-derived granzyme B. Nature 1995; 377:446–448.
76. Sarin A, Williams MS, Alexander-Miller MA, et al. Target cell lysis by CTL granule exocytosis is independent of ICE/CED-3 family proteases. Immunity 1997; 6: 209–215.
77. Trapani JA, Jans DA, Jans PJ, et al. Efficient nuclear targeting of granzyme B and the nuclear consequences of apoptosis induced by granzyme B and perforin are caspase-dependent, but cell death is caspase-independent. J Biol Chem 1998; 273:27934–27938.
78. Beresford PJ, Xia Z, Greenberg AH, et al. Granzyme A loading induces rapid cytolysis and a novel form of DNA damage independently of caspase activation. Immunity 1999; 10:585–594.
79. Zhou Q, Salvesen GS. Activation of pro-caspase-7 by serine proteases includes a non-canonical specificity. Biochem J 1997; 324:361–364.
80. Wright SC, Wei QS, Zhong J, et al. Purification of a 24-kD protease from apoptotic tumor cells that activates DNA fragmentation. J Exp Med 1994; 180:2113–2123.
81. Park IC, Park MJ, Choe TB, et al. TNF-alpha induces apoptosis mediated by AEBSF-sensitive serine protease(s) that may involve upstream caspase-3/CPP32 protease activation in a human gastric cancer cell line. Int J Oncol 2000; 16: 1243–1248.
82. Cauwels A, Janssen B, Waeytens A, et al. Caspase inhibition causes hyperacute tumor necrosis factor-induced shock via oxidative stress and phospholipase A2. Nat Immunol 2003; 4:387–393.

7

Autophagic Cell Death in Mammalian Cells

David T. Madden
Buck Institute for Age Research, Novato, California, U.S.A.

INTRODUCTION

Depending upon the particular conditions encountered by a cell, any of a number of programmed cell death pathways may be triggered. Using strictly morphological descriptions to segregate one form of programmed cell death from another, Peter Clarke parsed programmed cell death into three principle forms (1): apoptosis (Type I, see Refs. 2 and 3), autophagic (Type II, see Refs. 4–6), and cytoplasmic cell death (Type III, with multiple subtypes, see Refs. 2,3,7). With the definition of a set of genes in *Saccharomyces cerevisiae* required for autophagy in the mid-1990s [genes that were later dubbed Atg genes], a wealth of information has been gathered about the role of autophagy in various biological processes in metazoans by studying the mammalian homologs of yeast Atg genes (5,8,9). Autophagy, a cellular mechanism for bulk delivery of organelles and cytosol to the lysosome for degradation, has been classically studied as a cellular response to amino acid deprivation. More recently, autophagy is being studied in the context of trophic factor withdrawal, insulin signaling, longevity, innate immunity, neurodegeneration, cancer, and programmed cell death (5). Despite the availability of RNA interference (RNAi) and knockout mice, the debate over whether autophagy can actively participate in the destruction of a mammalian cell has yet to be settled (10). In this review, I summarize the mechanics of autophagy, review important historical (pre-RNAi) literature on autophagic cell death, and highlight some of the most recently published studies that have contributed to our understanding of autophagic cell death in mammalian cells.

First, it should be emphasized that autophagy can occur without cell death. For clarity, the term "autophagy" will not be used synonymously with "autophagic cell death" in this review. Rather, "autophagy" will be reserved for the process in which cytosolic content is delivered to the lysosome, whereas "autophagic cell death" will be used when discussing programmed cell death pathways that require autophagy.

AUTOPHAGY

Autophagy is a major degradative pathway of the cell in which a double-membrane structure surrounds a large portion of the cytosol, including organelles, and delivers these contents to the lysosome to be degraded (Fig. 1) (5,8,11). A set of cytosolic proteins and integral membrane proteins constituting the autophagy machinery produce the double-membrane structure, which is thought to begin as a vesicle, termed the preautophagosomal structure (PAS), the origin of which remains unknown (9). As a result of the activity of the autophagy machinery, this vesicle is elongated and curvature is imparted on the membrane to produce a pancake-like structure, which curves into the shape of a bowl. This structure, termed the phagophore, is further curved and elongated until the leading edges eventually fuse. The resulting large, double-membrane structure is known as the autophagic vacuole or, alternatively, the autophagosome. The limiting membrane of the autophagic vacuole fuses with a lysosome to produce an autolysosome, where lysosomal hydrolases degrade the entire contents of the delivered autophagic vacuole (Fig. 1).

Autophagy Requires Two Ubiquitin-like Conjugation Systems

In order to shape the PAS into a phagophore, and ultimately, an autophagic vacuole, the proteins of the autophagic machinery bind as protein complexes and as protein-lipid conjugates to the surface of the PAS (9). In order to function as autophagic components, many autophagy proteins require covalent modification; the activity of two ubiquitin-like conjugation systems directs the production of mature autophagy protein complexes (Fig. 2). The first conjugation system uses the E1- and E2-like activities of Atg7 and Atg10, respectively, to covalently attach the ubiquitin-like molecule Atg12 to the E3-like protein Atg5, producing a stable Atg5-Atg12 adduct. The second conjugation system covalently attaches the ubiquitin-like molecule microtubule-associated protein light chain 3 (LC3) (the mammalian homolog of yeast Atg8) to the amine head-group of phosphatidylethanolamine (PE) via the activities of Atg7 (E1-like) and Atg3 (E2-like) (Fig. 2) (9).

Besides the products of the two ubiquitin-like conjugation systems (Atg5-Atg12 and LC3-PE), the production of autophagic vacuoles requires the function of several other proteins including Atg9, an integral membrane protein required for Atg protein recycling, and Atg16, a peripheral membrane protein

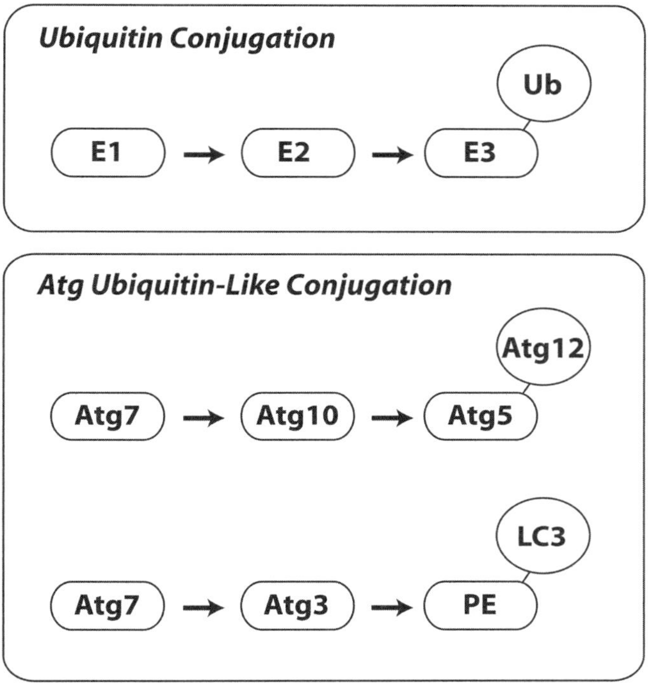

Figure 1 Two ubiquitin-like conjugation systems are required for autophagy. The production of an autophagic vacuole requires the activity of two ubiquitin-like conjugation reactions. Ubiquitin conjugation to substrate proteins requires the ubiquitin-activating enzyme E1, wherein ubiquitin is transiently covalently attached to E1. Subsequently, ubiquitin is passed on to an E2 ubiquitin ligase and on to an E3 ubiquitin ligase. Ubiquitin is attached to substrate proteins via the activity of the E3 ubiquitin ligase. During autophagy, two analogous conjugation systems work to produce Atg5-Atg12 and PE-LC3. Both Atg12 and LC3 are ubiquitin-like proteins that are first activated by the E1 ubiquitin-activating enzyme-like Atg7. Next, Atg12 is passed on to Atg10 (the E2-like enzyme) and on to Atg5 (the E3-like enzyme). Unlike ubiquitin, Atg12 is not passed on to another protein, but is maintained as a complex with Atg5. Similarly, LC3 is passed from Atg7 to Atg3 (the E2-like enzyme) and on to the phospholipid PE. The PE-LC3 product is membrane bound. *Abbreviations*: Atg, autophagy; PE, phosphatidylethanolamine; LC3, microtubule-associated protein light chain 3.

that forms a complex with Atg5-Atg12 (9). To produce an autophagic vacuole, Atg5-Atg12 (along with Atg16) and LC3-PE are recruited to the PAS, where they act to elongate the PAS into a phagophore and produce an autophagic vacuole (12). Upon completion of autophagic vacuole formation, the autophagy components uncoat, with the exception of the pool of LC3-PE, which is bound to the inner membrane of the autophagic vacuole; this fraction of LC3 is degraded in the lysosome (9).

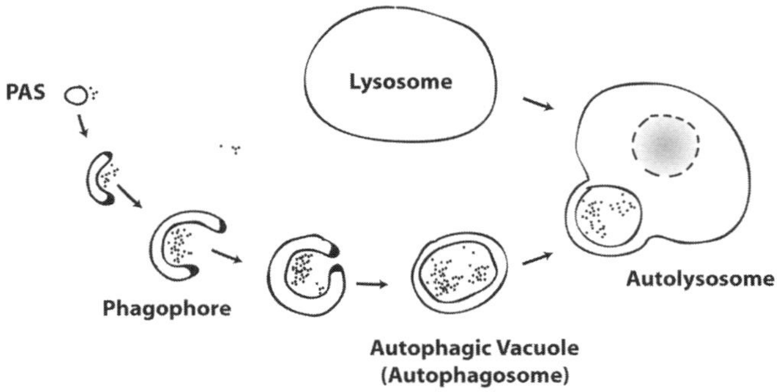

Figure 2 Autophagy. The nascent vesicle required for the production of autophagic vacuoles is the PAS, the origin of which is unknown. As a consequence of the activity of autophagy proteins, the PAS is elongated into a phagophore, which encircles large portions of cytosol, including organelles like mitochondria and endoplasmic reticulum. When the leading edges of the phagophore meet and fuse, a double-membrane autophagic vacuole (or autophagosome) is produced. Fusion of the limiting membrane of the autophagic vacuole with the lysosome (the autolysosome) results in the delivery of the sequestered cytosolic material to the lysosome, where it is degraded. *Abbreviations*: PAS, preautophagosomal structure.

Beclin 1 Multiprotein Complex

One of the early biochemical events necessary for autophagy is the production of phosphatidylinositol-3-phosphate (PI3P) from phosphatidylinositol, which is mediated by class III phosphatidylinositol 3-kinase (PI3K) (9). PI3P is required for the recruitment of proteins with PI3P-binding domains such as FYVE domain, ENTH domain, or Phox-homology (PH) domain–containing proteins (9). In yeast, two proteins required for cytosol-to-vacuolar targeting (a pathway that requires many of the Atg genes), Atg20p and Atg24p, contain PX domains, bind to PI3P, and localize to the PAS (9). Likewise, it is expected that PI3P-binding proteins required for autophagy will be identified in mammalian cells, but none have been found thus far.

The catalytic subunit of class III PI3K is hVps34 (9,13,14). In yeast, Vps34p is associated with two functionally distinct protein complexes; the subunits of the autophagy-promoting PI3K complex include Vps34p, Vps15p, Atg6p (yeast homolog of Beclin 1), and Atg14p (9). To date, the mammalian class III PI3K also consists of four subunits: Beclin 1, hVps34, UVRAG, and Bcl-2 (15–17). Elucidation of this Beclin 1 multiprotein complex began in the mid-1990s when the human homolog of yeast Vps34p was cloned by Michael Waterfield's group, and the recombinant hVps34 protein was demonstrated to have PI3K activity in vitro (13). Independently, Beth Levine's group determined that Beclin 1, identified as a Bcl-2-interacting protein by yeast-2-hybrid (16), contained significant sequence similarity to yeast *ATG6/VPS30* (18). Beclin 1 shares 24% identity with yeast Atg6p, can complement an *atg6Δ* yeast strain, and functions to upregulate autophagy in mammalian cells (17,18). The fourth component, UVRAG, was discovered in Jae Jung's laboratory using a biochemical search for proteins that interact with a viral homolog of Bcl-2 (15). Of the proteins from HEK293T lysate that bound to GST-vBcl-2 were Beclin 1, hVps34, and UVRAG (15).

These four proteins, hVps34, Beclin 1, Bcl-2, and UVRAG, were established as comprising a bona fide protein complex after it was shown that manipulation of subunits within this complex has profound effects on the activity of PI3K and autophagy (15,17,18). Beclin 1 interacts directly with Bcl-2, hVps34, and UVRAG, and the interactions between Bcl-2, hVps34, and UVRAG are Beclin 1 dependent (15,17). Beclin 1 and UVRAG were shown to be necessary for hVps34's PI3K activity, and furthermore, it was demonstrated that Beclin 1, UVRAG, and hVps34 are necessary for autophagy (15,17–19). Counter to the pro-autophagic activity of Beclin 1, UVRAG, and hVps34, overexpression of Bcl-2 [as well as Bcl-X_L and myeloid cell leukemia-1 (Mcl-1)] inhibits autophagy, and conversely, decreased Bcl-2 expression levels increase the autophagic activity of cells (17,20,21). In addition, overexpression of Bcl-2 mutants that fail to interact with Beclin 1 do not repress autophagy, and expression of Beclin 1 mutants that fail to bind to Bcl-2 results in constitutive autophagy (17,20,21). Amino acid starvation has been shown to cause disruption of the Bcl-2-Beclin 1 interaction, thereby derepressing the Beclin 1 multiprotein complex's PI3K activity to produce PI3P and induce autophagy (17,19–21).

INITIAL STUDIES OF AUTOPHAGIC CELL DEATH

3-methyladenine: Proceed with Caution

Initially, a role of autophagy in autophagic cell death was hypothesized based on morphological criteria alone (1). However, as morphological characterization does not definitively link autophagy to programmed cell death, subsequent studies tested the requirement for autophagy by examining programmed cell death pathways in the presence of autophagy inhibitors. Early studies relied upon 3-methyladenine (3-MA), a nucleotide analog that was discovered to inhibit starvation-induced autophagy in rat liver hepatocytes (22,23). Subsequently, 3-MA was shown to inhibit autophagy by blocking the activity of PI3K (14,24). However, many other enzymes are likely to be inhibited at the concentrations of 3-MA typically required to inhibit autophagy (10 mM) (25), and Aviva Tolkovsky's group showed that 3-MA has the capacity to interfere with mitogen-activated protein kinase (MAPK) signaling (26) as well as mitochondrial permeability transition (27). Because 3-MA has so many off-target effects, experiments relying upon 3-MA should be interpreted with caution. Nevertheless, prior to the availability of RNAi technology in mammalian cells to deplete autophagy genes, 3-MA was widely used to test the role of autophagy in programmed cell death.

TNFα-Induced Cell Death

One of the first studies to characterize the role of autophagy in cell death was one in which T-lymphoblastic leukaemic cells were treated with TNFα (28). Upon TNFα treatment, Jia et al. discovered by electron microscopy (EM) that a massive accumulation of autophagic vacuoles occurred prior to death (28). When 3-MA was co-incubated with TNFα, autophagic vacuole numbers were reduced, and cells were rescued from TNFα-induced toxicity (28). At about the same time, Bursch et al. published that autophagic vacuoles were frequently observed in MCF-7 cells treated with tamoxifen prior to the onset of cell death (29). Dying cells also became pyknotic, and by using this as a criterion for death, Bursch et al. showed that 3-MA could block tamoxifen-induced autophagic vacuole formation and cell death (29). Later, Tolkovsky's group described a programmed cell death pathway in sympathetic neurons, which was induced by withdrawal of nerve growth factor (NGF) (26). Autophagic vacuoles accumulated in response to withdrawal of NGF, nuclei became pyknotic, DNA appeared as a smear when resolved on an agarose gel, caspases were active, and cells died (26). All of these cell death events were blocked in the presence of the autophagy inhibitor 3-MA, suggesting, at least in this model of cell death, that autophagy may work upstream of apoptotic effector molecules to cause cell death (26).

Death-Associated Protein Kinases

In an elegant screen for molecules required for INFγ-induced cell death in HeLa cells, Adi Kimchi and colleagues identified two novel death-signaling kinases: Death-associated protein kinase (DAPk) and DAPk-related protein kinase (DRP)-1 (30). In a paper by Inbal et al., DAPk was shown to cause plasma membrane blebbing, autophagic vacuole formation, and death in multiple cell types (31). DAPk-induced death proceeds via a caspase-independent, Bcl-2 and Bcl-X_L nonblockable pathway in which cytochrome c is not released (31). Despite the features of DAPk-induced cell death that suggest autophagy is a required component of this pathway, this has yet to be demonstrated. Nevertheless, the discovery of a signaling molecule that caused features consistent with autophagic cell death bolstered the notion that cells might have a dedicated programmed cell death pathway requiring autophagy.

Tumor Necrosis Factor–Related Apoptosis-Inducing Ligand

Programmed cell death is required for many processes during mammalian development (1,32) and recently, attention has focused on the mechanism of hollow lumen formation in glandular structures (33,34). Joan Brugge's group took advantage of an in vitro 3D cell culture model using immortalized human mammary epithelial MCF-10A cells to study the signaling pathways that govern elimination of centrally located cells to produce an acinar-like structure (34). Earlier, Debnath et al. discovered that overexpression of Bcl-2 or Bcl-X_L in this system only delays cavitation, suggesting that Bcl-2 nonblockable programmed cell death pathways also participate (33). Furthermore, dying cells accumulate autophagic vacuoles even in the presence of Bcl-2 or Bcl-X_L, pointing to the possibility that autophagy may play a role in lumen formation (33). Mills et al. identified a tumor necrosis factor–related apoptosis-inducing ligand (TRAIL)-mediated signaling pathway, which participated with apoptosis in the cavitation process (34). Inhibition of endogenous TRAIL signaling blocks the formation of autophagic vacuoles in dying cells, and exogenously added TRAIL causes autophagic vacuole production (34). Although a direct connection between autophagy and death of cells within the lumen was not made, the fact that the combination of Bcl-X_L overexpression and inhibition of the autophagy-promoting ligand TRAIL blocks cavitation in acinar-like structures is highly suggestive that autophagy plays an active role in this process (33,34).

TWO SEMINAL AUTOPHAGIC CELL DEATH PAPERS

z-Val-Ala-Asp(OMe)-Fluoromethylketone-Induced Cell Death

Ultimately, it was not until 2004 that the first example of Atg gene–dependent cell death was described in mammalian cells (35). In a paper by Michael Lenardo's group, the authors used the broad-spectrum caspase inhibitor z-Val-Ala-Asp(OMe)-fluoromethylketone (zVAD) to induce cell death in the mouse fibrosarcoma cell line L929. When either Atg7 or Beclin 1 was depleted in cells by RNAi, zVAD-induced cell death was greatly reduced (35). In addition to killing L929 cells, zVAD was shown to induce cell death in macrophages including RAW264.7 cells, mouse peritoneal macrophages, and U937 human leukemic monocyte lymphoma cells. Furthermore, a signal transduction cascade was elucidated in which inhibition of basal caspase-8 activity by zVAD resulted in the accumulation of an active form of receptor-interacting protein (RIP), which could then activate autophagic cell death in an mitogen-activated protein kinase kinase 7 (MKK7), c-Jun Nterminal kinase (JNK), and c-Jun–dependent manner (35). In a subsequent report by the same group, Yu et al. showed that in response to zVAD, the cytosolic form of catalase is destroyed in an autophagy-dependent manner, and that the signaling molecules RIP, MKK7, JNK, and c-Jun are also required for catalase depletion (36). As a consequence of decreased catalase levels, free radicals accumulate in cells, and oxidative damage ensues [scavengers of reactive oxygen species (ROS) rescued cells from zVAD-induced cell death] (36). Not only did this work demonstrate the first autophagy gene–dependent programmed cell death pathway, but it also defined a dedicated signaling cascade that results in Type II autophagic cell death (35,36).

Apoptotic Stimuli Induce Autophagic Cell Death in *Bax*$^{-/-}$/*Bak*$^{-/-}$ MEFs

Also, in 2004, Yoshihide Tsujimoto's group followed the viability of *Bax*$^{-/-}$/*Bak*$^{-/-}$ MEFs (mouse embryonic fibroblasts) in response to the classic apoptotic stimuli etoposide and staurosporine (37). Bax and Bak are proapoptotic so-called "multi-domain" Bcl-2 family members, which cause mitochondrial outer membrane permeabilization when left unchecked by Bcl-2 or other antiapoptotic Bcl-2 family members (38,39). The proteins Bax and Bak serve redundant functions and have been proposed to constitute a requisite gateway to apoptosis (40); many reports have indicated that $Bax^{-/-}/Bak^{-/-}$ MEFs are refractory to most apoptotic triggers including etoposide and staurosporine (38,40–42). It was surprising, then, when Shimizu et al. reported that extended incubation with etoposide or staurosporine killed $Bax^{-/-}/Bak^{-/-}$ MEFs, because it suggested that alternative, nonapoptotic forms of cell death were at play (37). Shimizu et al. found that within 24 hours of etoposide or staurosporine treatment, cells accumulated autophagic vacuoles and were killed. Importantly, small interfering RNAs (siRNAs) directed against Beclin 1 and Atg5 reduced the number of dead cells in response to both etoposide and staurosporine indicating that the production of autophagic vacuoles was a prerequisite for cell death. Curiously, $Bax^{-/-}/Bak^{-/-}$ MEFs were sensitized to etoposide and staurosporine when the antiapoptotic proteins Bcl-2 or Bcl-X_L were overexpressed, and this augmented death required Beclin 1 (37). From this report, it seemed that Bcl-2 family members regulate both apoptosis and autophagic cell death, but with opposite signs (Bax/Bak promote apoptosis, but seem to be inhibitory to autophagic cell death, whereas Bcl-2 blocks apoptosis, but stimulates autophagic cell death) (37). Although the model proposed by Shimizu et al. is enticing, it has been difficult to reconcile the differences between the results of this report with others that indicate Bcl-2, Bcl-X_L, and Mcl-1 are inhibitory to autophagy (17,20,21).

RECENT STUDIES OF AUTOPHAGIC CELL DEATH PATHWAYS

A Fetal Alcohol Syndrome–Associated Protein with Death Domain- and Atg5-Requiring Autophagic Cell Death Pathway

In an effort to better understand the role of fetal alcohol syndrome (Fas)–associated protein with death domain (FADD) in programmed cell death pathways, Yong-Keun Jung and colleagues used the yeast-2-hybrid system to identify novel protein interactors with FADD, and from their screen identified Atg5 (43). Having pulled out an autophagy gene from their screen, Pyo et al. tested the role of Atg5 and FADD in INFγ-induced programmed cell death, a model of cell death featuring autophagy (43) (as described by Kimchi's group) (30,31). Antisense oligonucleotides directed toward Atg5 blocked INFγ-induced cell death, but had no effect on etoposide, staurosporine, or cisplatin-induced cell death (43) [although this finding conflicts with that reported by Yousefi et al. (see below) (44)]. INFγ-induced cell death was also blocked by antisense FADD, and could be inhibited by zVAD [unlike that reported by Kimchi's group (31)] and 3-MA, suggesting that multiple cell death pathways might cooperate to dismantle the cell in response to INFγ (43). Interestingly, a point mutant of Atg5 in which the lysine used to link Atg12 to Atg5 (K130) is mutated also blocked

INFγ-induced cell death, suggesting that this mutant can be used as a dominant-negative mutant to block autophagy and autophagy-requiring processes (43). Overexpression of Atg5 and FADD were shown to be sufficient to kill HeLa cells, and testing Atg5- and FADD-induced cell death for reliance on autophagy, caspases, FADD, and Atg5 led to a model proposed by Pyo et al. in which INFγ induces a signaling cascade that flows from Atg5 to autophagy, FADD, caspases, and ultimately, cell death (43).

Calpains, Autophagy, and Death

Recently, several reports described the role of calpains in autophagy and autophagic cell death. In one study from Claudio Schneider's laboratory, MEFs lacking calpain small subunit 1 [(CAPNS1), a subunit shared between calpain-1 and calpain-2] were shown to have reduced levels of long-lived protein degradation in response to amino acid starvation (45). From this and other experiments, Demarchi et al. proposed that calpain activity is necessary for autophagy (45). Furthermore, CAPNS1$^{-/-}$ MEFs–induced apoptosis when starved of amino acids for 20 hours (45), consistent with studies that showed that knockdown of autophagy genes lead to apoptosis when cells were deprived of amino acids (46).

However, data in other reports have lead to the opposite conclusion that calpain activity blocks autophagy (44,47). In a study that followed up on the mechanism of zVAD-induced autophagic cell death, Dale Bredesen's group discovered that the autophagic cell death–promoting activity of zVAD could be parsed into two activities: a caspase inhibiting activity and a calpain inhibiting activity (47). Off-target effects of zVAD have been described previously, including cathepsin and calpain inhibition (30,48–50), but a second-generation broad-spectrum caspase inhibitor, Q-VD-OPh, is far more selective for caspases (51,52). Madden et al. showed that by using Q-VD-OPh to block caspase activity, which alone was not toxic to L929s, caspases and calpains needed to be inhibited simultaneously in order for mouse L929 fibroblasts to be killed (47). The results from Madden et al. are suggestive that calpains are inhibitory to autophagy and autophagic cell death (47).

In another report describing an inhibitory role of calpain on autophagy, Yousefi et al. made the observation [as did Pyo et al. (43)] that Atg5 overexpression resulted in increased sensitivity to proapoptotic triggers (44). Typically, the 33-kD form of Atg5 cannot be visualized by western blot because the vast majority is covalently attached to Atg12 and runs as a 55-kD complex by sodium dodecyl sulfate polyacrylamide gel electrophoresis (SDS-PAGE) (53). However, Yousefi et al. discovered that, with many proapoptotic insults, a 24-kD fragment recognized by an anti-Atg5 antibody could be visualized by Western blot (44). By surveying a number of proteases for those that could cleave recombinant Atg5 in vitro, they discovered that calpain-1 and calpain-2 were not only active in that regard, but also produced the 24-kDa fragment seen in earlier experiments. Subsequently, the calpain cleavage site on Atg5 was mapped, and it was shown that the 24-kD species of Atg5-induced apoptosis when overexpressed. Furthermore, 24-kD Atg5 is incapable of supporting autophagy compared to full-length Atg5 (44). The authors proposed a mechanism of cell death in which 24-kD Atg5 translocates to the mitochondria where it induces cytochrome c release and apoptosis. In this example, the autophagy-promoting activity of Atg5 is lost upon cleavage, but a resulting fragment gains a proapoptotic function (44). From this study came two findings of outstanding interest: (*i*) Calpain activity inhibits autophagy and (*ii*) programmed cell death pathways that are Atg5-dependent may not be autophagy-dependent. Because of these findings, studies demonstrating a requirement for Atg5 in a programmed cell death pathway may have to be revisited to definitively show that autophagy per se, and not pro-death autophagy protein derivatives, is responsible for programmed cell death.

HIV-Induced Killing of Noninfected T Cells

It has long been recognized that HIV infection results in the elimination of noninfected T cells (54), but the precise mechanism of bystander killing by HIV is unknown. Recently, a report by Martine Biard-Piechaczyk's group demonstrated that bystander killing occurs as a consequence of ligating the chemokine receptor CXCR4 localized on noninfected cells with HIV-encoded Env from an infected cell (55). As a result of this interaction, a signaling cascade leads to accumulation of autophagic vacuoles and cell death (55). In this elegant study, Espert et al. were able to reconstitute this process in three different settings using populations of "effector cells," those which express Env, and "target cells," those which express CXCR4 and the necessary coreceptor CD-4 (55). With these systems, the authors demonstrated that, when presented with Env-expressing effector cells, CXCR4-expressing target cells produce autophagic vacuoles, upregulate levels of Beclin 1, and lose viability. 3-MA and siRNAs directed against either Beclin 1 or Atg7 blocked Env/CXCR4-mediated cell death, demonstrating that autophagy is required for cell death. The authors also showed that during the course of cell death, an increase in caspase-3 activity could be detected, and that this activity was inhibited by zVAD as well as RNAi against Beclin 1 or Atg7 (55). The fact that autophagy gene knockdown blocked the Env/CXCR4-induced caspase-3 activity indicates that autophagy acts upstream of caspase activation in this system. Whereas some caspase activation might be expected as an epiphenomenon during autophagic cell death, zVAD also diminished death induced by Env/CXCR4 signaling, suggesting that caspase-mediated pathways might also participate in promoting cell death (55). If caspases are truly major players in Env/CXCR4-mediated programmed cell death, one would expect to see signs of apoptotic morphology, including chromatin or nuclear condensation in the dying cells. However, the EMs of dying cells presented in this paper show nuclei that are very well preserved (55). Thus, it seems that Env/CXCR4-mediated programmed cell death might be better characterized as autophagic cell death, rather than autophagy-requiring apoptosis. Nevertheless, the work presented by Espert et al. is an outstanding contribution to our understanding of programmed cell death pathways that require autophagy.

AUTOPHAGY AND CANCER

One mechanism of tumor suppression utilized by metazoans is apoptosis, which can be activated when proliferation rates exceed a certain threshold (56). However, many cancers are refractory to apoptotic stimuli, meaning that drugs that activate nonapoptotic programmed cell death pathways, including autophagic cell death, are likely to have greater therapeutic impact (57,58).

Autophagy Sensitizes Cells to Radiation

One common cancer treatment is radiation, which likely induces multiple forms of programmed cell death (59). However, Bo Lu's group made the surprising observation that $Bax^{-/-}/Bak^{-/-}$ MEFs are more sensitive to radiation than wild-type MEFs (60). In response to radiation, $Bax^{-/-}/Bak^{-/-}$ MEFs produce autophagic vacuoles and upregulate Beclin 1 and Atg5-Atg12 by inactivating the mammalian target of rapamycin (mTor) pathway (60). The possibility that $Bax^{-/-}/Bak^{-/-}$ MEFs were being killed by autophagic cell death was tested by inhibiting autophagy with 3-MA and RNAi against either Atg5 or Beclin 1. Doing so demonstrated that the sensitivity of $Bax^{-/-}/Bak^{-/-}$ MEFs to radiation was reduced to that of wild-type when autophagy was inhibited (60). Additionally, the authors showed that induction of autophagy by inhibition of mTor or co-overexpression of Atg5 and Beclin 1 was sufficient to sensitize wild-type MEFs to radiation (60). From this work it seems that in some settings, induction of autophagy could promote destruction of cancer cells when combined with other chemotherapeutic agents such as radiation.

Damage-Regulated Autophagy Modulator

One of the major obstacles to tumor formation is the tumor suppressor p53, the inactivation of which is required for genome instability, wherein inactivation of other tumor suppressors is made possible (56).

To identify novel targets of p53, Kevin Ryan's group used a doxycicline-inducible p53 system in a cell line that lacks p53 and performed microarray analyses (61). One transcript not previously characterized was one that encoded a multi-spanning transmembrane protein the authors dubbed damage-regulated autophagy modulator (DRAM) (61). Crighton et al. discovered that p53 and DRAM expression induce autophagy, and furthermore, that upregulation of autophagy by p53 is DRAM-dependent (61). Whereas DRAM expression is not sufficient to kill cells, about 60% of p53-induced cell death is DRAM-dependent and the same fraction of p53-induced cell death is Atg5-dependent. Given that autophagy induced by p53 is DRAM-dependent, it is a reasonable conclusion that DRAM governs the autophagy-requiring arm of p53-induced cell death; however, many details of DRAM's function have yet to be elucidated. Crighton et al. also surveyed human tumors for DRAM expression and found that 50% of squamous cancers had reduced DRAM mRNA levels. Even more striking, a significant inverse correlation exists between mutant p53 status and loss of DRAM expression: Tumors with mutant p53 were more likely to express DRAM and vice versa (61). The exciting prospect from this work is that small molecules that can complement the activity of DRAM may restore the tumor suppressor activity of p53 in cancers that express wild-type p53.

Autophagy and Cancer In Vivo

That Bcl-2 overexpression in vivo results in cancer has been interpreted thus far with the assumption that Bcl-2 is inhibitory to apoptosis only (38). However, it is plausible that given Bcl-2's capacity to inhibit autophagy, Bcl-2 overexpression may prevent autophagic cell death as well. If this were true, one would expect that downregulation of the other Beclin 1 multiprotein complex subunits would also result in the formation of tumors. Consistent with this hypothesis, Beclin 1 has been found to be monoallelically deleted in 40% to 75% of sporadic breast cancers and ovarian and prostate cancers (18,62–64), and UVRAG is monoallelically deleted in colon cancer (15,65). In light of these data, in addition to those by Crighton et al. on DRAM (61), could it be that autophagic cell death is a physiological mechanism of tumor suppression?

Although this is a tantalizing possibility, this interpretation has been questioned based on a number of findings. First, MCF-7 cells are not killed by Beclin 1 overexpression unless a Beclin 1 mutant that fails to bind to Bcl-2 is overexpressed (17). Furthermore, enforced expression of Beclin 1 resulted in slowed growth rates, but no loss in cell viability over seven days of expression (18) and overexpression of Beclin 1 in other systems does not lead to cell death (66). Similarly, enforced expression of UVRAG in UVRAG-deficient HCT116 cells resulted in reduced growth rates and suppressed tumorigenicity, but neither an increase in cell death nor a decrease in viability was observed (15). Likewise for DRAM, enforced expression does not result in cell death, and clonogenicity is unchanged (61).

One possible connection between diminished autophagic activity and cancer may be related to mitophagy—the selective degradation of mitochondria via autophagy (67,68). Evidence has emerged that mitophagy may not treat all mitochondria alike, and that those mitochondria that are losing their membrane potential or producing ROS are selectively incorporated into autophagic vacuoles and destroyed (67,68). If this process was slowed, then ROS could accumulate resulting in mutations, genome instability, and cancer. If and when mitophagy-specific proteins are identified, one would predict that this protein(s) will be a tumor suppressor.

AUTOPHAGY AND Bcl-2 FAMILY MEMBERS: A CLOSER LOOK

Recently, two groups independently reported that Beclin 1 contains a Bcl-2 homology 3 (BH3) domain within its coding region and is responsible for Beclin 1's interaction with Bcl-2, Bcl-X_L, and Mcl-1 (20,21,69). Yigong Shi's group solved the crystal structure of the BH3 peptide from Beclin 1 complexed to Bcl-X_L (69). The structure of the complex was remarkably similar to that of Bim's BH3 peptide bound to Bcl-X_L, proving that Beclin 1 is a genuine BH3-only protein (69). Since BH3-only proteins are proapoptotic, is it possible that the presence of a BH3 domain within Beclin 1 would confer proapoptotic properties on Beclin 1, and that this activity could explain the tumor suppressive function of Beclin 1? To date, no evidence supports this notion and to the contrary, Guido Kroemer's group demonstrated that the function of the Beclin 1 BH3 domain–Bcl-2 interaction is to regulate autophagy (20,21). Maiuri et al. showed that ABT-737, a BH3 peptide mimetic with a Bad-like profile [Bad is a BH3-only protein that selectively targets Bcl-2, Bcl-X_L, and Bcl-w (38)], disrupts the Beclin 1–Bcl-2 protein interaction, thereby mimicking the cellular response to amino acid starvation (20). What is more, disruption of the interaction between Beclin 1 and Bcl-2 takes place only at the surface of the endoplasmic reticulum (ER); Beclin 1–Bcl-2 complexes immunoprecipitated from heavy membrane fractions rich in mitochondria are not disrupted by starvation or ABT-737, whereas those Beclin 1–Bcl-2 immunoprecipitates isolated from an ER-enriched microsomal preparation are disrupted (20). Using mitochondria- or ER-restricted forms of Bcl-2, Maiuri et al. confirmed that amino acid starvation as well as ABT-737 only disrupted the ER-localized complexes between Beclin 1 and Bcl-2 (20). These findings suggest that some hitherto unknown posttranslational modification, such as phosphorylation, of Bcl-2 or the Beclin 1 protein complex may regulate this interaction, and that the modification is mediated by an activity restricted to the ER.

Interestingly, Thomas Roberts's group found that the protein phosphatase PP2A co-purified with Bcl-2 isolated from an ER-enriched membrane fraction (70). In this case, depletion of $PP2A_C$ by RNAi sensitized MEFs to cell death induced by various toxic insults (70). This and other data presented by Lin et al. suggested that phosphorylation of ER-localized Bcl-2 led to proteasome-mediated degradation of Bcl-2 and that the role of $PP2A_C$ is to maintain Bcl-2 in a stable dephosphorylated state. This report contradicts other publications that indicate phosphorylation of Bcl-2 prevents proteasome-mediated degradation (71–74). One report in particular implicated a mitochondrion-localized protein kinase C alpha (PKCα) in the stabilization of Bcl-2 by phosphorylation (74), suggesting that the stability of phosphorylated Bcl-2 depends upon the subcellular localization of Bcl-2; phospho-Bcl-2 is stable at the mitochondria, yet degraded at the ER.

That Bcl-2, belonging to a class of potent antiapoptotic proteins, would also play such an instrumental role in controlling autophagy may not seem surprising considering that both autophagic cell death and apoptosis result in cellular destruction. From this perspective, it is tempting to speculate that the production of PI3P by the Beclin 1 multiprotein complex serves as a "gateway" to autophagic cell death, much as Bax and Bak serve as a gateway to apoptosis. Whereas the capacity of Bcl-2 to block a particular cell death paradigm has always implicated the involvement of mitochondria, it now seems formally possible that the capacity of Bcl-2 to block cell death could be the result of Bcl-2 blunting the autophagy-promoting activity of the Beclin 1 multiprotein complex at the ER. Further studies are required to determine whether "anti-death" Bcl-2 family members may differentially regulate either autophagic cell death or apoptosis, depending upon subcellular localization and phosphorylation status.

ACKNOWLEDGMENTS

The author would like to thank Dale E. Bredesen for maintaining a productive research environment as well as financial and mentor support. I thank Fiona Pickford, Brigit Riley, Philipp Jaeger, Sebastian Schuck, Malene Hansen, Nicole Meyer-Morse, Jayanta Debnath, William Jackson, and Matthew Taylor for many insightful discussions. David Madden is supported by a fellowship from the Larry L. Hillblom Foundation.

REFERENCES

1. Clarke PG. Developmental cell death: morphological diversity and multiple mechanisms. Anat Embryol (Berl) 1990; 181(3):195–213.
2. Yuan J, Lipinski M, Degterev A. Diversity in the mechanisms of neuronal cell death. Neuron 2003; 40(2):401–413.
3. Bredesen DE, Rao RV, Mehlen P. Cell death in the nervous system. Nature 2006; 443(7113):796–802.
4. Gozuacik D, Kimchi A. Autophagy and cell death. Curr Top Dev Biol 2007; 78:217–245.
5. Levine B, Klionsky DJ. Development by self-digestion: molecular mechanisms and biological functions of autophagy. Dev Cell 2004; 6(4):463–477.
6. Levine B, Yuan J. Autophagy in cell death: an innocent convict? J Clin Invest 2005; 115(10):2679–2688.
7. Leist M, Jaattela M. Four deaths and a funeral: from caspases to alternative mechanisms. Nat Rev Mol Cell Biol 2001; 2(8):589–598.
8. Lum JJ, DeBerardinis RJ, Thompson CB. Autophagy in metazoans: cell survival in the land of plenty. Nat Rev Mol Cell Biol 2005; 6(6):439–448.
9. Yorimitsu T, Klionsky DJ. Autophagy: molecular machinery for self-eating. Cell Death Differ 2005; 12(suppl 2):1542–1552.
10. Debnath J, Baehrecke EH, Kroemer G. Does autophagy contribute to cell death? Autophagy 2005; 1(2):66–74.
11. Kroemer G, Jaattela M. Lysosomes and autophagy in cell death control. Nat Rev Cancer 2005; 5(11):886–897.
12. Suzuki K, Kubota Y, Sekito T, et al. Hierarchy of Atg proteins in pre-autophagosomal structure organization. Genes Cells 2007; 12(2):209–218.

13. Volinia S, Dhand R, Vanhaesebroeck B, et al. A human phosphatidylinositol 3-kinase complex related to the yeast Vps34p-Vps15p protein sorting system. EMBO J 1995; 14(14):3339–3348.
14. Petiot A, Ogier-Denis E, Blommaart EF, et al. Distinct classes of phosphatidylinositol 3′-kinases are involved in signaling pathways that control macroautophagy in HT-29 cells. J Biol Chem 2000; 275(2):992–998.
15. Liang C, Feng P, Ku B, et al. Autophagic and tumour suppressor activity of a novel Beclin1-binding protein UVRAG. Nat Cell Biol 2006; 8(7):688–699.
16. Liang XH, Kleeman LK, Jiang HH, et al. Protection against fatal Sindbis virus encephalitis by beclin, a novel Bcl-2-interacting protein. J Virol 1998; 72(11): 8586–8596.
17. Pattingre S, Tassa A, Qu X, et al. Bcl-2 antiapoptotic proteins inhibit Beclin 1-dependent autophagy. Cell 2005; 122(6):927–939.
18. Liang XH, Jackson S, Seaman M, et al. Induction of autophagy and inhibition of tumorigenesis by beclin 1. Nature 1999; 402(6762):672–676.
19. Furuya N, Yu J, Byfield M, et al. The evolutionarily conserved domain of Beclin 1 is required for Vps34 binding, autophagy and tumor suppressor function. Autophagy 2005; 1(1):46–52.
20. Maiuri MC, Le Toumelin G, Criollo A, et al. Functional and physical interaction between Bcl-X(L) and a BH3-like domain in Beclin-1. EMBO J 2007; 26(10): 2527–2539.
21. Maiuri MC, Criollo A, Tasdemir E, et al. BH3-only proteins and BH3 mimetics induce autophagy by competitively disrupting the interaction between Beclin 1 and Bcl-2/Bcl-X(L). Autophagy 2007; 3(4):374–376.
22. Kopitz J, Kisen GO, Gordon PB, et al. Nonselective autophagy of cytosolic enzymes by isolated rat hepatocytes. J Cell Biol 1990; 111(3):941–953.
23. Seglen PO, Gordon PB. 3-Methyladenine: specific inhibitor of autophagic/lysosomal protein degradation in isolated rat hepatocytes. Proc Natl Acad Sci U S A 1982; 79(6):1889–1892.
24. Blommaart EF, Krause U, Schellens JP, et al. The phosphatidylinositol 3-kinase inhibitors wortmannin and LY294002 inhibit autophagy in isolated rat hepatocytes. Eur J Biochem 1997; 243(1–2):240–246.
25. Caro LH, Plomp PJ, Wolvetang EJ, et al. 3-Methyladenine, an inhibitor of autophagy, has multiple effects on metabolism. Eur J Biochem 1988; 175(2): 325–329.
26. Xue L, Fletcher GC, Tolkovsky AM. Autophagy is activated by apoptotic signalling in sympathetic neurons: an alternative mechanism of death execution. Mol Cell Neurosci 1999; 14(3):180–198.
27. Xue L, Borutaite V, Tolkovsky AM. Inhibition of mitochondrial permeability transition and release of cytochrome c by anti-apoptotic nucleoside analogues. Biochem Pharmacol 2002; 64(3):441–449.
28. Jia L, Dourmashkin RR, Allen PD, et al. Inhibition of autophagy abrogates tumour necrosis factor alpha induced apoptosis in human T-lymphoblastic leukaemic cells. Br J Haematol 1997; 98(3):673–685.
29. Bursch W, Ellinger A, Kienzl H, et al. Active cell death induced by the anti-estrogens tamoxifen and ICI 164 384 in human mammary carcinoma cells (MCF-7) in culture: the role of autophagy. Carcinogenesis 1996; 17(8):1595–1607.

30. Deiss LP, Feinstein E, Berissi H, et al. Identification of a novel serine/threonine kinase and a novel 15-kD protein as potential mediators of the gamma interferon-induced cell death. Genes Dev 1995; 9(1):15–30.
31. Inbal B, Bialik S, Sabanay I, et al. DAP kinase and DRP-1 mediate membrane blebbing and the formation of autophagic vesicles during programmed cell death. J Cell Biol 2002; 157(3):455–468.
32. Baehrecke EH. How death shapes life during development. Nat Rev Mol Cell Biol 2002; 3(10):779–787.
33. Debnath J, Mills KR, Collins NL, et al. The role of apoptosis in creating and maintaining luminal space within normal and oncogene-expressing mammary acini. Cell 2002; 111(1):29–40.
34. Mills KR, Reginato M, Debnath J, et al. Tumor necrosis factor-related apoptosis-inducing ligand (TRAIL) is required for induction of autophagy during lumen formation in vitro. Proc Natl Acad Sci U S A 2004; 101(10):3438–3443.
35. Yu L, Alva A, Su H, et al. Regulation of an ATG7-beclin 1 program of autophagic cell death by caspase-8. Science 2004; 304(5676):1500–1502.
36. Yu L, Wan F, Dutta S, et al. Autophagic programmed cell death by selective catalase degradation. Proc Natl Acad Sci U S A 2006; 103(13):4952–4957.
37. Shimizu S, Kanaseki T, Mizushima N, et al. Role of Bcl-2 family proteins in a non-apoptotic programmed cell death dependent on autophagy genes. Nat Cell Biol 2004; 6(12):1221–1228.
38. Adams JM, Cory S. The Bcl-2 apoptotic switch in cancer development and therapy. Oncogene 2007; 26(9):1324–1337.
39. Adams JM, Cory S. The Bcl-2 protein family: arbiters of cell survival. Science 1998; 281(5381):1322–1326.
40. Wei MC, Zong WX, Cheng EH, et al. Proapoptotic BAX and BAK: a requisite gateway to mitochondrial dysfunction and death. Science 2001; 292(5517):727–730.
41. Ruiz-Vela A, Opferman JT, Cheng EH, et al. Proapoptotic BAX and BAK control multiple initiator caspases. EMBO Rep 2005; 6(4):379–385.
42. Danial NN, Korsmeyer SJ. Cell death: critical control points. Cell 2004; 116(2): 205–219.
43. Pyo JO, Jang MH, Kwon YK, et al. Essential roles of Atg5 and FADD in autophagic cell death: dissection of autophagic cell death into vacuole formation and cell death. J Biol Chem 2005; 280(21):20722–20729.
44. Yousefi S, Perozzo R, Schmid I, et al. Calpain-mediated cleavage of Atg5 switches autophagy to apoptosis. Nat Cell Biol 2006; 8(10):1124–1132.
45. Demarchi F, Bertoli C, Copetti T, et al. Calpain is required for macroautophagy in mammalian cells. J Cell Biol 2006; 175(4):595–605.
46. Boya P, Gonzalez-Polo RA, Casares N, et al. Inhibition of macroautophagy triggers apoptosis. Mol Cell Biol 2005; 25(3):1025–1040.
47. Madden DT, Egger L, Bredesen DE. A calpain-like protease inhibits autophagic cell death. Autophagy 2007; 3(5):519–522.
48. Wolf BB, Goldstein JC, Stennicke HR, et al. Calpain functions in a caspase-independent manner to promote apoptosis-like events during platelet activation. Blood 1999; 94(5):1683–1692.
49. Schotte P, Declercq W, Van Huffel S, et al. Non-specific effects of methyl ketone peptide inhibitors of caspases. FEBS Lett 1999; 442(1):117–121.

50. Rozman-Pungercar J, Kopitar-Jerala N, Bogyo M, et al. Inhibition of papain-like cysteine proteases and legumain by caspase-specific inhibitors: when reaction mechanism is more important than specificity. Cell Death Differ 2003; 10(8): 881–888.
51. Caserta TM, Smith AN, Gultice AD, et al. Q-VD-OPh, a broad spectrum caspase inhibitor with potent antiapoptotic properties. Apoptosis 2003; 8(4):345–352.
52. Chauvier D, Ankri S, Charriaut-Marlangue C, et al. Broad-spectrum caspase inhibitors: from myth to reality? Cell Death Differ 2007; 14(2):387–391.
53. Mizushima N, Yamamoto A, Hatano M, et al. Dissection of autophagosome formation using Apg5-deficient mouse embryonic stem cells. J Cell Biol 2001; 152(4):657–668.
54. Ahr B, Robert-Hebmann V, Devaux C, et al. Apoptosis of uninfected cells induced by HIV envelope glycoproteins. Retrovirology 2004; 1:12.
55. Espert L, Denizot M, Grimaldi M, et al. Autophagy is involved in T cell death after binding of HIV-1 envelope proteins to CXCR4. J Clin Invest 2006; 116(8): 2161–2172.
56. Hanahan D, Weinberg RA. The hallmarks of cancer. Cell 2000; 100(1):57–70.
57. Kondo Y, Kondo S. Autophagy and cancer therapy. Autophagy 2006; 2(2):85–90.
58. Kondo Y, Kanzawa T, Sawaya R, et al. The role of autophagy in cancer development and response to therapy. Nat Rev Cancer 2005; 5(9):726–734.
59. Verheij M, Bartelink H. Radiation-induced apoptosis. Cell Tissue Res 2000; 301 (1):133–142.
60. Kim KW, Mutter RW, Cao C, et al. Autophagy for cancer therapy through inhibition of pro-apoptotic proteins and mammalian target of rapamycin signaling. J Biol Chem 2006; 281(48):36883–36890.
61. Crighton D, Wilkinson S, O'Prey J, et al. DRAM, a p53-induced modulator of autophagy, is critical for apoptosis. Cell 2006; 126(1):121–134.
62. Yue Z, Jin S, Yang C, et al. Beclin 1, an autophagy gene essential for early embryonic development, is a haploinsufficient tumor suppressor. Proc Natl Acad Sci U S A 2003; 100(25):15077–15082.
63. Qu X, Yu J, Bhagat G, et al. Promotion of tumorigenesis by heterozygous disruption of the beclin 1 autophagy gene. J Clin Invest 2003; 112(12):1809–1820.
64. Aita VM, Liang XH, Murty VV, et al. Cloning and genomic organization of beclin 1, a candidate tumor suppressor gene on chromosome 17q21. Genomics 1999; 59(1): 59–65.
65. Ionov Y, Nowak N, Perucho M, et al. Manipulation of nonsense mediated decay identifies gene mutations in colon cancer Cells with microsatellite instability. Oncogene 2004; 23(3):639–645.
66. Furuya D, Tsuji N, Yagihashi A, et al. Beclin 1 augmented cis-diamminedichloroplatinum induced apoptosis via enhancing caspase-9 activity. Exp Cell Res 2005; 307(1):26–40.
67. Kim I, Rodriguez-Enriquez S, Lemasters JJ. Selective degradation of mitochondria by mitophagy. Arch Biochem Biophys 2007; 462(2):245–253.
68. Devenish RJ. Mitophagy: growing in intricacy. Autophagy 2007; 3(4):293–294.
69. Oberstein A, Jeffrey PD, Shi Y. Crystal structure of the Bcl-XL-Beclin 1 peptide complex: Beclin 1 is a novel BH3-only protein. J Biol Chem 2007; 282(17): 13123–13132.

70. Lin SS, Bassik MC, Suh H, et al. PP2A regulates BCL-2 phosphorylation and proteasome-mediated degradation at the endoplasmic reticulum. J Biol Chem 2006; 281(32):23003–23012.
71. Breitschopf K, Haendeler J, Malchow P, et al. Posttranslational modification of Bcl-2 facilitates its proteasome-dependent degradation: molecular characterization of the involved signaling pathway. Mol Cell Biol 2000; 20(5):1886–1896.
72. Dimmeler S, Breitschopf K, Haendeler J, et al. Dephosphorylation targets Bcl-2 for ubiquitin-dependent degradation: a link between the apoptosome and the proteasome pathway. J Exp Med 1999; 189(11):1815–1822.
73. Deng X, Gao F, Flagg T, et al. Mono- and multisite phosphorylation enhances Bcl2's antiapoptotic function and inhibition of cell cycle entry functions. Proc Natl Acad Sci U S A 2004; 101(1):153–158.
74. Ruvolo PP, Deng X, Carr BK, et al. A functional role for mitochondrial protein kinase Calpha in Bcl2 phosphorylation and suppression of apoptosis. J Biol Chem 1998; 273(39):25436–25442.

8

The Cellular Decision Between Apoptosis and Autophagy

Yongjun Fan
Department of Molecular Genetics & Microbiology,
Stony Brook University, Stony Brook, New York, U.S.A.

Erica Ullman
Graduate Program of Molecular & Cellular Biology,
Stony Brook University, Stony Brook, New York, U.S.A.

Wei-Xing Zong
Department of Molecular Genetics & Microbiology,
Stony Brook University, Stony Brook, New York, U.S.A.

INTRODUCTION

Programmed cell death was originally described as a cell self-destruction process that plays a crucial role during the development of metazoans. It is responsible for the structural reorganization of embryos and for maintaining tissue homeostasis, by eliminating unwanted or damaged cells. The term "apoptosis" was invented to describe the type of programmed cell death that displays specific microscopic features such as chromatin condensation, nuclear fragmentation, and plasma membrane blebbing (1). For a good period of time, the terms "apoptosis" and "programmed cell death" were used synonymously, until "autophagic cell death" or "type II programmed cell death" was brought to the attention of the research community. Accordingly, apoptosis is currently referred to as "type I programmed cell death" (2,3).

Like apoptosis, autophagic cell death was initially described by the microscopic features of dying/dead cells. Autophagic cells often lack the typical apoptotic features and almost always contain multi- or double-layer intracellular membrane structures enclosing a bulk of cytoplasmic materials or subcellular organelles. These structures are termed autophagosomes, which are also observed in yeast cells upon nutrient starvation. Autophagosomes fuse with lysosomes to form "autolysosomes." Contents enclosed in autolysosomes are then digested by the lysosomal proteases. This process allows the recycling of intracellular organelles and proteins and benefits cells by (*i*) removing damaged or unwanted organelles and protein molecules, and (*ii*) providing an energy source and building blocks for cellular functions and biosynthesis. These functions clearly have a positive effect on cell survival. Yeast strains harboring mutations in autophagy genes die rapidly during starvation, suggesting that autophagy serves as a survival mechanism (4,5). This survival-promoting function of autophagy in yeast is contradictory to the theory that autophagy is a death mechanism. This raises the question as to whether autophagy can really serve as an active form of cell death or whether it is a cell-responsive event whose features are observed when cells die of irreparable damage (6). Indeed, the situation in multicellular organisms is much more complicated than it is in yeast, which lacks well-defined apoptosis machinery as seen in metazoans. It is becoming increasingly apparent that processes such as apoptosis and autophagy may not be mutually exclusive but rather they may be closely linked events, whose activation and inhibition involve complementary pathways. It has been observed that morphological features characteristic of both autophagy and apoptosis occur simultaneously in the same tissue. Furthermore, the ability of autophagy to elicit a cell survival or cell death response seems to be significantly impacted by whether or not the apoptotic machinery is functional. The interplay between autophagy and apoptosis is a seemingly complex process whose outcome depends on cell type and environmental conditions. Situations have been described in which autophagy acts to antagonize apoptosis, whereas in other cases, autophagy acts as an agonist of apoptosis and yet there are cases where autophagy seemingly induces cell death irrespective of apoptosis.

Autophagy Promotes Cell Survival

Autophagy's ability to provide energy and nutrients to starved cells has been well documented in yeast and metazoans. In this scenario, autophagy functions as an adaptive mechanism, aiding to circumvent cell death. When murine cells deficient in apoptosis are deprived of growth factors, they are observed to atrophy; however cell death is not induced. Inhibition of autophagy in these cells results in cell death with no obvious apoptotic features (7). Another example of the protective role autophagy plays is observed in an apoptosis-deficient xenograft mouse tumor model. Autophagy plays a protective role in these tumor cells, as inhibition of autophagy results in accelerated necrotic death (8). The shortage of nutrients to certain areas of the expanding tumor, due to the lack of vascularization, leads to bioenergetic failure and subsequent cell death.

The other well-documented role for autophagy is degradation of long-lived proteins and damaged organelles. This function of autophagy may actually serve to antagonize apoptosis by preventing the release of death-inducing components from damaged organelles (9).

Autophagy Promotes Cell Death

Several studies have pointed to autophagy as a cell death pathway both in apoptosis-competent and apoptosis-deficient cells. One of the more established ideas of how autophagy leads to cell death is that autophagy, though initially functioning to overcome the cellular stress, reaches a point when so much, or vital components, of the cell has been consumed, thus leading to the collapse of cell function and death. However, it may be possible that proteins involved in autophagy take a more active approach to death induction. It is plausible that certain stimuli activate autophagy proteins that lead directly to a cell death pathway. In some instances, autophagy appears to act as an agonist of other forms of cell death, whereby its activation leads to the induction of apoptosis or necrosis.

Several observations have acknowledged autophagy as a cell death pathway distinct from apoptosis and necrosis. Some of these observations are based solely on morphological features characteristic of autophagic cell death accumulation of autophagic vacuoles, and bulk degradation of cytoplasm followed by nuclear collapse, as well as the absence of apoptotic features. Using these morphological distinctions, it was determined that autophagy, rather than apoptosis, plays an important role during specific types of development, such as death of central cells of the intestine during cavity formation, regression of mammary gland, or atrophy of prostate (2,10). Since these are purely morphological observations, it is difficult to state with absolute certainty that autophagy is acting without apoptosis to kill cells. More significant is the biochemical data using inhibitors of autophagy or RNAi directed against autophagy genes. For example, it is observed that mouse L929 cells treated with zVAD undergo autophagic cell death. These cells were protected when RNAi against autophagy proteins Atg7 and Beclin 1 were used (11). Cells deficient in apoptosis, due to the absence of Bax and Bak, were observed to undergo autophagic cell death when treated with etoposide and staurosporine. This cell death was blocked by RNAi knockdown of Beclin 1 and Atg5 (12).

In response to certain stimuli, autophagy may not be able to efficiently carry out cell death on its own but rather it may act as an agonist of other forms of cell death. For example, in some experimental systems, execution of apoptosis is preceded by, and even depends on the occurrence of autophagy. In these settings, inhibition of autophagy delays apoptosis. This has been observed in primary sympathetic neurons exposed to neural growth factor deprivation or c-Ara treatment (13), in TNF-α-induced apoptosis of T-lymphoblastic leukemia cell lines (14), as well as in the cell death of $CD4^+$ T cells triggered by the HIV-1 envelope (Env) glycoproteins (15). Furthermore, it has also been discovered that the effectiveness of a chemotherapeutic vitamin D analog, EB1089, lies in its ability to activate autophagy leading to nuclear apoptosis (16). Depletion of Beclin 1 in this instance confers resistance to apoptosis. Importantly, Beclin 1 recently was identified as a BH3-only protein that can interact with anti-apoptotic proteins Bcl-2 and Bcl-X_L (17,18). This interaction is most likely significant in the ability of autophagy to activate apoptosis. Another autophagy-related gene, Atg5, when overexpressed in multiple cell lines, is able to sensitize cells to death induced by ceramide or the genotoxins etoposide and doxorubicin. The resulting apoptotic cell death is associated with calpain-mediated Atg5 cleavage (19). Autophagy induction may also lead to necrotic cell death, which appears to be the case in an experimental system where apoptosis-deficient $bax^{-/-}bak^{-/-}$ cells were treated with ER stressors (20).

MOLECULES REGULATING BOTH APOPTOSIS AND AUTOPHAGY

Apoptosis and autophagy are closely connected, both being a cellular response to stress and damage. The mechanisms underlying the interplay between apoptosis and autophagy are currently obscure. Nevertheless, several molecules have been reported to affect both autophagy and apoptosis, suggesting that they may act as molecular switches of these two cellular processes in response to cell-damaging signalings.

Bcl-2 Family of Proteins

The Bcl-2 family of proteins is critical in controlling mitochondrial membrane permeabilization (MMP) and apoptosis. These proteins are distinguished by the presence of up to four Bcl-2 homology domains (BH1–4 domains) and are usually grouped into three distinct subfamilies based on their function and the BH domains: (*i*) the anti-apoptotic proteins including Bcl-2, Bcl-X_L, Mcl-1, Bcl-W, and A1/Bfl-1; (*ii*) the "multidomain" pro-apoptotic proteins including Bax, Bak, and Bok; and (*iii*) the "BH3-only" proteins such as Bad, Bim, Noxa, Puma, Bmf, and Bnip3. The multidomain pro-apoptotic proteins Bax and Bak are the key molecules necessary for inducing apoptosis. In response to apoptosis stimuli, these proteins undergo conformational changes leading to their oligomerization on the mitochondrial outer membrane, followed by cytochrome c release and activation of the caspase cascade. The anti-apoptotic Bcl-2 proteins block this process by interacting with Bax and Bak. The BH3-only proteins are responders to different cell death stimuli. All BH3-only proteins identified so far are pro-apoptotic and function by either activating Bax and Bak, or inhibiting the anti-apoptotic Bcl-2 proteins (21).

Anti-apoptotic Bcl-2 Proteins

The anti-apoptotic Bcl-2 proteins Bcl-2 and Bcl-X_L can inhibit autophagy via interaction with an autophagy essential factor Beclin 1. In fact, Beclin 1 was discovered in a yeast two-hybrid screening for Bcl-2 interacting proteins (22). A similar interaction was found in *C. elegans* between Beclin 1 (BEC-1) and the Bcl-2 homolog CED-9 (23). Other anti-apoptotic Bcl-2 family members, for example, Bcl-X_L, Mcl-1, and a herpesvirus encoded Bcl-2 homolog (vBcl-2), were also found to interact with Beclin 1 (18,24). These anti-apoptotic Bcl-2 proteins suppress starvation-induced autophagy by inhibiting the formation of the Beclin 1/Vps34 PI-3 kinase (class III) complex (25). The subcellular localization of Bcl-2 appears to affect its function as an autophagy regulator. While the majority of cellular Bcl-2 is localized to mitochondria in most cells, a portion of Bcl-2 is localized to the ER membrane (26). It has been found that the ER-targeted Bcl-2, but not mitochondrial Bcl-2, inhibits autophagy (25). A possible explanation for this is that Bcl-2 may require some factors such as IP3 receptors on the ER membrane to function as an autophagy modulator.

Another theory for the inhibitory effect of Bcl-2 on autophagy is through modulation of intracellular Ca^{2+} homeostasis. The increase of free cytosolic Ca^{2+} can induce autophagy. This may be mediated by the ability of calmodulin-dependent kinase kinase-β and AMP-activated protein kinase (AMPK) to inhibit the target of rapamycin (TOR) signaling pathway (27). Bcl-2 can reduce the steady-state level of ER Ca^{2+}, preventing the stress-induced increase of cytosolic Ca^{2+} (28,29), thus suppressing the induction of autophagy (27).

Contrary to the above conditions, where Bcl-2 functions to inhibit autophagy, it has been found that Bcl-2 and Bcl-X_L may enhance autophagy under circumstances such as in cells treated with etoposide and staurosporine (12). The underlying mechanism whereby Bcl-2/Bcl-X_L can promote autophagy is currently unclear. It is possible that different stress signals, for example, starvation versus genotoxins, may cause Bcl-2 family proteins to localize to different subcellular compartments, or to interact with different sets of molecules, to impose the distinctive roles on autophagy induction.

Bax and Bak

Bax and Bak are key molecules in mediating the intrinsic apoptosis pathway. Bax/Bak doubly deficient cells are resistant to a wide range of apoptotic stimuli (30). Nevertheless, these cells can still undergo nonapoptotic cell death, namely autophagic and necrotic cell death (12,31). Long-term treatment of $bax^{-/-}bak^{-/-}$ cells with the genotoxin etoposide leads to a decrease in cell viability, which can be blocked by the inhibition of autophagy, suggesting that autophagy is critical in this Bax/Bak-independent cell death (12). Similar results were found in irradiated $bax^{-/-}bak^{-/-}$ cells (32,33). Noting that Bax and Bak directly interact with the anti-apoptotic Bcl-2 proteins, it is reasonable to consider that Bax and Bak may displace Bcl-2/Bcl-X_L from Beclin 1, whereby affecting autophagy. Another hypothesis suggests that Bax and Bak may play a role in keeping the phosphatase and tension homolog (PTEN) activity low, which can result in a constitutively active AKT-mammalian TOR (AKT-mTOR) signaling and the inhibition of autophagy (34). However, there exists no direct evidence to support these theories. Indeed, in certain settings, Bax and Bak proteins do not seem to affect the induction of autophagy (20). It is more likely that the deficiency in apoptosis due to the lack of Bax and Bak enables autophagy to accumulate to an extent that the cells die of damage resulting from excessive autophagy.

BH3-Only Proteins

The pro-apoptotic BH3-only proteins are categorized into two types: one includes Bim and Bid, which contain an α-helical BH3 domain that can directly induce oligomerization of Bax and Bak on the mitochondrial outer membrane, leading to cytochrome c release. The other includes Bad and Bik, whose BH3 domain binds and antagonizes the anti-apoptotic Bcl-2 proteins and sensitizes cells to apoptosis. In both cases, the BH3 peptides interact with a hydrophobic groove formed by the BH1, BH2, and BH3 domains, and upon doing so, induce apoptosis either by activating Bax and Bak, or by neutralizing Bcl-2 and Bcl-X_L (35). A small molecule compound that mimics the BH3 peptide has been shown to induce apoptosis in cancer cells and can be explored as a potential anticancer drug (36).

In addition to their pro-apoptotic function, BH3-only proteins may also regulate autophagy. BNIP3, a hypoxia-inducible BH3 domain containing protein, leads to accumulation of autophagosomes and nonapoptotic cell death when overexpressed (37). Arsenic trioxide (As_2O_3), which can kill hematological tumor cells via apoptosis, induces autophagic cell death in solid tumor cells through upregulating the expression of BNIP3 (38).

Recently, a chemical BH3 peptide mimetic ABT737, as well as Bad and the *C. elegans* BH3-only homolog EGL-1, have been found to enhance basal level and starvation-induced autophagy. This autophagy enhancement is through the competitive interaction of BH3 peptides with Bcl-2/Bcl-X_L, which displaces Bcl-2/Bcl-X_L from Beclin 1 (18). Thus, the BH3-only proteins may have dual functions serving as inducers for both apoptosis and autophagy. It remains to be determined whether the Bid/Bim-like and Bad/Bik-like BH3-only proteins have the same or different effects in regulating autophagy. Nevertheless, in the above settings, autophagy acts as a survival mechanism, as the inhibition of autophagy enhanced apoptosis induced by the BH3 mimetics and BH3-only proteins (18). Moreover, while Beclin 1/Bcl-2 interaction requires the BH1 and BH2 domains in Bcl-2 (25), a putative BH3 domain has been found in Beclin 1 (17), suggesting a retrospect role of Beclin 1 in regulating apoptosis.

Calpains

Calpains are a family of Ca^{2+}-dependent cysteine proteases. There are at least 14 members that can be categorized into two subfamilies: μ-calpains and m-calpains, named as such to reflect their requirements for Ca^{2+}. μ-calpains are activated by micromolar and m-calpains by millimolar concentrations of Ca^{2+} (39,40). The two subfamilies share a common 30 kDa regulatory subunit and contain a distinct ~80 kDa catalytic subunit (41). Calpains reside in the cytosol in an inactive form. In response to increased levels of cytosolic Ca^{2+}, calpains translocate to the intracellular membranes and are activated by autocatalytic hydrolysis (40). Calpains are involved in both apoptotic and necrotic cell death. Apoptosis is promoted by calpain-mediated cleavage of anti-apoptotic proteins such as Bcl-2, or by the cleavage of certain pro-caspases thereby leading to their activation (42–44). Calpain-mediated cleavage of the Na^{+}/Ca^{2+} exchanger in the plasma membrane has been shown to result in sustained secondary intracellular Ca^{2+} overload and subsequent necrotic cell death (45). Calpains can also contribute to the activation of cathepsins by causing lysosomal membrane permeability (LMP), which leads to the release of lysosomal enzymes and cell death (46).

Recent work suggests that calpains are also involved in autophagy and thus act as a molecular switch to regulate cell fate between apoptosis and autophagy. Calpains are activated by several stimuli that trigger apoptosis and autophagy, such as starvation, ceramide, and etoposide. One recent report showed that calpains are required for autophagy induced by these agents. Genetic deletion or RNAi knockdown of the calpain small 1 regulatory subunit (CAPNS1), which is required for the activation of both μ-calpain and m-calpain, results in a block of autophagy induced by rapamycin, starvation, and ceramide (47). Inhibition of autophagy due to calpain deprivation makes cells more sensitive to apoptosis, suggesting that calpain-mediated autophagy plays a protective role against apoptosis. Another report, however, showed that in response to similar insults, Atg5 is cleaved by calpain at its C-terminus, generating a 24 kDa fragment from its 33 kDa proform. This 24 kDa fragment translocates to mitochondria where it antagonizes Bcl-X_L thereby inducing apoptosis (19). Thus, in this setting, calpain activation switches cells from autophagy to apoptosis by cleaving Atg5. These reports indicate that calpains are an important regulator in apoptosis and autophagy, although the discrepancy between these reports needs to be further investigated.

p53

p53 is a major cellular stress-sensing molecule. Its activation leads to cell cycle arrest or apoptosis (48,49). Recently p53 has also been shown to induce autophagy. Treatment with genotoxins including etoposide, doxorubicin, and actinomycin D resulted in p53-dependent induction of autophagy (50,51). One mechanism for p53-mediated autophagy is through the activation of AMPK, and subsequent activation of TSC1 and TSC2 kinases, leading to the inhibition of mTOR (50). In addition to this rapidly induced inhibition of mTOR, p53 activation also leads to the upregulation of PTEN and TSC2 at the transcriptional level, which may contribute to a long-term suppression of mTOR (50). Another mechanism for p53-induced autophagy is through transcriptional activation of damage-regulated autophagy modulator (DRAM), a p53 transcriptional target (51). While DRAM is required for p53-mediated autophagy, it is not required for amino acid starvation-induced autophagy. This DRAM-dependent autophagy appears to facilitate p53-dependent apoptosis. While it is known that DRAM is a lysosomal protein, how it activates autophagy remains unclear.

While p53 is well recognized as a DNA damage responsive factor, its ability to activate AMPK and inhibit mTOR signifies that it is also capable of signaling to nutrient-sensing pathway. Additional support for this role of p53 is illustrated by its activation in response to nutrient starvation. Glucose starvation leads to the activation of p53 through the phosphorylation of serine 15 on human p53 (serine 18 in mouse). This phosphorylation is mediated by AMPK (50,52). In normal cells, phosphorylation of serine 15 is quickly reversed by a phosphatase composed of the $\alpha 4$ regulatory subunit and the PP2A catalytic subunit, which is in turn activated via phosphorylation by TOR kinase (53). Thus TOR forms a negative feedback loop in the phosphorylation of serine 15 of p53, while AMPK positively acts on serine 15 phosphorylation and activates p53.

The fact that p53 interconnects with the nutrient-sensing molecules is of significant importance. On one hand, genotoxic stress activates p53, which can signal to the nutrient-sensing pathway to switch the cellular bioenergetic status from anabolic to catabolic in order to halt cell growth. On the other hand, nutrient poor conditions can activate p53, causing it to halt DNA replication and cell cycle progression. p53-mediated autophagy may play an important role in regulating these processes, as well as act as a switch to determine cell fate. Autophagy may facilitate the selective degradation of damaged molecules or organelles in order to provide an energy source for the damage repair process. Alternatively, when the extent of damage is beyond repair, autophagy may act to accelerate apoptotic cell death in response to p53 activation.

$p27^{Kip1}$

$p27^{Kip1}$ is a member of the Cip/Kip family of Cdk inhibitors, which binds to Cdk2 (and to other Cdks) and potently inhibits Cdk2 kinase activity (54). $p27^{Kip1}$ overexpression in human cells leads to cell cycle arrest in the G1 phase. It has been reported that $p27^{kip1}$ mediates the decision to enter autophagy or apoptosis when cells are under metabolic stress (55). The stability and function of $p27^{kip1}$ is regulated by phosphorylation by different kinases. Activation of the LKB1-AMPK pathway results in the phosphorylation of Thr198 of $p27^{kip1}$; the phosphorylation promotes $p27^{kip1}$ stability. Overexpression of a stable $p27^{kip1}$ mutant (Thr198-Asp)-induced autophagy even in the presence of serum. In serum-starved cells, both Thr198 phosphorylation and $p27^{kip1}$ stability increased, concurrent with autophagy induction, suggesting that $p27^{kip1}$ plays a role in serum deprivation–induced autophagy. While serum deprivation, inhibition of

PI(3)K by LY294002, or inhibition of mTOR by rapamycin induced autophagy in cells with normal $p27^{kip1}$ levels, these treatments induced apoptotic cell death in $p27^{kip1}$-RNAi knockdown cells (55). Thus, the $p27^{kip1}$ can function as a determinant of whether quiescent cells enter the autophagy cell survival pathway or undergo apoptosis. Since tumor cells often encounter metabolic stress imposed by their microenvironment and certain chemotherapeutic treatments, $p27^{kip1}$ may function to promote tumor cell survival through the maintenance of autophagy.

smARF

The mouse p19ARF (p14ARF in humans) is a tumor suppressor protein. It activates p53 by antagonizing a p53 inhibitor Mdm2 (Hdm2 in human) (56). ARF initiates p53-dependent cell cycle arrest and enhances apoptosis. However, in mice lacking functional Mdm2 and p53, deficiency in p19ARF makes these animals more tumor prone, suggesting a Mdm2- and p53-independent tumor suppressive function of p19ARF (57). While several explanations have been proposed for p53-independent effects of p19ARF (58), it has been recently found that a short form of ARF (smARF) can induce autophagic cell death in a p53-independent manner. This may also contribute to its p53-independent tumor suppressor function (59).

smARF protein is translationally initiated from the single internal methionine codon in both mouse and human ARF mRNA (Met45 in mouse p19ARF and Met48 in human p14ARF). Mutation of the canonical p19ARF initiation codon (Met1) or deletion of sequences encoding ARF's 40 N-terminal amino acids leads to increased production of smARF. Unlike the full-length ARF, smARF is localized to mitochondria. It causes mitochondrial morphological changes, loss of mitochondrial membrane potential, and eventually cell death. These effects are observed in murine embryonic fibroblasts (MEFs) lacking functional p53, or the pro-apoptotic proteins Bax and Bak, and could not be circumvented by the anti-apoptotic factors Bcl-2 and Bcl-X_L. Moreover, smARF does not trigger cytochrome c release and caspase activation, all pointing to a nonapoptotic form of cell death. Indeed, smARF was found to stimulate the accumulation of autophagic vesicles. RNAi knockdown of Atg5 and BEC-1 attenuated smARF-induced cell death. Thus, it appears that smARF can induce autophagy as a mechanism to promote cell death (59). Therefore, there may exist different tumor suppressing mechanisms mediated by the two ARF isoforms: the full-length p19ARF triggers a rapid, p53-dependent nuclear response, and the short form smARF activates a slowly evolving, mitochondria-based autophagy program that gains primacy when p53 is dysfunctional.

Death-Associated Protein Kinase Family

Death-associated protein kinase (DAPK) is a calcium/calmodulin-regulated serine/threonine kinase, initially isolated as a mediator of apoptosis induced by interferon-γ (60). Further studies revealed that DAPK is a founder member of a protein kinase family, which includes two other proteins, ZIPk and DRP-1. These proteins share a high degree of homology in their kinase catalytic domains and function to promote cell death triggered by various death stimuli including interferon-γ, cell death receptors, TGF-β, oncogene expression, and anoikis (loss of adherence to extracellular matrix) (61,62). Interestingly, the cell death type induced by ectopic expression of the DAPK family members depends on cellular and experimental conditions. In primary fibroblasts, overexpression of DAPKs leads to apoptosis, whereas in tumor cell lines such as HeLa, MCF-7, or 293T cells, overexpression of DAPKs leads to cell death with autophagic features (63). A dominant-negative DRP-1 reduced the level of starvation and tamoxifen-induced autophagy in MCF-7 breast carcinoma cells, whereas the reduction of DAPK expression by antisense RNA attenuated interferon-γ-induced autophagy in HeLa cells (63), suggesting that these kinases may be necessary for autophagy processes. Interestingly, DRP-1 is localized to the lumen of autophagic vesicles. This raises the possibility that DRP-1 may have a direct role in autophagic vesicle formation, possibly by phosphorylating factors involved in this process (63).

DNA-PK

DNA-PK is a nuclear serine/threonine kinase and a member of the phosphatidylinositol 3-kinase-like family. It consists of the catalytic subunit (DNA-PKcs), and a DNA binding component Ku70/Ku80 (64). Its activity is stimulated by double-stranded DNA ends. DNA-PK plays a central role in the nonhomologous end joining (NHEJ) pathway for double-stranded break (DSB) repair, and in V(D)J recombination. The phosphorylation targets of the DNA-PKcs include DNA-PKcs (65), both Ku subunits (66), XRCC4 (67), p53 (68), Mdm2, (69) and c-Abl (70). The phosphorylation of DNA-PKcs, Ku subunits, and XRCC4 is associated with DNA repair, whereas that of p53, Mdm2, and c-Abl induces apoptosis. The different functions of these targets are, therefore, consistent with the notion that DNA-PK has dual roles in responding to DNA damage: one is to sense DNA damage and facilitate the repair processes, and the other is to induce apoptosis (71). DNA damage does not induce apoptosis in DNA-PKcs$^{-/-}$ thymocytes (72,73). Further studies found that when treated with γ-irradiation, DNA-PKcs-deficient M059J cells die with autophagic features. Inhibition of DNA-PKcs-induced autophagy and sensitized malignant glioma cells to γ-irradiation (74). Thus, DNA-PKcs may also multitask by inducing apoptosis while inhibiting autophagy.

CONCLUSION

Autophagy, as a cellular response to nutrient deprivation, is an evolutionarily more ancient event than apoptosis. As unicellular organisms such as yeast developed a complex molecular pathway to regulate autophagy, apoptosis appeared later in metazoans to ensure tissue and organismal homeostasis. Both processes play critical roles in controlling cell death/survival, thus they need to be delicately regulated. It is thus not surprising that certain canonical apoptotic molecules can regulate autophagy, and vice versa. Multitasking by certain molecules offers advantages to regulate these important biological events more rapidly, efficiently, and precisely. Autophagy clearly has both pro-survival and pro-death functions, and this is largely dependent on the cell's ability to die by apoptosis. In response to stress or damage, autophagy is induced to shut down the anabolic processes and keep cells viable in a metabolic inert status, so that cells can do the repair and escape further damage. When the cellular damage is too severe, apoptosis kicks in to eliminate irreparable cells in order to maintain tissue homeostasis. In cells with defective apoptosis, autophagy may serve as a backup strategy for cellular demise, or as a mechanism to promote other forms of cell death, such as necrosis. This is of significant interest in cancer cells since the apoptosis machinery is often mutated during tumorigenesis. Thus, targeting autophagy should be an important consideration in developing novel strategies for cancer therapy.

REFERENCES

1. Kerr JF, Wyllie AH, Currie AR. Apoptosis: a basic biological phenomenon with wide-ranging implications in tissue kinetics. Br J Cancer 1972; 26(4):239–257.
2. Clarke PG. Developmental cell death: morphological diversity and multiple mechanisms. Anat Embryol (Berl) 1990; 181(3):195–213.
3. Lockshin RA, Zakeri Z. Apoptosis, autophagy, and more. Int J Biochem Cell Biol 2004; 36(12):2405–2419.
4. Tsukada M, Ohsumi Y. Isolation and characterization of autophagy-defective mutants of *Saccharomyces cerevisiae*. FEBS Lett 1993; 333(1–2):169–174.
5. Otto GP, Wu MY, Kazgan N, et al. Macroautophagy is required for multicellular development of the social amoeba *Dictyostelium discoideum*. J Biol Chem 2003; 278(20):17636–17645.
6. Levine B, Yuan J. Autophagy in cell death: an innocent convict? J Clin Invest 2005; 115(10):2679–2688.
7. Lum JJ, Bauer DE, Kong M, et al. Growth factor regulation of autophagy and cell survival in the absence of apoptosis. Cell 2005; 120(2):237–248.
8. Degenhardt K, Mathew R, Beaudoin B, et al. Autophagy promotes tumor cell survival and restricts necrosis, inflammation, and tumorigenesis. Cancer Cell 2006; 10(1):51–64.
9. Lemasters JJ, Nieminen AL, Qian T, et al. The mitochondrial permeability transition in cell death: a common mechanism in necrosis, apoptosis and autophagy. Biochim Biophys Acta 1998; 1366(1–2):177–196.
10. Bursch W. The autophagosomal-lysosomal compartment in programmed cell death. Cell Death Differ 2001; 8(6):569–581.
11. Yu L, Alva A, Su H, et al. Regulation of an ATG7-beclin 1 program of autophagic cell death by caspase-8. Science 2004; 304(5676):1500–1502.

12. Shimizu S, Kanaseki T, Mizushima N, et al. Role of Bcl-2 family proteins in a nonapoptotic programmed cell death dependent on autophagy genes. Nat Cell Biol 2004; 6(12):1221–1228.
13. Xue L, Fletcher GC, Tolkovsky AM. Autophagy is activated by apoptotic signaling in sympathetic neurons: an alternative mechanism of death execution. Mol Cell Neurosci 1999; 14(3):180–198.
14. Jia L, Dourmashkin RR, Allen PD, et al. Inhibition of autophagy abrogates tumour necrosis factor alpha induced apoptosis in human T-lymphoblastic leukaemic cells. Br J Haematol 1997; 98(3):673–685.
15. Espert L, Denizot M, Grimaldi M, et al. Autophagy is involved in T cell death after binding of HIV-1 envelope proteins to CXCR4. J Clin Invest 2006; 116(8): 2161–2172.
16. Hoyer-Hansen M, Bastholm L, Mathiasen IS, et al. Vitamin D analog EB1089 triggers dramatic lysosomal changes and Beclin 1-mediated autophagic cell death. Cell Death Differ 2005; 12(10):1297–1309.
17. Oberstein A, Jeffrey P, Shi Y. Crystal structure of the BCL-XL-beclin 1 peptide complex: Beclin 1 is a novel BH3-only protein. J Biol Chem 2007; 282(17): 13123–13132.
18. Maiuri MC, Le Toumelin G, Criollo A, et al. Functional and physical interaction between Bcl-X(L) and a BH3-like domain in Beclin-1. EMBO J 2007; 26(10): 2527–2539.
19. Yousefi S, Perozzo R, Schmid I, et al. Calpain-mediated cleavage of Atg5 switches autophagy to apoptosis. Nat Cell Biol 2006; 8(10):1124–1132.
20. Ullman E, Fan Y, Stawowczyk M, et al. Autophagy promotes necrosis in apoptosis-deficient cells in response to ER stress. Cell Death Differ 2008; 16:422–425.
21. Danial NN, Korsmeyer SJ. Cell death: critical control points. Cell 2004; 116(2): 205–219.
22. Liang XH, Kleeman LK, Jiang HH, et al. Protection against fatal Sindbis virus encephalitis by beclin, a novel Bcl-2-interacting protein. J Virol 1998; 72(11): 8586–8596.
23. Takacs-Vellai K, Vellai T, Puoti A, et al. Inactivation of the autophagy gene bec-1 triggers apoptotic cell death in *C. elegans*. Curr Biol 2005; 15(16):1513–1517.
24. Liang C, Feng P, Ku B, et al. Autophagic and tumour suppressor activity of a novel Beclin1-binding protein UVRAG. Nat Cell Biol 2006; 8(7):688–699.
25. Pattingre S, Tassa A, Qu X, et al. Bcl-2 antiapoptotic proteins inhibit Beclin 1-dependent autophagy. Cell 2005; 122(6):927–939.
26. Ferri KF, Kroemer G. Organelle-specific initiation of cell death pathways. Nat Cell Biol 2001; 3(11):E255–E263.
27. Hoyer-Hansen M, Bastholm L, Szyniarowski, P et al. Control of macroautophagy by calcium, calmodulin-dependent kinase kinase-beta, and Bcl-2. Mol Cell 2007; 25(2):193–205.
28. Pinton P, Ferrari D, Magalhaes P, et al. Reduced loading of intracellular Ca(2+) stores and downregulation of capacitative Ca(2+) influx in Bcl-2-overexpressing cells. J Cell Biol 2000; 148(5):857–862.
29. Foyouzi-Youssefi R, Arnaudeau S, Borner C, et al. Bcl-2 decreases the free Ca2+ concentration within the endoplasmic reticulum. Proc Natl Acad Sci U S A 2000; 97(11):5723–5728.

30. Wei MC, Zong WX, Cheng EH, et al. Proapoptotic BAX and BAK: a requisite gateway to mitochondrial dysfunction and death. Science 2001; 292(5517):727–730.
31. Zong WX, Ditsworth D, Bauer DE, et al. Alkylating DNA damage stimulates a regulated form of necrotic cell death. Genes Dev 2004; 18(11):1272–1282.
32. Kim KW, Mutter RW, Cao C, et al. Autophagy for cancer therapy through inhibition of pro-apoptotic proteins and mammalian target of rapamycin signaling. J Biol Chem 2006; 281(48):36883–36890.
33. Buytaert E, Callewaert G, Hendrickx N, et al. Role of endoplasmic reticulum depletion and multidomain proapoptotic BAX and BAK proteins in shaping cell death after hypericin-mediated photodynamic therapy. FASEB J 2006; 20(6): 756–758.
34. Moretti L, Attia A, Kim KW, et al. Crosstalk between Bak/Bax and mTOR signaling regulates radiation-induced autophagy. Autophagy 2007; 3(2):142–144.
35. Letai A, Bassik MC, Walensky LD, et al. Distinct BH3 domains either sensitize or activate mitochondrial apoptosis, serving as prototype cancer therapeutics. Cancer Cell 2002; 2(3):183–192.
36. Oltersdorf T, Elmore SW, Shoemaker AR, et al. An inhibitor of Bcl-2 family proteins induces regression of solid tumours. Nature 2005; 435(7042):677–681.
37. Vande Velde C, Cizeau J, Dubik D, et al. BNIP3 and genetic control of necrosis-like cell death through the mitochondrial permeability transition pore. Mol Cell Biol 2000; 20(15):5454–5468.
38. Kanzawa T, Zhang L, Xiao L, et al. Arsenic trioxide induces autophagic cell death in malignant glioma cells by upregulation of mitochondrial cell death protein BNIP3. Oncogene 2005; 24(6):980–991.
39. Suzuki K, Sorimachi H. A novel aspect of calpain activation. FEBS Lett 1998; 433(1–2):1–4.
40. Goll DE, Thompson VF, Li H, et al. The calpain system. Physiol Rev 2003; 83(3): 731–801.
41. Suzuki K, Hata S, Kawabata Y, et al. Structure, activation, and biology of calpain. Diabetes 2004; 53(suppl 1):S12–S18.
42. Chen M, He H, Zhan S, et al. Bid is cleaved by calpain to an active fragment in vitro and during myocardial ischemia/reperfusion. J Biol Chem 2001; 276(33): 30724–30728.
43. Nakagawa T, Yuan J. Cross-talk between two cysteine protease families. Activation of caspase-12 by calpain in apoptosis. J Cell Biol 2000; 150(4):887–894.
44. Wood DE, Thomas A, Devi LA, et al. Bax cleavage is mediated by calpain during drug-induced apoptosis. Oncogene 1998; 17(9):1069–1078.
45. Bano D, Young KW, Guerin CJ, et al. Cleavage of the plasma membrane Na+/Ca2+ exchanger in excitotoxicity. Cell 2005; 120(2):275–285.
46. Yamashima T. Ca2+-dependent proteases in ischemic neuronal death: a conserved 'calpain-cathepsin cascade' from nematodes to primates. Cell Calcium 2004; 36(3–4):285–293.
47. Demarchi F, Bertoli C, Copetti T, et al. Calpain is required for macroautophagy in mammalian cells. J Cell Biol 2006; 175(4):595–605.
48. Kastan MB, Zhan Q, el-Deiry WS, et al. A mammalian cell cycle checkpoint pathway utilizing p53 and GADD45 is defective in ataxia-telangiectasia. Cell 1992; 71(4):587–597.

49. Lowe SW, Ruley HE, Jacks T, et al. p53-dependent apoptosis modulates the cytotoxicity of anticancer agents. Cell 1993; 74(6):957–967.
50. Feng Z, Zhang H, Levine AJ, et al. The coordinate regulation of the p53 and mTOR pathways in cells. Proc Natl Acad Sci U S A 2005; 102(23):8204–8209.
51. Crighton D, Wilkinson S, O'Prey J, et al. DRAM, a p53-induced modulator of autophagy, is critical for apoptosis. Cell 2006; 126(1):121–134.
52. Jones RG, Plas DR, Kubek S, et al. AMP-activated protein kinase induces a p53-dependent metabolic checkpoint. Mol Cell 2005; 18(3):283–293.
53. Kong M, Fox CJ, Mu J, et al. The PP2A-associated protein alpha4 is an essential inhibitor of apoptosis. Science 2004; 306(5696):695–698.
54. Sherr CJ, Roberts JM. CDK inhibitors: positive and negative regulators of G1-phase progression. Genes Dev 1999; 13(12):1501–1512.
55. Liang J, Shao SH, Xu ZX, et al. The energy sensing LKB1-AMPK pathway regulates p27(kip1) phosphorylation mediating the decision to enter autophagy or apoptosis. Nat Cell Biol 2007; 9(2):218–224.
56. Kamijo T, Weber JD, Zambetti G, et al. Functional and physical interactions of the ARF tumor suppressor with p53 and Mdm2. Proc Natl Acad Sci U S A 1998; 95(14):8292–8297.
57. Weber JD, Jeffers JR, Rehg JE, et al. p53-independent functions of the p19(ARF) tumor suppressor. Genes Dev 2000; 14(18):2358–2365.
58. Sherr CJ, Bertwistle D, DEN Besten W, et al. p53-dependent and -independent functions of the Arf tumor suppressor. Cold Spring Harb Symp Quant Biol 2005; 70:129–137.
59. Reef S, Zalckvar E, Shifman O, et al. A short mitochondrial form of p19ARF induces autophagy and caspase-independent cell death. Mol Cell 2006; 22(4):463–475.
60. Deiss LP, Feinstein E, Berissi H, et al. Identification of a novel serine/threonine kinase and a novel 15-kD protein as potential mediators of the gamma interferon-induced cell death. Genes Dev 1995; 9(1):15–30.
61. Cohen O, Kimchi A. DAP-kinase: from functional gene cloning to establishment of its role in apoptosis and cancer. Cell Death Differ 2001; 8(1):6–15.
62. Jang CW, Chen CH, Chen CC, et al. TGF-beta induces apoptosis through Smad-mediated expression of DAP-kinase. Nat Cell Biol 2002; 4(1):51–58.
63. Inbal B, Bialik S, Sabanay I, et al. DAP kinase and DRP-1 mediate membrane blebbing and the formation of autophagic vesicles during programmed cell death. J Cell Biol 2002; 157(3):455–468.
64. Smith GC, Jackson SP. The DNA-dependent protein kinase. Genes Dev 1999; 13(8): 916–934.
65. Douglas P, Sapkota GP, Morrice N, et al. Identification of in vitro and in vivo phosphorylation sites in the catalytic subunit of the DNA-dependent protein kinase. Biochem J 2002; 368(pt 1):243–251.
66. Chan DW, Ye R, Veillette CJ, et al. DNA-dependent protein kinase phosphorylation sites in Ku 70/80 heterodimer. Biochemistry 1999; 38(6):1819–1828.
67. Yu Y, Wang W, Ding Q, et al. DNA-PK phosphorylation sites in XRCC4 are not required for survival after radiation or for V(D)J recombination. DNA Repair (Amst) 2003; 2(11):1239–1252.
68. Woo RA, McLure KG, Lees-Miller SP, et al. DNA-dependent protein kinase acts upstream of p53 in response to DNA damage. Nature 1998; 394(6694):700–704.

69. Mayo LD, Turchi JJ, Berberich SJ. Mdm-2 phosphorylation by DNA-dependent protein kinase prevents interaction with p53. Cancer Res 1997; 57(22):5013–5016.
70. Kharbanda S, Pandey P, Jin S, et al. Functional interaction between DNA-PK and c-Abl in response to DNA damage. Nature 1997; 386(6626):732–735.
71. Bernstein C, Bernstein H, Payne CM, et al. DNA repair/pro-apoptotic dual-role proteins in five major DNA repair pathways: fail-safe protection against carcinogenesis. Mutat Res 2002; 511(2):145–178.
72. Wang S, Guo M, Ouyang H, et al. The catalytic subunit of DNA-dependent protein kinase selectively regulates p53-dependent apoptosis but not cell-cycle arrest. Proc Natl Acad Sci U S A 2000; 97(4):1584–1588.
73. Woo RA, Jack MT, Xu Y, et al. DNA damage-induced apoptosis requires the DNA-dependent protein kinase, and is mediated by the latent population of p53. EMBO J 2002; 21(12):3000–3008.
74. Daido S, Yamamoto A, Fujiwara K, et al. Inhibition of the DNA-dependent protein kinase catalytic subunit radiosensitizes malignant glioma cells by inducing autophagy. Cancer Res 2005; 65(10):4368–4375.

9

Parthanatos: PARP- and AIF-Dependent Programmed Cell Death

Valina L. Dawson and Ted M. Dawson
Institute for Cell Engineering, Johns Hopkins University School of Medicine, Baltimore, Maryland, U.S.A.

INTRODUCTION

In the nervous system, cell death is important during development, replacement of mitotic cells, and for remodeling. Cell death in response to traumatic, ischemic, or intrinsic events can lead to loss of normal function in diseases of the nervous system ranging from traumatic brain injury, stroke, neurodegenerative diseases, demyelinating diseases, and cancer. The pathways toward cell death in the nervous system have been of scientific and medical interest. During development, there is programmed cell death that follows criteria originally described by Lockshin and colleagues in insect cells (1) and subsequently extend into understanding the development of the mammalian nervous system. Outside of development, it is not clear if programmed cell death continues to play a role. Apoptosis and necrosis were originally defined morphologically (2), and subsequently, biochemical pathways for apoptosis have been intensively studied and defined. Activation of caspases is the defining biochemical feature of apoptosis. Apoptosis can occur in the nervous system in restricted settings, but in most cases, cell death does not conform to the morphologic criteria of apoptosis and is caspase independent. Autophagy, originally defined as a mechanism to recycle organelles and proteins, has emerged as a potentially important response to diseases involving misfolded proteins and thus may be important in certain neurodegenerative diseases. Interference with the autophagic process can lead to neurodegenerative phenotypes in mice (3–5).

Apoptosis, necrosis, and autophagy do not fully cover the range of cell death processes in the nervous system, leaving many forms of cell death unaccounted. Recently, we have found that the molecule poly(ADP-ribose) (PAR), which is generated by the nuclear enzyme poly(ADP-ribose) polymerase-1 (PARP-1), is a cell death signal (6,7). PARP-1-dependent cell death can be triggered by excitotoxicity in neuronal systems or by DNA nicks and breaks in neuronal and non-neuronal cells due to DNA-alkylating agents. This form of cell death results morphologically in a rapidly shrunken nucleus and subsequent reduction in cell size, annexin-V positivity, and membrane permeability (8,9). The protein apoptosis inducing factor (AIF) translocates to the nucleus shortly after PAR formation, which likely accounts for the rapid reduction in nuclear size. This form of cell death is caspase independent in that broad-spectrum inhibition of caspases does not rescue against toxicity (6–9).

POLY(ADP-RIBOSE) POLYMERASE-1

While it has been over 40 years since the observation that PAR polymer could be generated from nicotinamide adenine dinucleotide (NAD) (10), it was not until 1987 that the gene for PARP-1 was identified (11–13). For several years, it was thought to be the only enzyme capable of conducting poly-(ADP-ribosyl)ation reactions, but clearly, PARP-1 belongs to a family of proteins that currently stands at 17 putative isoforms. Of these 17 PARPs, to date, six isoforms have been characterized as functionally capable of carrying out poly-ADP-ribosylation reactions. These include PARP-1, PARP-2 and PARP-3, vault PARP (PARP-4) and tankyrase 1 (PARP-5), and tankyrase 2a (PARP-6a) (14). Of these isoforms, it is thought that PARP-1 has the primary ability to synthesize large, complex, branched-chain PAR (15). Of all the PARP family members, PARP-1 is the enzyme largely responsible for generating cytotoxic PAR, based on the observations of robust cell survival in PARP-1 knockout cells and tissues that are exposed to a variety of toxic insults.

The biologic significance of PARP-1 is mediated through its direct interaction with other proteins, the poly(ADP-ribosyl)ation of target proteins, and nonenzymatic modification of proteins by PAR generated by PARP-1. The physical association of PARP-1 with transcription factors and cofactors such as p300/CBP, NF-kB, and AP-1 is important in the acute inflammatory response (14,16). While these activities have direct relevance to tissue injury, they have not been observed to be directly involved in PARP-1-dependent, caspase-independent cell death. The enzymatic activation of PARP-1 and the production of PAR appears to be the more proximal event in cell death (6,7). PARP-1 generates PAR through the catabolism of NAD^+ in response to DNA-strand nicks and breaks (Fig. 1). PAR can be a linear or branched-chain polyanion and can be several ribose units to several hundred ribose units in size

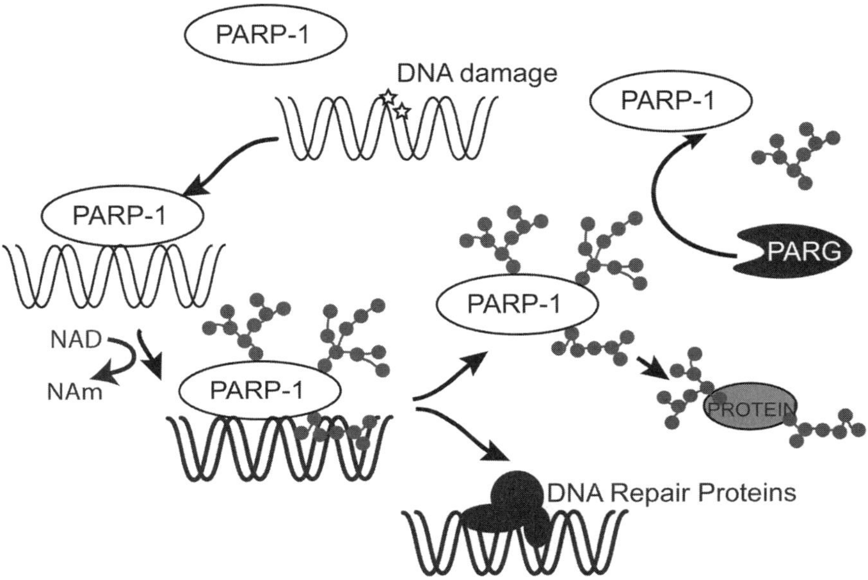

Figure 1 PARP-1 cycle in response to DNA damage. The longest studied action of PARP-1 is its role in maintenance of genomic integrity. PARP-1 senses DNA-strand nicks and breaks. This activates the enzymatic activity of PARP-1 leading to the generation of large, branched-chain PAR. PARP-1 ribosylates itself as well as other nuclear proteins such as histones. PARP-1 dissociates from the damaged DNA allowing DNA repair enzymes access. The glycolytic enzyme, PARG, degrades PAR, removing PAR from PARP-1 and other ribosylated proteins. *Abbreviations*: PARP-1, poly(ADP-ribose) polymerase-1; PAR, poly(ADP-ribose); PARG, poly(ADP-ribose) glycohydrolase.

(14,16). It is not yet known what events regulate the size or complexity of PAR generated by PARP-1.

The catabolism of PAR is mediated by poly(ADP-ribose) glycohydrolase (PARG) (16,17). In mammalian cells, there appears to be only one gene for PARG that generates several proteins of difference sizes by alternative splicing. The biologic roles and cellular distribution of these splice variants are not yet known. PARG appears capable of fully degrading PAR through exoglycosidase and endoglycosidase reactions, resulting in free PAR and free ADP-ribose (16,17). However, genetic deletion results in an embryonic lethal in both drosophila (18) and mice (19), suggesting a key role of PARG in regulating PAR homeostasis in cells. Recently, PARG activity has been assigned to ARH3, a structurally unrelated protein (20). The glycohydrolase activity of this protein is much lower than that of PARG and endogenous activity of this protein has yet to be determined. Since there does not appear to be compensation for PARG by ARH3 in the PARP gene–deleted animals, it would appear that the glycohydrolase functions of ARH3 may not contribute significantly to cell viability.

Poly(ADP-ribosyl)ation is a common and ancient biochemical pathway. The rapid synthesis of PAR following DNA nicks and damage and the subsequent degradation of PAR by PARG indicate a tightly coordinated action between these two proteins to regulate PAR metabolism and the levels of PAR. Disruption of PAR homeostasis can result in cell injury and death.

PARP-1 IN MODELS OF DISEASE

The profound protection afforded by inhibition of PARP or genetic deletion of PARP-1 across organ systems has sparked intense interest. Through these experimental approaches, PARP-1 activity has been implicated in ischemia-reperfusion injury in the brain (21–27), spinal cord (28,29), retina (30–32), cochlea (33), heart (34–40), lung (41–44), liver (45–49), skeletal muscle (38,50,51), kidney (52–57), intestine (58), and testes (59,60). Activation of PARP-1 has been implicated in experimental models of neurodegeneration including the 1-methyl-4-phenyl-1,2,3,6-tetrahydropyridine (MPTP) intoxication model of Parkinson's disease (61–64) and Alzheimer's disease (65,66). PARP inhibition or gene deletion provides protection against models of traumatic brain injury (67–73) and spinal cord injury (28,29,74–77). A role of PARP activation in models of diabetes has been long appreciated and extensively studied. Inhibition or gene deletion of PARP-1 can protect against streptozotozin-induced loss of pancreatic β-islet cells and development of experimental diabetes (78–80). Inhibition of PARP in type 2 diabetic mice and nonobese diabetic mouse models of type 1 diabetes protects and reverses the cellular complications of diabetes (81,82). The broad and diverse organ systems and tissue types affected by PARP in injury and disease suggest that PARP-1 and PAR signaling in cell death is an important cell death cascade to understand and control.

PARP-1 AND EXCITOTOXICITY

The signal cascade for PARP-1- and PAR-mediated cell death has been explored in cell culture models of excitotoxicity in primary neuronal cultures and DNA damage in primary mouse fibroblast and HeLa cell cultures. Ischemia-reperfusion injury in the brain is caused in large part by activation of glutamate receptors of the N-methyl-D-aspartate (NMDA) receptor subtype. This injury can be modeled in cell culture by exposure to oxygen-glucose deprivation (83) or to brief excitotoxic concentrations of glutamate or NMDA (84). Because of rapid and efficient transport of glutamate, NMDA is the toxin of choice, although it is not a compound naturally produced by mammalian cells. In a manner similar to ischemic and excitotoxic injury to the brain, short exposure to glutamate or glutamate analogs does not kill neurons immediately, but activates a signaling cascade, resulting in cell death many hours later (85). The influx of calcium and loss of calcium homeostasis is a key event in glutamate excitotoxicity.The next key step in the cascade is the activation of nitric oxide synthase and the production of nitric oxide and peroxynitrite (86–89). Peroxynitrite is a potent oxidant with many cellular targets (90). During excitotoxicity, the primary target is oxidative damage of DNA leading to DNA nicks and breaks (8). This DNA damage triggers the activation of PARP-1 and subsequent neurotoxicity (21,91).

It was presumed that the causal event for PARP-1-mediated cell death was the consumption of NAD leading to significant decrements in cellular NAD and ATP. However, direct evidence for energy depletion playing a role in PARP-1-dependent cell death is difficult to obtain. Increasing the energy reserves in neurons clearly provides protection (92,93), however, these experiments are difficult to control. Because of a plethora of targets for energy molecules in a cell, simply increasing the availability of bioenergetic substrates or adding NAD to the cell can have many off-target signaling events (94) that may be independent of the events due to catabolism of NAD by PARP-1. In the nervous system, following PARP-1 activation, NAD^+ and ATP depletion are early events; yet, neurons die hours later, suggesting that other downstream mediators might be required for cell death. While inhibition and genetic deletion of PARP-1 preserves NAD^+ and ATP levels, recent data indicate that preservation of energy stores in PARP-1 knockout mice is not the mechanism for the reduced infarct volume following experimental stroke (95). The time course and severity of the apparent diffusion coefficient (ADC), an in vivo measure of cellular energy stores, was not altered in PARP-1 knockout brains compared to wild-type brains (95). In primary neuronal cultures, activation of PARP-1 leads to cell death, but energy collapse is not the primary mediator of cytotoxicity (95). Recent evidence suggests that energy depletion alone might not be sufficient to mediate PARP-1-dependent cell death (92,96). Identification of the translocation of AIF as a downstream event in PARP-1-mediated cell death (7–9,24) provides a tool to dissect the importance of bioenergetics in PARP-1-dependent cell death (Fig. 2).

PARP-1- AND PAR-MEDIATED CELL DEATH

The enzymatic activation of PARP-1 results in a reduction in NAD due to catabolism to generate PAR. PARP-1 ribosylates itself and other nuclear proteins. Recent work indicates that there are proteins that can form strong noncovalent links with PAR, thus affecting protein function (16,97). A protein-binding motif composed of hydrophobic and basic amino acid residues is remarkably specific and selective (15,98). It is thus possible that PAR can act as a cell signaling molecule independent of the localization of PARP-1. Synthetic PAR is directly toxic to neurons in a complexity and dose-dependent manner (6). Degradation of the PAR eliminates its neurotoxic effects. Experimental manipulations to decrease cellular PAR with neutralizing antibodies to PAR or by overexpression of PARG reduces PARP-1-dependent NMDA excitotoxicity of cortical neurons. Conversely, neuronal cultures with reduced levels of PARG

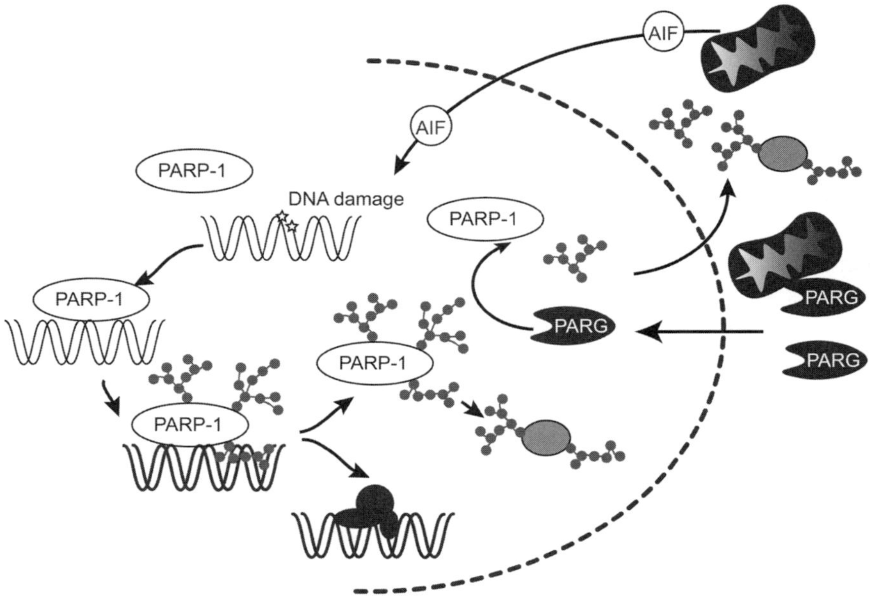

Figure 2 PAR signaling. The generation of PAR by PARP-1 leads to ribosylated proteins. PARG is largely localized to the cytosol but can shuttle between the cytosol and nucleus and is recruited to the nucleus by as yet unknown signals when PARP-1 is activated. PAR can leave the nucleus to act at mitochondria. It is not yet known if this PAR is free or bound to a protein. PAR itself is sufficient to induce the release of AIF from mitochondria. On release from mitochondria, AIF translocates to the nucleus. Subsequently, chromatin condenses and there is large DNA fragmentation and nuclear condensation. This is likely a commitment point for cell death. *Abbreviations*: PAR, poly(ADP-ribose); PARP-1, poly(ADP-ribose) polymerase-1; PARG, poly(ADP-ribose) glycohydrolase; AIF, apoptosis inducing factor.

and increased levels of PAR are more sensitive to NMDA excitotoxicity then wild-type cultures. These observations are reproducible in vivo in an experimental stroke model. Transgenic mice overexpressing PARG have significantly reduced infarct volumes following focal ischemia, while heterozygotic PARG-deficient mice have significantly increased infarct volumes following focal ischemia, compared to wild-type littermate controls (6). Fractionation studies and immunohistochemical studies show that PAR is found in the cytosol following PARP activation and that it partially associates with mitochondria (7). PAR is capable of triggering the release of AIF from mitochondria (7). Taken together, the published literature indicates that PARP-1 cell death is a ubiquitous form of cell death occurring in many different cell types. It is mediated, in large part, by the release of AIF from the nucleus. In Harlequin mice, AIF is knocked down by 80% (99). Cultures from these mice are resistant to NMDA excitotoxicity and PAR-dependent cell death (7,100). Harlequin mice are also resistant to experimental stroke injury (24).

The role of AIF in PAR-dependent cell death adds another level of complexity to the actions of AIF in cell death. AIF translocates from mitochondria following mitochondrial membrane permeability and participates in apoptotic cell death in both caspase-dependent and caspase-independent manners (101). The protease calpain, along with Bax, can also activate AIF translocation and subsequent cell death (102,103). PAR is another route to activate release of this death effector, which appears as a common element in diverse cell death signaling cascades.

PARTHANATOS: CELLULAR FEATURES

Given that PARP-1-dependent cell death occurs in multiple tissue types and organ systems and that it is morphologically and biochemically distinct from previously defined forms of cell death, it is practical to provide a name to distinguish this form of cell death. The features of Parthanatos include a rapid activation of PARP-1 resulting in high concentrations of complex PAR. Within minutes this PAR is observed in the cytosol, phosphotidyl serine is exposed (annexin-V positive cells), mitochondrial membranes are depolarized, and AIF translocates to the nucleus. The nucleus shrinks to ~30% of its original size without the formation of

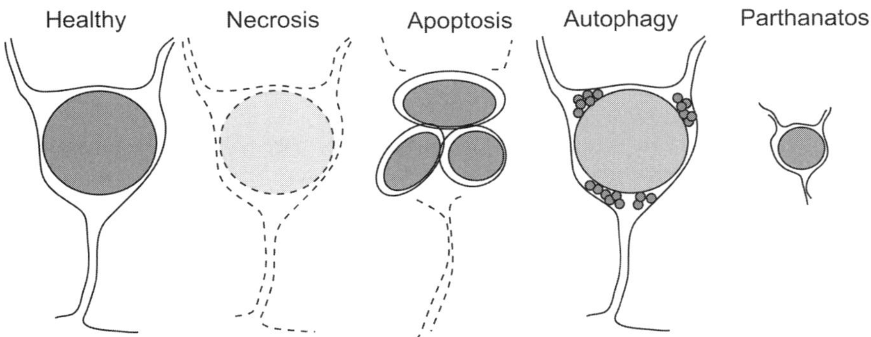

Figure 3 Parthanatos and cell death patterns. Parthanatos is a distinct form of cell death, both morphologically and biochemically. Necrosis is defined as uncontrolled cellular dissolution. The cellular membranes are disrupted leading to cell death. Apoptosis is a tightly choreographed cell death, biochemically characterized by activation of caspases. Morphologically, apoptosis is characterized by apoptotic bodies that comprise membrane-bound cytosol and fragmented nuclei. Autophagy is a self-digestion and is characterized by an increase in autophagic vesicles in the cytoplasm. Parthanatos is biochemically characterized by the generation of PAR and translocation of PAR to the cytosol. AIF translocation leads to rapid nuclear condensation. The overall size of the cell also shrinks with the nucleus. The nucleus remains intact but pyknotic. Eventually cellular membranes become permeable. *Abbreviations*: PAR, poly(ADP-ribose); AIF, apoptosis inducing factor.

apoptotic bodies. Within hours measurable cytochrome c is released from the mitochondria and caspase-3 is activated. The nucleus continues to shrink and condense. Within 6–12 hours, cell membrane integrity is lost and the cells are propidium iodide–positive. At the end stage, dead cells have a very small pyknotic nuclei surrounded by cellular membrane and debris (Fig. 3).

REFERENCES

1. Lockshin R, Williams C. Programmed cell death. II. Endocrine potentiation of the breakdown of the intersegmental muscles of silkmoths. J Insect Physiol 1964; 10:643–649.
2. Kerr JF, Wyllie AH, Currie AR. Apoptosis: a basic biological phenomenon with wide-ranging implications in tissue kinetics. Br J Cancer 1972; 26(4):239–257.
3. Hara T, Nakamura K, Matsui M, et al. Suppression of basal autophagy in neural cells causes neurodegenerative disease in mice. Nature 2006; 441(7095):885–889.
4. Komatsu M, Waguri S, Chiba T, et al. Loss of autophagy in the central nervous system causes neurodegeneration in mice. Nature 2006; 441(7095):880–884.
5. Shacka JJ, Klocke BJ, Young C, et al. Cathepsin D deficiency induces persistent neurodegeneration in the absence of Bax-dependent apoptosis. J Neurosci 2007; 27(8):2081–2090.
6. Andrabi SA, Kim NS, Yu SW, et al. Poly(ADP-ribose) (PAR) polymer is a death signal. Proc Natl Acad Sci U S A 2006; 103(48):18308–18313.
7. Yu SW, Andrabi SA, Wang H, et al. Apoptosis-inducing factor mediates poly(ADP-ribose) (PAR) polymer-induced cell death. Proc Natl Acad Sci U S A 2006; 103(48):18314–18319.
8. Wang H, Yu SW, Koh DW, et al. Apoptosis-inducing factor substitutes for caspase executioners in NMDA-triggered excitotoxic neuronal death. J Neurosci 2004; 24(48):10963–10973.
9. Yu SW, Wang H, Poitras MF, et al. Mediation of poly(ADP-ribose) polymerase-1-dependent cell death by apoptosis-inducing factor. Science 2002; 297(5579): 259–263.
10. Chambon P, Weill JD, Mandel P. Nicotinamide mononucleotide activation of new DNA-dependent polyadenylic acid synthesizing nuclear enzyme. Biochem Biophys Res Commun 1963; 11:39–43.
11. Alkhatib HM, Chen DF, Cherney B, et al. Cloning and expression of cDNA for human poly(ADP-ribose) polymerase. Proc Natl Acad Sci U S A 1987; 84(5): 1224–1228.
12. Kurosaki T, Ushiro H, Mitsuuchi Y, et al. Primary structure of human poly(ADP-ribose) synthetase as deduced from cDNA sequence. J Biol Chem 1987; 262(33): 15990–15997.
13. Uchida K, Morita T, Sato T, et al. Nucleotide sequence of a full-length cDNA for human fibroblast poly(ADP-ribose) polymerase. Biochem Biophys Res Commun 1987; 148(2):617–622.
14. Schreiber V, Dantzer F, Ame JC, et al. Poly(ADP-ribose): novel functions for an old molecule. Nat Rev Mol Cell Biol 2006; 7(7):517–528.
15. D'Amours D, Desnoyers S, D'Silva I, et al. Poly(ADP-ribosyl)ation reactions in the regulation of nuclear functions. Biochem J 1999; 342(pt 2):249–268.

16. Gagne JP, Hendzel MJ, Droit A, et al. The expanding role of poly(ADP-ribose) metabolism: current challenges and new perspectives. Curr Opin Cell Biol 2006; 18(2):145–151.
17. Davidovic L, Vodenicharov M, Affar EB, et al. Importance of poly(ADP-ribose) glycohydrolase in the control of poly(ADP-ribose) metabolism. Exp Cell Res 2001; 268(1):7–13.
18. Hanai S, Kanai M, Ohashi S, et al. Loss of poly(ADP-ribose) glycohydrolase causes progressive neurodegeneration in Drosophila melanogaster. Proc Natl Acad Sci U S A 2004; 101(1):82–86.
19. Koh DW, Lawler AM, Poitras MF, et al. Failure to degrade poly(ADP-ribose) causes increased sensitivity to cytotoxicity and early embryonic lethality. Proc Natl Acad Sci U S A 2004; 101(51):17699–17704.
20. Oka S, Kato J, Moss J. Identification and characterization of a mammalian 39-kDa poly(ADP-ribose) glycohydrolase. J Biol Chem 2006; 281(2):705–713.
21. Eliasson MJ, Sampei K, Mandir AS, et al. Poly(ADP-ribose) polymerase gene disruption renders mice resistant to cerebral ischemia. Nat Med 1997; 3(10): 1089–1095.
22. Abdelkarim GE, Gertz K, Harms C, et al. Protective effects of PJ34, a novel, potent inhibitor of poly(ADP-ribose) polymerase (PARP) in in vitro and in vivo models of stroke. Int J Mol Med 2001; 7(3):255–260.
23. Chiarugi A, Meli E, Calvani M, et al. Novel isoquinolinone-derived inhibitors of poly(ADP-ribose) polymerase-1: pharmacological characterization and neuroprotective effects in an in vitro model of cerebral ischemia. J Pharmacol Exp Ther 2003; 305(3):943–949.
24. Culmsee C, Zhu C, Landshamer S, et al. Apoptosis-inducing factor triggered by poly (ADP-ribose) polymerase and Bid mediates neuronal cell death after oxygen-glucose deprivation and focal cerebral ischemia. J Neurosci 2005; 25(44): 10262–10272.
25. Endres M, Scott G, Namura S, et al. Role of peroxynitrite and neuronal nitric oxide synthase in the activation of poly(ADP-ribose) synthetase in a murine model of cerebral ischemia-reperfusion. Neurosci Lett 1998; 248(1):41–44.
26. Endres M, Wang ZQ, Namura S, et al. Ischemic brain injury is mediated by the activation of poly(ADP-ribose)polymerase. J Cereb Blood Flow Metab 1997; 17(11): 1143–1151.
27. Kabra DG, Thiyagarajan M, Kaul CL, et al. Neuroprotective effect of 4-amino-1, 8-napthalimide, a poly(ADP ribose) polymerase inhibitor in middle cerebral artery occlusion-induced focal cerebral ischemia in rat. Brain Res Bull 2004; 62(5): 425–433.
28. Casey PJ, Black JH, Szabo C, et al. Poly(adenosine diphosphate ribose) polymerase inhibition modulates spinal cord dysfunction after thoracoabdominal aortic ischemia-reperfusion. J Vasc Surg 2005; 41(1):99–107.
29. Maier C, Scheuerle A, Hauser B, et al. The selective poly(ADP)ribose-polymerase 1 inhibitor INO1001 reduces spinal cord injury during porcine aortic cross-clamping-induced ischemia/reperfusion injury. Intensive Care Med 2007; 33(5):845–850.
30. Chiang SK, Lam TT. Post-treatment at 12 or 18 hours with 3-aminobenzamide ameliorates retinal ischemia-reperfusion damage. Invest Ophthalmol Vis Sci 2000; 41(10):3210–3214.
31. Lam TT. The effect of 3-aminobenzamide, an inhibitor of poly-ADP-ribose polymerase, on ischemia/reperfusion damage in rat retina. Res Commun Mol Pathol Pharmacol 1997; 95(3):241–252.

32. Weise J, Isenmann S, Bahr M. Increased expression and activation of poly(ADP-ribose) polymerase (PARP) contribute to retinal ganglion cell death following rat optic nerve transection. Cell Death Differ 2001; 8(8):801–807.
33. Tabuchi K, Ito Z, Tsuji S, et al. Poly(adenosine diphosphate-ribose) synthetase inhibitor 3-aminobenzamide alleviates cochlear dysfunction induced by transient ischemia. Ann Otol Rhinol Laryngol 2001; 110(2):118–121.
34. Bowes J, Ruetten H, Martorana PA, et al. Reduction of myocardial reperfusion injury by an inhibitor of poly (ADP-ribose) synthetase in the pig. Eur J Pharmacol 1998; 359(2–3):143–150.
35. Gilad E, Zingarelli B, Salzman AL, et al. Protection by inhibition of poly (ADP-ribose) synthetase against oxidant injury in cardiac myoblasts In vitro. J Mol Cell Cardiol 1997; 29(9):2585–2597.
36. Grupp IL, Jackson TM, Hake P, et al. Protection against hypoxia-reoxygenation in the absence of poly (ADP-ribose) synthetase in isolated working hearts. J Mol Cell Cardiol 1999; 31(1):297–303.
37. Szabo G, Soos P, Mandera S, et al. INO-1001 a novel poly(ADP-ribose) polymerase (PARP) inhibitor improves cardiac and pulmonary function after crystalloid cardioplegia and extracorporal circulation. Shock 2004; 21(5):426–432.
38. Thiemermann C, Bowes J, Myint FP, et al. Inhibition of the activity of poly(ADP ribose) synthetase reduces ischemia-reperfusion injury in the heart and skeletal muscle. Proc Natl Acad Sci U S A 1997; 94(2):679–683.
39. Yang Z, Zingarelli B, Szabo C. Effect of genetic disruption of poly (ADP-ribose) synthetase on delayed production of inflammatory mediators and delayed necrosis during myocardial ischemia-reperfusion injury. Shock 2000; 13(1):60–66.
40. Yeh CH, Chen TP, Lee CH, et al. Inhibition of poly(adp-ribose) polymerase reduces cardiomyocytic apoptosis after global cardiac arrest under cardiopulmonary bypass. Shock 2006; 25(2):168–175.
41. Cuzzocrea S, McDonald MC, Mazzon E, et al. Effects of 5-aminoisoquinolinone, a water-soluble, potent inhibitor of the activity of poly (ADP-ribose) polymerase, in a rodent model of lung injury. Biochem Pharmacol 2002; 63(2):293–304.
42. Farivar AS, Woolley SM, Fraga CH, et al. Intratracheal poly (ADP) ribose synthetase inhibition ameliorates lung ischemia reperfusion injury. Ann Thorac Surg 2004; 77(6):1938–1943.
43. Koksel O, Yildirim C, Cinel L, et al. Inhibition of poly(ADP-ribose) polymerase attenuates lung tissue damage after hind limb ischemia-reperfusion in rats. Pharmacol Res 2005; 51(5):453–462.
44. Woolley SM, Farivar AS, Naidu BV, et al. Role of poly (ADP) ribose synthetase in lung ischemia-reperfusion injury. J Heart Lung Transplant 2004; 23(11): 1290–1296.
45. Black JH, Casey PJ, Albadawi H, et al. Poly adenosine diphosphate-ribose polymerase inhibitor PJ34 abolishes systemic proinflammatory responses to thoracic aortic ischemia and reperfusion. J Am Coll Surg 2006; 203(1):44–53.
46. Bowes J, Thiemermann C. Effects of inhibitors of the activity of poly (ADP-ribose) synthetase on the liver injury caused by ischaemia-reperfusion: a comparison with radical scavengers. Br J Pharmacol 1998; 124(6):1254–1260.
47. Chen CF, Wang D, Hwang CP, et al. The protective effect of niacinamide on ischemia-reperfusion-induced liver injury. J Biomed Sci 2001; 8(6):446–452.

48. Roesner JP, Vagts DA, Iber T, et al. Protective effects of PARP inhibition on liver microcirculation and function after haemorrhagic shock and resuscitation in male rats. Intensive Care Med 2006; 32(10):1649–1657.
49. Yadav SS, Sindram D, Perry DK, et al. Ischemic preconditioning protects the mouse liver by inhibition of apoptosis through a caspase-dependent pathway. Hepatology 1999; 30(5):1223–1231.
50. Conrad MF, Albadawi H, Stone DH, et al. Local administration of the Poly ADP-Ribose Polymerase (PARP) inhibitor, PJ34 during hindlimb ischemia modulates skeletal muscle reperfusion injury. J Surg Res 2006; 135(2):233–237.
51. Hua HT, Albadawi H, Entabi F, et al. Polyadenosine diphosphate-ribose polymerase inhibition modulates skeletal muscle injury following ischemia reperfusion. Arch Surg 2005; 140(4):344–351; discussion 51–52.
52. Avlan D, Unlu A, Ayaz L, et al. Poly (adp-ribose) synthetase inhibition reduces oxidative and nitrosative organ damage after thermal injury. Pediatr Surg Int 2005; 21(6):449–455.
53. Chatterjee PK, Cuzzocrea S, Thiemermann C. Inhibitors of poly (ADP-ribose) synthetase protect rat proximal tubular cells against oxidant stress. Kidney Int 1999; 56(3):973–984.
54. Chatterjee PK, Zacharowski K, Cuzzocrea S, et al. Inhibitors of poly (ADP-ribose) synthetase reduce renal ischemia-reperfusion injury in the anesthetized rat in vivo. FASEB J 2000; 14(5):641–651.
55. Martin DR, Lewington AJ, Hammerman MR, et al. Inhibition of poly(ADP-ribose) polymerase attenuates ischemic renal injury in rats. Am J Physiol Regul Integr Comp Physiol 2000; 279(5):R1834–R1840.
56. Stone DH, Al-Badawi H, Conrad MF, et al. PJ34, a poly-ADP-ribose polymerase inhibitor, modulates renal injury after thoracic aortic ischemia/reperfusion. Surgery 2005; 138(2):368–374.
57. Zheng J, Devalaraja-Narashimha K, Singaravelu K, et al. Poly(ADP-ribose) polymerase-1 gene ablation protects mice from ischemic renal injury. Am J Physiol Renal Physiol 2005; 288(2):F387–F398.
58. Liaudet L, Szabo A, Soriano FG, et al. Poly (ADP-ribose) synthetase mediates intestinal mucosal barrier dysfunction after mesenteric ischemia. Shock 2000; 14(2): 134–141.
59. Bozlu M, Coskun B, Cayan S, et al. Inhibition of poly(adenosine diphosphate-ribose) polymerase decreases long-term histologic damage in testicular ischemia-reperfusion injury. Urology 2004; 63(4):791–795.
60. Bozlu M, Eskandari G, Cayan S, et al. The effect of poly (adenosine diphosphate-ribose) polymerase inhibitors on biochemical changes in testicular ischemia-reperfusion injury. J Urol 2003; 169(5):1870–1873.
61. Cosi C, Marien M. Implication of poly (ADP-ribose) polymerase (PARP) in neurodegeneration and brain energy metabolism. Decreases in mouse brain NAD+ and ATP caused by MPTP are prevented by the PARP inhibitor benzamide. Ann N Y Acad Sci 1999; 890:227–239.
62. Iwashita A, Yamazaki S, Mihara K, et al. Neuroprotective effects of a novel poly(ADP-ribose) polymerase-1 inhibitor, 2-[3-[4-(4-chlorophenyl)-1-piperazinyl] propyl]-4(3H)-quinazolinone (FR255595), in an in vitro model of cell death and in mouse 1-methyl-4-phenyl-1,2,3,6-tetrahydropyridine model of Parkinson's disease. J Pharmacol Exp Ther 2004; 309(3):1067–1078.

63. Mandir AS, Przedborski S, Jackson-Lewis V, et al. Poly(ADP-ribose) polymerase activation mediates 1-methyl-4-phenyl-1, 2,3,6-tetrahydropyridine (MPTP)-induced parkinsonism. Proc Natl Acad Sci U S A 1999; 96(10):5774–5779.
64. Outeiro TF, Grammatopoulos TN, Altmann S, et al. Pharmacological inhibition of PARP-1 reduces alpha-synuclein- and MPP(+)-induced cytotoxicity in Parkinson's disease in vitro models. Biochem Biophys Res Commun 2007; 357(3):596–602.
65. Love S, Barber R, Wilcock GK. Increased poly(ADP-ribosyl)ation of nuclear proteins in Alzheimer's disease. Brain 1999; 122(pt 2):247–253.
66. Strosznajder JB, Jesko H, Strosznajder RP. Effect of amyloid beta peptide on poly (ADP-ribose) polymerase activity in adult and aged rat hippocampus. Acta Biochim Pol 2000; 47(3):847–854.
67. Besson VC, Croci N, Boulu RG, et al. Deleterious poly(ADP-ribose)polymerase-1 pathway activation in traumatic brain injury in rat. Brain Res 2003; 989(1): 58–66.
68. Besson VC, Zsengeller Z, Plotkine M, et al. Beneficial effects of PJ34 and INO-1001, two novel water-soluble poly(ADP-ribose) polymerase inhibitors, on the consequences of traumatic brain injury in rat. Brain Res 2005; 1041(2):149–156.
69. Joashi UC, Greenwood K, Taylor DL, et al. Poly(ADP ribose) polymerase cleavage precedes neuronal death in the hippocampus and cerebellum following injury to the developing rat forebrain. Eur J Neurosci 1999; 11(1):91–100.
70. LaPlaca MC, Zhang J, Raghupathi R, et al. Pharmacologic inhibition of poly(ADP-ribose) polymerase is neuroprotective following traumatic brain injury in rats. J Neurotrauma 2001; 18(4):369–376.
71. Simbulan-Rosenthal CM, Rosenthal DS, Luo R, et al. Inhibition of poly(ADP-ribose) polymerase activity is insufficient to induce tetraploidy. Nucleic Acids Res 2001; 29(3):841–849.
72. Whalen MJ, Clark RS, Dixon CE, et al. Traumatic brain injury in mice deficient in poly-ADP(ribose) polymerase: a preliminary report. Acta Neurochir Suppl 2000; 76:61–64.
73. Whalen MJ, Clark RS, Dixon CE, et al. Reduction of cognitive and motor deficits after traumatic brain injury in mice deficient in poly(ADP-ribose) polymerase. J Cereb Blood Flow Metab 1999; 19(8):835–842.
74. Genovese T, Mazzon E, Muia C, et al. Attenuation in the evolution of experimental spinal cord trauma by treatment with melatonin. J Pineal Res 2005; 38(3):198–208.
75. Genovese T, Mazzon E, Muia C, et al. Inhibitors of poly(ADP-ribose) polymerase modulate signal transduction pathways and secondary damage in experimental spinal cord trauma. J Pharmacol Exp Ther 2005; 312(2):449–457.
76. Mao J, Price DD, Zhu J, et al. The inhibition of nitric oxide-activated poly(ADP-ribose) synthetase attenuates transsynaptic alteration of spinal cord dorsal horn neurons and neuropathic pain in the rat. Pain 1997; 72(3):355–366.
77. Scott GS, Szabo C, Hooper DC. Poly(ADP-ribose) polymerase activity contributes to peroxynitrite-induced spinal cord neuronal cell death in vitro. J Neurotrauma 2004; 21(9):1255–1263.
78. Masiello P, Cubeddu TL, Frosina G, et al. Protective effect of 3-aminobenzamide, an inhibitor of poly (ADP-ribose) synthetase, against streptozotocin-induced diabetes. Diabetologia 1985; 28(9):683–686.
79. Masutani M, Suzuki H, Kamada N, et al. Poly(ADP-ribose) polymerase gene disruption conferred mice resistant to streptozotocin-induced diabetes. Proc Natl Acad Sci U S A 1999; 96(5):2301–2304.

80. Pieper AA, Brat DJ, Krug DK, et al. Poly(ADP-ribose) polymerase-deficient mice are protected from streptozotocin-induced diabetes. Proc Natl Acad Sci U S A 1999; 96(6):3059–3064.
81. Mabley JG, Suarez-Pinzon WL, Hasko G, et al. Inhibition of poly (ADP-ribose) synthetase by gene disruption or inhibition with 5-iodo-6-amino-1,2-benzopyrone protects mice from multiple-low-dose-streptozotocin-induced diabetes. Br J Pharmacol 2001; 133(6):909–919.
82. Szabo C, Biser A, Benko R, et al. Poly(ADP-ribose) polymerase inhibitors ameliorate nephropathy of type 2 diabetic Leprdb/db mice. Diabetes 2006; 55(11): 3004–3012.
83. Monyer H, Giffard RG, Hartley DM, et al. Oxygen or glucose deprivation-induced neuronal injury in cortical cell cultures is reduced by tetanus toxin. Neuron 1992; 8(5):967–973.
84. Koh JY, Choi DW. Vulnerability of cultured cortical neurons to damage by excitotoxins: differential susceptibility of neurons containing NADPH-diaphorase. J Neurosci 1988; 8(6):2153–2163.
85. Choi DW. Glutamate neurotoxicity and diseases of the nervous system. Neuron 1988; 1(8):623–634.
86. Dawson VL, Dawson TM, Bartley DA, et al. Mechanisms of nitric oxide-mediated neurotoxicity in primary brain cultures. J Neurosci 1993; 13(6):2651–2661.
87. Dawson VL, Dawson TM, London ED, et al. Nitric oxide mediates glutamate neurotoxicity in primary cortical cultures. Proc Natl Acad Sci U S A 1991; 88(14):6368–6371.
88. Dawson VL, Kizushi VM, Huang PL, et al. Resistance to neurotoxicity in cortical cultures from neuronal nitric oxide synthase-deficient mice. J Neurosci 1996; 16(8): 2479–2487.
89. Huang Z, Huang PL, Panahian N, et al. Effects of cerebral ischemia in mice deficient in neuronal nitric oxide synthase. Science 1994; 265(5180):1883–1885.
90. Pacher P, Beckman JS, Liaudet L. Nitric oxide and peroxynitrite in health and disease. Physiol Rev 2007; 87(1):315–424.
91. Zhang J, Dawson VL, Dawson TM, et al. Nitric oxide activation of poly(ADP-ribose) synthetase in neurotoxicity. Science 1994; 263(5147):687–689.
92. Fossati S, Cipriani G, Moroni F, et al. Neither energy collapse nor transcription underlie in vitro neurotoxicity of poly(ADP-ribose) polymerase hyper-activation. Neurochem Int 2007; 50(1):203–210.
93. Zeng J, Yang GY, Ying W, et al. Pyruvate improves recovery after PARP-1-associated energy failure induced by oxidative stress in neonatal rat cerebrocortical slices. J Cereb Blood Flow Metab 2007; 27(2):304–315.
94. Berger F, Ramirez-Hernandez MH, Ziegler M. The new life of a centenarian: signalling functions of NAD(P). Trends Biochem Sci 2004; 29(3):111–118.
95. Goto S, Xue R, Sugo N, et al. Poly(ADP-ribose) polymerase impairs early and long-term experimental stroke recovery. Stroke 2002; 33(4):1101–1106.
96. Chiarugi A. Poly(ADP-ribose) polymerase: killer or conspirator? The 'suicide hypothesis' revisited. Trends Pharmacol Sci 2002; 23(3):122–129.
97. Gagne JP, Hunter JM, Labrecque B, et al. A proteomic approach to the identification of heterogeneous nuclear ribonucleoproteins as a new family of poly(ADP-ribose)-binding proteins. Biochem J 2003; 371(pt 2):331–340.

98. Pleschke JM, Kleczkowska HE, Strohm M, et al. Poly(ADP-ribose) binds to specific domains in DNA damage checkpoint proteins. J Biol Chem 2000; 275(52): 40974–40980.
99. Klein JA, Longo-Guess CM, Rossmann MP, et al. The harlequin mouse mutation downregulates apoptosis-inducing factor. Nature 2002; 419(6905):367–374.
100. Cheung EC, Melanson-Drapeau L, Cregan SP, et al. Apoptosis-inducing factor is a key factor in neuronal cell death propagated by BAX-dependent and BAX-independent mechanisms. J Neurosci 2005; 25(6):1324–1334.
101. Modjtahedi N, Giordanetto F, Madeo F, et al. Apoptosis-inducing factor: vital and lethal. Trends Cell Biol 2006; 16(5):264–272.
102. Moubarak RS, Yuste VJ, Artus C, et al. Sequential activation of poly(ADP-ribose) polymerase 1, calpains, and Bax is essential in apoptosis-inducing factor-mediated programmed necrosis. Mol Cell Biol 2007; 27(13):4844–4862.
103. Polster BM, Basanez G, Etxebarria A, et al. Calpain I induces cleavage and release of apoptosis-inducing factor from isolated mitochondria. J Biol Chem 2005; 280(8): 6447–6454.

10

Paraptosis

Sabina Sperandio and Ian de Belle
Centre de Recherche du CHUL, Université Laval, Québec, Canada

INTRODUCTION

Naturally occurring cell death was first recognized toward the middle of the 19th century, shortly after the establishment of the cell theory by Schleiden and Schwann (1). In 1842, the German scientist Carl Vogt produced a monograph wherein he clearly describes the disappearance or destruction of embryonal cells preceding the formation or change of determinate structures during the development of the midwife toad (2). Significant investigation on this topic was carried out throughout the second half of the 19th century.

Between the end of the 19th century and the 1960s, elegant work extensively described morphological features of cell death (1,3). At least three main types of naturally occurring cell death were recognized: type 1 or nuclear cell death today known as apoptosis, type 2 corresponding to autophagic cell death, and type 3 or cytoplasmic cell death (4). Type 3 cell death was further distinguished as type 3A, or nonlysosomal disintegration, and type 3B, morphologically similar to 3A, but distinct in the lack of total cellular fragmentation. In addition, cells undergoing type 3B cell death appear to trigger phagocytosis by neighboring cells. The term "programmed cell death" (PCD) was introduced to describe the demise of cells occurring at a determined time and place during insect development—thus these cell deaths were predictable (5). In 1966, Tata demonstrated that the cell death observed during tadpole metamorphosis was blocked by cycloheximide, and therefore required production of new proteins by the cell (6). Today, PCD has acquired the more generalized meaning of a process by which the cell participates actively in its own demise (cell suicide), as opposed to necrosis or accidental cell death. In 1972, the term apoptosis was introduced to describe the cell death previously known as "shrinkage necrosis" or "nuclear cell death" (7). In their seminal work, Kerr et al. first linked this specific morphology occurring during development to the cell death observed in physiological adult tissue turnover and that induced by toxins and hormones (8,9). The first mechanistic tool to define apoptosis arrived with the association of this cell death with specific products of DNA degradation (ladders) (10–12). Major interest in apoptosis was ignited in the late eighties upon the discovery of the first cellular participant in the apoptotic network, Bcl-2 (13). Following the dissection of the genetic elements of apoptosis (14–16), the study of this type of cell death underwent a greater expansion witnessed by the exponential growth of publications on this subject.

A wealth of literature has been produced over the past two decades, largely regarding apoptosis, due in part to the availability of many specific tools for its identification. A variety of stimuli have now been recognized as inducers of apoptosis and apoptosis specific inhibitors have been identified and extensively used as well. A growing number of apoptosis modulators have now entered the clinical trial phase to treat conditions due to either excessive or reduced apoptosis (17,18).

Forms of PCD distinct from apoptosis, although known for many years, have somehow been disregarded during the period of major progress in the understanding of apoptosis, but their importance has resurfaced slowly in more recent years. These studies are uncovering the existence of a multiplicity of mechanisms leading to cell death in both normal and pathological conditions that lead to distinct morphological presentations of cell death (19,20). Sometimes cell death pathways present hybrid morphologies that complicate their classification. For example, dying chondrocytes observed in chick embryo and rabbit growth plates and in articular cartilage, also called "dark chondrocytes," present some of the typical features of apoptosis without the production of apoptotic bodies. This is combined with the presence of autophagic vacuoles that are thought to be essential to the autodestruction of the chondrocytes. This peculiar morphology named chondroptosis could result from an adaptation of an apoptotic pathway to a particular cell type or it may represent a distinct form of cell death (21).

Because the classification of different types of PCD is based largely on morphological criteria, it is not always possible to correlate these definitions with specific molecular mechanisms and therefore define what constitutes a bona fide cell death pathway.

Here we will follow the aforementioned classification adopted by Clarke (4) based largely on the original distinction proposed by Schweichel and Merker (22). We distinguish necrosis as the cell destruction derived from passive cell death in which the cell's own suicide machinery is not actively engaged.

WHAT IS PARAPTOSIS?

A few years ago, we reported that the expression of a construct encoding a membrane targeted intracytoplasmic domain of the insulin-like growth factor-I receptor (IGF-IR-IC) induced a form of cell death that we could not easily classify (23). Our interest in the IGF-IR was triggered by previous reports demonstrating apoptosis induction following the expression of a C-terminal peptide of the receptor (24). Because of the known prosurvival activity of its ligands, IGF-I and IGF-II, the IGF-IR appeared to be a good candidate to function as a dependence receptor (i.e., a receptor that can promote survival when expressed in combination with its trophic ligand but induce cell death when expressed in the unbound state) (25). Typically, cleavage by a caspase of an unligated form of a dependence receptor triggers apoptosis. Such receptors may have important roles in regulating cell homeostasis and in cancer suppression, in particular in preventing metastasis, by linking the survival of cells expressing the dependence receptor to the availability of its specific ligand. This hypothesis has been proven true at least in the case of the dependence receptor DCC (deleted in colorectal cancer) (26).

Consistent with this hypothesis, we found that the expression of the membrane-targeted IGF-IR-IC induced cell death in all cell lines that were tested. This cell death is programmatic in nature as it requires new protein synthesis for its execution, and therefore is distinct from necrosis (23). We found that membrane-bound IGF-IR-IC was also able to form a complex with the proapoptotic protease caspase-9 and that the cell death was partially inhibited by a dominant negative mutant of caspase-9. Surprisingly, though, this cell death activity was not sensitive to caspase inhibitors and was apparently independent of Apaf-1 recruitment as well as of cytochrome c release form mitochondria.

The paradox was explained by the demonstration that caspase-9 itself, when expressed by transfection, could induce both apoptosis and a cell death similar to paraptosis, and that only the apoptotic component could be blocked by caspase inhibitors (23). Interestingly, recruitment of caspase-9 and independence from Apaf-1 is also a feature of apoptosis induced by the dependence receptor DCC. Despite this similarity, and for reasons that we will explain below, in our experimental system, IGF-IR was neither acting as a dependence receptor nor inducing apoptosis.

Due to these peculiar features and a possible link to the apoptotic pathway, we dubbed this form of cell death paraptosis, meaning “next to,” or “related to,” apoptosis.

Subsequently, our colleagues reported that the binding of the peptide neurotransmitter substance P to its receptor, neurokinin-1 receptor, also induces a nonapoptotic form of PCD with similarities to paraptosis in its morphology, caspase independence and requirement for gene transcription and translation. However, substance P–induced cell death also displays some similarity to autophagic cell death (27). Other groups have reported the observation of cell deaths similar to paraptosis including the death of 293T cells in response to overexpression of the TAJ/TROY tumor necrosis factor (TNF) receptor (28), cerebellar granule neurons in response to intracellular acidification (29), U251MG or T9-C glioma cells in response to macrophage-colony-stimulating factor (M-CSF) transduction or transfection (30,31), and Jurkat cells following calcium influx mediated by transfection of the vanilloid receptor calcium channel (32). Another important receptor, the epidermal growth factor receptor (EGFR), has been reported to mediate cell death with features similar to paraptosis in pituitary somato-lactotrope cells (33,34).

CHARACTERISTICS OF PARAPTOSIS

Paraptosis most resembles the cell death defined as type 3B cell death, notably a mild necrosis-like type of morphology. Owing to their ultrastructural similarities, we are currently working under the assumption that paraptosis corresponds to this type of cell death. However, it should be underscored that a definite classification awaits the availability of molecular markers that can link a defined cell death pathway to a morphological description.

Ultrastructural features of paraptosis include, in its early stages, rounding of the cell without formation of membrane blebbing, extensive cytoplasmic vacuolation derived mostly from the ER, and a relatively unperturbed nuclear structure (Fig. 1). Increased electron density in the remaining cytoplasm and nucleus is also observed and there seems to be a certain degree of mitochondrial swelling, though in the absence of cytochrome c release.

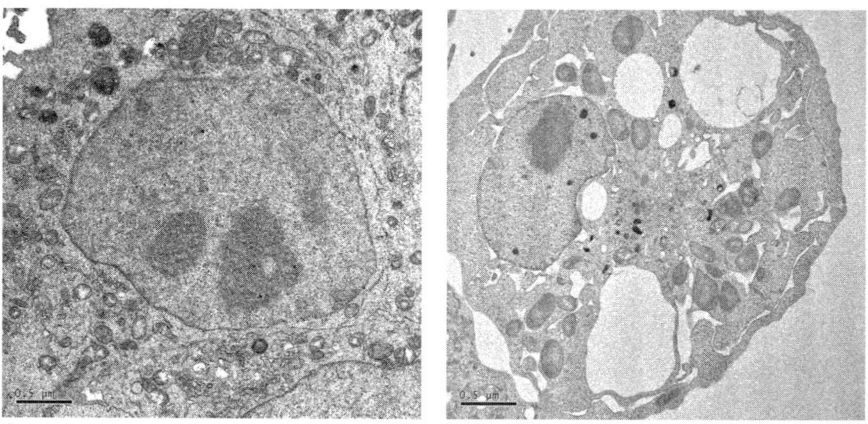

A B

Figure 1 Ultrastructural characteristics of cells undergoing paraptosis. 293T cells (human embryonic kidney) were transfected with either an empty vector (**A**) or the IGF-IR-IC construct (**B**). *Abbreviation*: IGF-IR-IC, insulin-like growth factor-I receptor intracytoplasmic domain.

As mentioned earlier, paraptosis is not sensitive to either pharmacological or biochemical inhibitors of apoptosis and cells undergoing paraptosis are generally negative for apoptotic markers such as terminal deoxynucleotidyl-transferase-mediated dUTP nick-end labeling (TUNEL) staining, annexinV reactivity, and the presence of internucleosomal DNA fragmentation (at least during the first 48 hours after induction) (23). Autophagic cell death inhibitors such as 3-methyladenine are also ineffective in preventing paraptosis.

Kinetically, paraptosis is slower than apoptosis. Figure 2 shows the comparison between the cell death rate measured over a five-day period following transfection of either the proapoptotic protein Bax or IGF-IR-IC. In the Bax transfected sample, the majority of cell death occurs within the first 24 hours

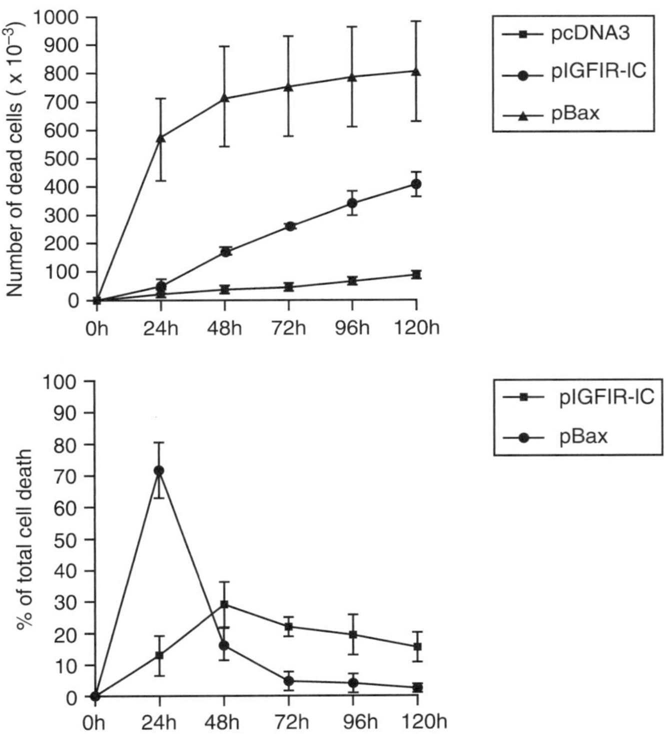

Figure 2 Kinetics of paraptotic cell death. IGF-IR-IC is compared with apoptotic cell death induced by Bax following transfection in 293T cells. Cell death kinetics are compared with a control empty vector (pcDNA3). Floating cells were collected from the culture medium at 24-hour intervals over a period of five days, and the cell death was scored by trypan exclusion shown in upper graph. The lower graph shows the percentage of cell death occurring at each time point with respect to the sum of all cell death over the five-day period. *Abbreviation*: IGF-IR-IC, insulin-like growth factor-I receptor intracytoplasmic domain.

following transfection, while in the paraptotic sample, a peak of cell death is observed at 48 hours. After this time the cell death rate slows down but remains significantly higher than in control cells over the entire five days. In a similar way, most of the cell death induced by transfection of caspase-9 into mammalian cells occurs within the first 24 hours, whereas the nonapoptotic cell death resulting from simultaneous administration of caspase inhibitors displays a peak at about 48 hours (23).

In an attempt to identify its characteristics, we undertook a proteomic analysis of cells undergoing paraptosis following the expression of the IGF-IR-IC. From this study, no evidence was found to indicate proteolytic processing of specific substrates, as observed in apoptotic cell death, although detection of small protein fragments could have been precluded by sensitivity factors. However, alterations or intracellular redistribution was found during paraptosis in several proteins including some cytoskeletal components. While several cytoskeletal and structure-related proteins are targeted by caspases during apoptosis (35), in autophagic cell death cytoskeletal filaments appear to be redistributed but preserved and are presumably required for the initial formation of the autophagosomes (36). In a similar way, it is conceivable that a functional reorganization of cytoskeletal components might take place in paraptosis to allow for the formation and expansion of the cytoplasmic vacuoli. Other proteins found altered during paraptosis included signal transduction proteins, tumor suppressors/transcription regulators, and some metabolic proteins.

The involvement of mitochondria in paraptosis has not yet been fully established. However, preliminary studies suggest that paraptosis requires ATP for its execution since treatment with either oligomycin B or antimycin can suppress the paraptotic morphology (S. Sperandio, unpublished observations). This requirement may be linked to an increase in mitochondrial levels of the ATP synthase β subunit, which is part of the oxidative phosphorylation complex V, found in cells transfected with IGF-IR-IC. Another mitochondrial protein found increased in paraptosis is prohibitin, which has also been previously implicated in apoptosis. A comparison of some of the known features between different cell death pathways discussed above is presented in Table 1.

OCCURRENCE OF PARAPTOSIS

Is paraptosis a relevant biological process or just a product of cell culture manipulation? Clues from the study of naturally occurring cell death in both normal and genetically modified animals suggest that paraptosis is, in fact, a physiologically important process.

A cell death morphologically similar to paraptosis has been observed in the developing nervous system, indicating that this may contribute to neuronal selection in vivo. Pilar and Landmesser described two main morphologies in degenerating neurons; one consistent with apoptosis and the other defined as

Table 1 Features Comparison Between Some of the Known Cell Death Types

	Apoptosis	Autophagic cell death	Paraptosis
Morphology	Chromatin condensation and fragmentation, cell shrinkage, membrane blebbing, formation of apoptotic bodies	Presence of autophagic vacuoles with inclusions, expansion of the lysosomal compartment, late chromatin condensation	Swelling of the ER and organelles, cytoplasmic vacuolation,
Triggers	Trophic factor withdrawal, cytotoxic cytokines, DNA damage	Hormones, amino acids and nutrients starvation	Trophotoxicity, substance P, some death receptors, anion imbalance
Mediators	Caspases, proapoptotic Bcl-2 family members	DAPK, JNK, ATG proteins, lysosomal proteases	JNK1, MAPK, Nur77
Inhibitors	Caspase inhibitors, anti-apoptotic Bcl-2 family members	3-methyladenine	AIP1/Alix, UO126, PEBP
Cytoskeletal changes	Proteolytic processing	Preservation of cytoskeletal components	Nonproteolytic alterations
Energy requirement	+	+	+

Abbreviations: ER, endoplasmic reticulum; DAPK, death-associated protein kinase; JNK, jun N terminal kinase; ATG, a topographically graded protein; MAPK, mitogen-activated protein kinase; AIP1, Alg-2-interacting protein 1; PEBP, phosphatidylethanolamine binding protein.

cytoplasmic cell death and similar to paraptosis, with major involvement of rough endoplasmic reticulum (RER)-derived cytoplasmic vacuoles (37). In a detailed review on naturally occurring neuron death during development, Cunningham explained that there appears to be a specific pattern of distribution of these two distinct cell death pathways within the nervous system (38). For example, in the rat superior colliculus, both types are present, but the cytoplasmic type is prominent (39), whereas only the nuclear type of degeneration (apoptosis) is observed in the retina (40). In the duck trochlear nucleus, only the cytoplasmic type is present (41) and both morphologies had been reported in chick spinal cord (42). Pilar and Landmesser extended their observations to target-deprived neurons and suggested that the type of morphology of degenerating neurons reflected their state of maturity at the time of cell death. They found that naturally degenerating chick ciliary ganglion cells normally undergo cytoplasmic type cell death. However, if the target structures for the neurons were removed before the innervation had occurred, the cell death was not only increased, but it was also morphologically changed to resemble the nuclear type (37). The removal of the peripheral target would therefore prevent maturation of the innervating neurons and explain the shift in cell death morphology. An alternative, but not mutually exclusive, interpretation is that, when the target was removed, trophic factor withdrawal led to apoptosis; however, in the presence of the target, high trophic factor concentration induced paraptosis. Such an interpretation recognizes the emerging awareness that mismatches of trophic factors/receptor interactions may induce PCD not only due to insufficient trophic factor concentration, but also due to hyperactivation of trophic factor receptor signaling (in the later case, via paraptosis).

At the moment, no definitive proof is available that the cytoplasmic cell death described in neuronal development corresponds to paraptosis. However, there is evidence to support a role for the IGF-IR in neuronal selection. The IGF system, including the IGF-IR and its ligands, are expressed at very early developmental stages and are known to play important roles throughout development. Mice genetically deleted of the IGF-IR die invariably at birth weighing only about 45% of their wild-type littermates' weight and display muscle hypoplasia and delayed bone and skin development (43). However, a higher neuronal density is observed in the brain stem and spinal cord of mutant mice. The authors of this article favored an explanation of this anomalous phenotype as a possible increase in neuronal crowdedness due to reduction of the surrounding neuropil rather than an absolute difference in the neuronal number. However, in the light of the newly discovered ability of the IGF-IR to promote cell death, the latter appears to be the simplest explanation.

The fundamental role of the apoptotic pathway in mammalian development has been well established by genetic deletion of specific caspases or caspase adaptor proteins (44–48). However, subsequent studies have uncovered the existence of an underlying mechanism(s) of PCD after depletion or pharmacological inhibition of caspases. For example, it was noted that interdigital web remodeling believed to occur by an apoptotic mechanism, was only delayed in the Apaf-$1^{-/-}$ mice and displayed a necrotic morphology (49). A thorough analysis of PCD in different subpopulations of developing neurons was performed by Oppenheim and colleagues in caspase-deleted mice (50). These authors showed that despite defects in neural tube closure and severe brain malformations in caspase-3 and caspase-9 null mice, the rate of PCD is normal in many neuronal subpopulations from brain stem and spinal cord. This cell death is nonapoptotic and appears morphologically consistent with type 3B cytoplasmic PCD. Similar results were obtained in examining motoneuron PCD in the chick embryo cervical spinal cord upon pharmacological inhibition of caspases (50,51). A possible explanation for the differential requirement for caspase expression in correct neuronal selection may reside in their mitotic potential at the time of selection (50). Abnormalities in the forebrain of caspase-deleted mice could arise from a delay in the death of the mitotically active neurons in this area, which would allow more replication cycles than normal. In contrast, the brain stem and spinal cord neuronal populations, which typically undergo PCD postmitotically, would not be affected by this delay. If paraptosis induced by the

IGF-IR or some other stimulus is implicated in the developmental cell death observed in the caspase-9 null mice, then it must be assumed that caspase-9 is not a requisite mediator of paraptosis.

These observations indicate that a parallel cell death pathway(s), distinct from apoptosis, is active in mammalian development. As apoptosis appears to be a quicker and more efficient means of cell demise, it is likely that paraptosis or a similar cell death pathway originated earlier than apoptosis during evolution. The existence of forms of PCD in yeast and other unicellular eukaryotes has been well established. PCD has been reported in yeast and lower eukaryotes such as the slime mold *Dictyostelium discoideum*. The preservation of pathways of alternative cell death in higher eukaryotes may have an advantage in situations where the cell's own apoptotic pathways are ineffective. For example, it is known that viruses have developed the ability to prevent altruistic cell suicide triggered by the infected hosts. In fact, the first inhibitor of apoptosis identified, CrmA (from cowpox virus), is a caspase inhibitor. Many other apoptosis inhibitors are expressed by viruses, including the baculovirus IAPs and p35, among others. A recent report has revealed the presence of paraptosis-like features in the dinoflagellate *Amphidium carterae* during PCD induced by darkness and culture senescence indicating that paraptosis might be an ancient, evolutionarily conserved process (52).

Multiple cell death mechanisms are thought to participate in neurodegeneration (19). For example, cell death morphologies similar to paraptosis have been reported in mouse models for Huntington's disease and amyotrophic lateral sclerosis (53,54), as well as in neuronal cultures after excitotoxic stimulation (55).

Participation of multiple cell death pathways in neurodegeneration might explain the limitation in the efficacy of apoptosis-based therapeutic strategies. For example, apoptosis inhibition has only been partially effective in the rescue of animal models of neurodegeneration (56–58).

MODULATION OF PARAPTOSIS

As mentioned earlier, unlike dependence receptor–induced death, paraptosis triggered by the IGF-IR does not proceed via classical apoptosis, nor does it result from the expression of a receptor in the unbound state. In fact, in a subsequent study we found that positive signaling through the receptor was responsible for the induction of cell death by the IGF-IR (59). In this report, it was determined that overexpression of the full-length receptor limited cell survival, and that this effect was exacerbated by the simultaneous administration of IGF-I or insulin at supraphysiological concentrations (where it is known to bind IGF-IR and mimic IGF-I effects), but not by an inactive IGF-I analog. In addition, the expression of a membrane-targeted intracytoplasmic domain was shown to mimic constitutive activation of the receptor and to induce the consequent autophosphorylation and activation of signal transduction pathways.

Such an effect can be defined as trophotoxicity, meaning the toxicity associated with activation of a trophic factor receptor. It is conceivable that such a mechanism might have originated to prevent autocrine loop-induced neoplasia. This hypothesis would predict mutations of paraptosis-mediating genes in autocrine loop tumors.

From a broader prospective, it could be extrapolated that apoptosis and paraptosis are complementary cell death programs induced at the two opposite ends of an ideal trophic support range. In this hypothesis we would predict that withdrawal of a trophic factor such as IGF-I can induce cell death through apoptosis while hyperactivation of the receptor above a certain trophic threshold, whereas blocking apoptosis induces an alternative cell death program, paraptosis. Adequate stimulation of a growth factor receptor (like IGF-IR), promotes cell survival through the suppression of apoptosis, while failing to induce trophotoxicity. A model explaining this hypothesis is shown in Figure 3.

As our understanding of the molecular mechanisms and triggers of paraptosis expands, paraptosis-based therapies may soon enter clinical experimentation in cases where apoptosis-based therapies have failed. Studies from our and other groups have begun to elucidate the mechanisms of paraptosis and cell deaths reported to be similar to paraptosis, and have outlined some similarities at the molecular level. At least two signaling pathways triggered by the IGF-IR were found to participate in paraptosis, the Jun N terminal kinase (JNK) and the

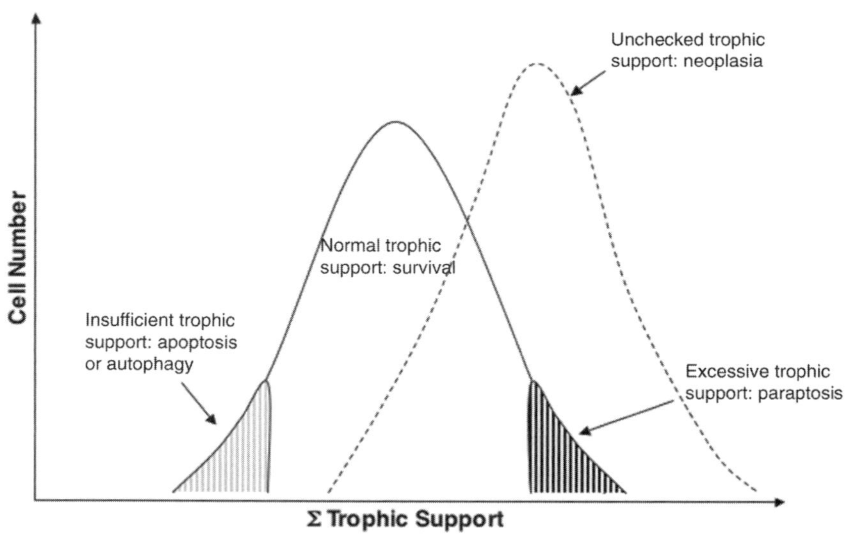

Figure 3 Model showing the hypothesis of cell number control mediated by trophic factor level. With low levels of trophic support, cells die by apoptosis or autophagic cell death, while with levels at the high end of the spectrum cells die by paraptosis. We propose that if high trophic support does not trigger paraptosis, then cell number would increase.

mitogen-activated protein kinase (MAPK) pathways. In particular, antisense oligonucleotide-mediated downregulation of JNK1 and pharmacological inhibition or RNA interference of MEK2 inhibited paraptosis induced by the IGF-IR (59). Similarly, the paraptosis-like cell death induced by substance P was preventable by the use of chemical inhibitors for MEK2 and by siRNA-mediated downregulation of extracellular signal-regulated kinase (ERK)2 and the ERK2 target Nurr77 (60).

Interestingly, it was recently shown that caspase-9 is also a direct substrate of ERK2 (61). Phosphorylation at threonine 125 by ERK2 has been demonstrated to inhibit caspase-9 induced apoptosis. One attractive hypothesis is that the induction of ERK2 during paraptosis may modify the activity of caspase-9 from proapoptotic to pro-paraptotic. While JNK activation has been known for a long time to participate in cell death, activation of the MAPK pathway more often mediates cell survival (62). However, ERK activation has been shown to be required for other paradigms of cell death including neuronal cell death induced by glutamate, okadaic acid, hemin, genistein, and 6-hydroxydopamine (63–67). In contrast, Fombonne et al. reported that the paraptosis-like cell death triggered in GH4C1 somato-lactotrope cells by epidermal growth factor (EGF) was not preventable by the MEK inhibitor UO126 despite its ability to inhibit ERK phosphorylation induced by the EGFR (34).

The protein PDCD5 (programmed cell death 5), has been proposed to be a modulator of paraptosis-like cell death induced by the TNF family member TAJ/TROY (28). As observed for paraptosis induced by the IGF-IR-IC, the cell death resulting from intracellular acidification of cerebellar granule neurons is not prevented by overexpression of Bcl-X_L, and requires new protein synthesis (29).

The first natural inhibitor to be found for paraptosis was ALG-2-interacting protein 1 (AIP1)/Alix, a protein cloned independently by two groups based on its interaction with the cell death-related calcium binding protein ALG2 (68,69). The mechanism of paraptosis inhibition by AIP1 was found to be at the level of the receptor autophosphorylation (59). Interestingly, AIP1 has also been shown to inhibit the paraptosis-like cell death induced by EGFR (33), suggesting that this protein may act as a general modulator of trophotoxicity, perhaps by reducing receptor tyrosine kinase phosphorylation.

Phosphatidylethanolamine binding protein (PEBP) is another inhibitor of paraptosis that has been recently identified. This protein was found to decrease during paraptosis, but if its levels are restored by means of transfection, the cell death induced by the IGF-IR-IC is prevented (Sperandio et al., submitted for publication). A member of the PEBP family, hPEBP4, was previously shown to induce cellular resistance to apoptosis mediated by TNF-α by inhibiting the activation of both MEK/ERK and JNK pathways (70). Thus, it appears that PEBP may be an important survival factor capable of preventing cell death induced by multiple pathways. Interestingly, PEBP appears to suppress paraptosis by inhibiting the activation of JNK but not MAPK induced by IGF-IR-IC, highlighting the specificity of activating and inhibitory signals in different cell death pathways.

PARAPTOSIS AND CANCER

The process by which a normal human cell becomes a cancer cell involves a complex series of genetic and epigenetic events, only some of which are currently understood. However, a wealth of recent studies has demonstrated that the genetic alterations occurring in cancer primarily affect the balance between cell proliferation and cell death. The precise mechanism(s) of interaction between cell growth and death pathways is not entirely understood but it is clear that identifying factors and events that tilt the balance between these pathways will be central to understanding carcinogenesis. Recognizing and understanding the specific cell death pathway(s) used by the cell will be of central importance in addressing the carcinogenic process. There is evidence of a paraptotic component in carcinogenesis suggesting that perturbations of this form of cell death may represent a significant goal for pharmacological intervention.

The involvement of the IGF axis in carcinogenesis has been studied for many years and there is ample documented evidence for its role in regulating the growth of a variety of tumor cell types (71). In contrast, as described above, the original description of paraptosis indicated that constitutive signaling through the IGF-IR could induce cell death and that there might exist an embedded signal capable of restraining uncontrolled growth. By extension, this might indicate that attenuating IGF-IR expression could be permissive for cancer cell growth. In accordance with this hypothesis, the work of Plymate and colleagues has demonstrated that, during prostate tumorigenesis, the expression of the IGF-IR is decreased in the progression from benign hyperplasia to the metastatic disease (72,73). Further, in cell-based models of different stages of prostate cancer, reintroduction of the IGF-IR in a prostate cancer cell line reduces its metastatic potential (74,75). In this same study, the authors described an increase in cell death associated with IGF-IR reexpression. The authors suggested that IGF-IR sensitizes cells to apoptosis based on the appearance of DNA ladders upon stimulation with 6-hydroxyurea. In addition, a decrease in IGF-IR expression has been reported in human prostate cancer bone metastases as well as in murine prostate cancer models (76,77). Furthermore, overexpression of fragments of the IGF-IR in cancer cells was shown to reduce tumorigenicity in nude mice and to induce cell death (23,24,78). Taken together, these studies provide evidence that the balance between cell growth and death, may in part, reside in alternate activities of growth factor receptors like the IGF-IR. These results would argue that the loss of receptor expression would facilitate tumorigenesis by compromising the cell's ability to commit suicide by eliminating the ability to activate cell death pathways including paraptosis.

Evidence for a potential role for paraptosis in cancer is not limited to the IGF axis and prostate cancer. For example, it has been demonstrated that human glioma cells lose their tumorigenicity in vivo when transfected with a membrane-bound form of M-CSF. These transfected cells are phagocytosed by macrophages and generate a lasting immunity in the host by inducing

CD3+ cells (30,31). Significantly, the M-CSF expressing glioma cells die by a process resembling paraptosis, including swelling and vacuolation with no apparent chromatin condensation. Apoptotic activation was not responsible for the cell death seen, although the authors did note evidence of late-stage DNA fragmentation as witnessed by TUNEL positive staining. The impact of these studies is highlighted by the authors' suggestion that the induction of paraptosis by M-CSF may provide a useful tool in the generation of tumor immunity and live tumor cell vaccines.

The ultimate goal for cancer researchers is to identify critical features of tumor cells that can be exploited for therapeutic intervention. It is well known that a number of chemotherapeutic drugs function through activating cell death pathways. One of the most widely used agents is the platinum-based drug cisplatin, which cross-links DNA and activates DNA damage repair and apoptosis. Some of the main drawbacks in the use of cisplatin include (*i*) its toxicity and (*ii*) cells often develop resistance. Recent efforts to develop new metal-based therapeutics with reduced toxicity have focused on using copper-based compounds, since it has been proposed that such compounds may be advantageous due to both reduced toxicity and alteration in the cellular redox balance. Marzano and colleagues have successfully synthesized a tris (hydroxymethyl) phosphine copper compound that is 2–18 times as effective as cisplatin in inhibiting cell growth depending upon the tumor cell line tested (79). Interestingly, the authors noted that, unlike cisplatin, their copper compound did not activate apoptosis but they did observe an increase in cellular granularity as determined by flow cytometry. This feature may implicate the appearance of large vacuoles reminiscent of the paraptotic phenotype. Perhaps the most significant finding from this study was the demonstration that the growth of cisplatin-resistant cells may be inhibited by their copper-based complex. This result has implications relating to treatment regimens for tumors with platinum-based drug resistance. In a separate study, Tardito and colleagues have studied the effects of another copper-based compound on the growth of a human tumor cell line (80). In this study, the authors used a copper (II) thioxotriazole complex to show antiproliferative effects comparable to those seen with cisplatin. This study also noted the presence of large vacuoles in the cells following treatment with the copper complex as well as maintenance of plasma membrane integrity, which was not seen in cisplatin treated cells. Based on morphological and biochemical criteria, the cell death associated with the copper compound treatment was suggested to be paraptotic. While cisplatin treatment resulted in caspase-3 activation, cell death associated with treatment with the copper complex was independent of caspase-3 activity. Further, coadministration of the copper complex and cisplatin resulted in an inhibition of the cisplatin-induced caspase-3 activity. This result was interpreted to signify that the copper compound can inhibit apoptosis and at the same time activate paraptosis. Whether the copper compound inhibits apoptosis, or paraptotic activation dominates, remains to be proven. Regardless of the above, these two studies using copper compounds represent the first suggestions for metal-based anticancer drugs that activate paraptosis, and at the same time, may provide useful adjuvants for tumors that resist current therapies. Together, these studies provide strong evidence that activation of paraptosis may constitute an important feature in targeting tumor cell death and should motivate great interest in dissecting the molecular features and mechanism of this PCD pathway.

CONCLUSION

Several distinct cell death morphologies have been identified, only a few of which have been intensively studied and characterized at the molecular level.

Our studies led to the partial characterization of paraptosis as a cell death pathway likely to correspond to the morphological type 3B previously described. Paraptosis appears to be triggered by trophotoxicity via activation of certain receptors including IGF-IR, EGFR, NK1R, and TAJ/TROY TNF receptor, but paraptosis-like cell death can also be induced by other cellular stimuli such as Ca^{++} influx and intracellular acidification.

As for other cell death pathways, paraptosis-like cell death appears to be present during development as well as in pathological conditions.

Future studies will likely lead to the recognition of novel and separate cell death pathways and further characterization of these pathways will lead to the identification of pathway-specific modulators. Accurate cell death regulation is essential to the homeostasis of higher organisms, and a more complete understanding of these processes will ultimately lead to improved therapeutic strategies to fight conditions in which this regulation has gone awry.

REFERENCES

1. Clarke PGH, Clarke S. Nineteenth century research on naturally occurring cell death and related phenomena. Anat Embryol (Berl) 1996; 193:81–99.
2. Vogt C. Untersuchungen uber die Entwicjlungsgeschichte der geburtschelerkroete (Alytes obstetrician), p130, Solothum. Jent und Gassman. 1842.
3. Lockshin RA. The early modern period in cell death. Cell Death Differ 1997; 4:347–351.
4. Clarke PG. Developmental cell death: morphological diversity and multiple mechanisms. Anat Embryol 1990; 181(3):195–213.
5. Lockshin RA, Williams CM. Programmed cell death. I. Cytology of degeneration in the intersegmental muscles of the Pernyi silkmoth. J Insect Physiol 1965; 11:123.
6. Tata JR. Requirement for RNA and protein synthesis for induced regression of the tadpole tail in organ culture. Dev Biol 1966; 13(1):77–94.
7. Kerr JF, Wyllie AH, Currie AR. Apoptosis: a basic biological phenomenon with wide-ranging implications in tissue kinetics. Br J Cancer 1972; 26(4):239–257.
8. Kerr JFR, Harmon B, Searle J. An electron microscope study of cell deletion in the anuran tadpole tail during spontaneous metamorphosis with special reference to apoptosis of striatal muscle fibres. J Cell Sci 1974; 14:571–585.

9. Wyllie AH, Kerr JFR, Currie AR. Cell death in the normal neonatal rat adrenal cortex. J Pathol 1973; 111:225.
10. Hewish DR, Burgoyne LA. Chromatin sub-structure. The digestion of chromatin DNA at regularly spaced sites by a nuclear deoxyribonuclease. Biochem Biophys Res Commun 1973; 52(2):504–510.
11. Williams JR, Little JB, Shipley WU. Association of mammalian cell death with a specific endonucleolytic degradation of DNA. Nature 1974; 252(5485):754–755.
12. Wyllie AH. Glucocorticoid-induced thymocyte apoptosis is associated with endogenous endonuclease activation. Nature 1980; 284(5756):555–556.
13. Vaux DL, Cory S, Adams JM. bcl-2 gene promotes haemopoietic cell survival and co-operates with c-myc to immortalise pre-B cells. Nature 1988; 335:440–442.
14. Hengartner MO, Ellis RE, Horvitz HR. *Caenorhabditis elegans* gene ced-9 protects cells from programmed cell death. Nature 1992; 356(6369):494–499.
15. Horvitz HR, Ellis HM, Sternberg PW. Programmed cell death in nematode development. Neurosci Comment 1982; 1:56–65.
16. Yuan JY, Horvitz HR. The Caenorhabditis elegans genes ced-3 and ced-4 act cell autonomously to cause programmed cell death. Dev Biol 1990; 138(1):33–41.
17. Reed JC. Apoptosis-based therapies. Nat Rev/Drug Discov 2002; 1:111–121.
18. Reed JC. Apoptosis-targeted therapies for cancer. Cancer Cell 2003; 3:17–22.
19. Bredesen DE, Rao RV, Mehlen P. Cell death in the nervous system. Nature 2006; 443:796–802.
20. Leist M, Jaattela M. Four deaths and a funeral: from caspases to alternative mechanisms. Nat Rev Mol Cell Biol 2001; 2:1–10.
21. Roach HI, Aigner T, Kourai JB. Chondroptosis: a variant of apoptotic cell death in chondrocytes? Apoptosis 2004; 9:265–277.
22. Schweichel JU, Merker HJ. The morphology of various types of cell death in prenatal tissues. Teratology 1973; 7(3):253–266.
23. Sperandio S, de Belle I, Bredesen DE. An alternative, non-apoptotic form of programmed cell death. Proc Natl Acad Sci U S A 2000; 97(26):14376–14381.
24. Liu Y, Lehar S, Corvi C, et al. Expression of the insulin-like growth factor I receptor C terminus as a myristylated protein leads to induction of apoptosis in tumor cells. Cancer Res 1998; 58(3):570–576.
25. Bredesen DE, Mehlen P, Rabizadeh S. Receptors that mediate cellular dependence. Cell Death Differ 2005; 12:1031–1043.
26. Mazelin L, Bernet A, Bonod-Bidaud C, et al. Netrin-1 controls colorectal tumorigenesis by regulating apoptosis. Nature 2004; 431:80–84.
27. Castro-Obregon S, Del Rio G, Chen SF, et al. A ligand-receptor pair that triggers a non-apoptotic form of programmed cell death. Cell Death Differ 2002; 9:807–817.
28. Wang Y, Li X, Wang L, et al. An alternative form of paraptosis-like cell death, triggered by TAJ/TROY and enhanced by PDCD5 overexpression. J Cell Sci 2004; 117:1525–1532.
29. Schneider D, Gerhardt E, Bock J, et al. Intracellular acidification by inhibition of the Na+/H+-exchanger leads to caspase-independent death of cerebellar granule neurons resembling paraptosis. Cell Death Differ 2004; 11:760–770.
30. Chen Y, Douglass T, Jeffes EW, et al. Living T9 glioma cells expressing membrane macrophage colony-stimulating factor produce immediate tumor destruction by polymorphonuclear leukocytes and macrophages via a "paraptosis"-induced pathway

that promotes systemic immunity against intracranial T9 gliomas. Blood 2002; 100(4): 1373–1380.
31. Jadus MR, Chen Y, Boldaji MT, et al. Human U251MG glioma cells expressing the membrane form of macrophage colony-stimulating factor (mM-CSF) are killed by human monocytes in vitro and are rejected within immunodeficient mice via paraptosis that is associated with increased expression of three different heat shock proteins. Cancer Gene Ther 2003; 10(5):411–420.
32. Jambrina E, Alonso R, Alcade M, et al. Calcium influx through receptor-operated channel induces mitochondria-triggered parpatotic cell death. J Biol Chem 2003; 278:14134–14145.
33. Fombonne J, Padron L, Enjalbert A, et al. A novel paraptosis pathway involving LEI/L-DNaseII for EGF-induced cell death in somato-lactotrope pituitary cells. Apoptosis 2006; 11:367–375.
34. Fombonne J, Reix S, Rasolonjanahary R, et al. Epidermal growth factor triggers an original, caspase-independent pituitary cell death with heterogeneous phenotype. Mol Biol Cell 2004; 15:4938–4948.
35. Fischer U, Janicke RU, Schulze-Osthoff K. Many cuts to ruin: a comprehensive update of caspase substrates. Cell Death Differ 2003; 10:76–100.
36. Bursch W, Hochegger K, Torok L, et al. Autophagic and apoptotic types of programmed cell death exhibit different fates of cytoskeletal filaments. J Cell Sci 2000; 113:1189–1198.
37. Pilar G, Landmesser L. Ultrastructural differences during embryonic cell death in normal and peripherally deprived ciliary ganglia. J Cell Biol 1976; 68(2):339–356.
38. Cunningham TJ. Naturally occurring neuron death and its regulation by developing neural pathways. Int Rev Cytol 1982; 74:163–186.
39. Giordano DL, Murray M, Cunningham TJ. Naturally occurring neuron death in the optic layers of superior colliculus of the postnatal rat. J Neurocytol 1980; 9(5): 603–614.
40. Cunningham TJ, Mohler IM, Giordano DL. Naturally occurring neuron death in the ganglion cell layer of the neonatal rat: morphology and evidence for regional correspondence with neuron death in superior colliculus. Brain Res 1981; 254(2): 203–215.
41. Sohal GS, Weidman TA. Ultrastructural sequence of embryonic cell death in normal and peripherally deprived trachlear nucleus. Exp Neurol 1978; 61(1):53–64.
42. Chu-Wang IW, Oppenheim RW. Cell death of motoneurons in the chick embryo spinal cord. I. A light and electron microscopic study of naturally occurring and induced cell loss during development. J Comp Neurol 1978; 177(1):33–57.
43. Liu J-P, Baker J, Perkins AS, et al. Mice carrying null mutations of the genes encoding insulin-like growth factor I (Igf-1) and type 1 IGF receptor (Igf1r). Cell 1993; 75:59–72.
44. Cecconi F, Alvarez-Bolado G, Meyer BI, et al. Apaf1 (CED-4 homolog) regulates programmed cell death in mammalian development. Cell 1998; 94(6):727–737.
45. Hakem R, Hakem A, Duncan GS, et al. Differential requirement for caspase 9 in apoptotic pathways in vivo. Cell 1998; 94(3):339–352.
46. Kuida K, Haydar TF, Kuan CY, et al. Reduced apoptosis and cytochrome c-mediated caspase activation in mice lacking caspase 9. Cell 1998; 94(3):325–337.
47. Kuida K, Zheng TS, Na S, et al. Decreased apoptosis in the brain and premature lethality in CPP32-deficient mice. Nature 1996; 384(6607):368–372.

48. Yoshida H, Kong YY, Yoshida R, et al. Apaf1 is required for mitochondrial pathways of apoptosis and brain development. Cell 1998; 94(6):739–750.
49. Chautan M, Chazal G, Cecconi F, et al. Interdigital cell death can occur through a necrotic and caspase- independent pathway. Curr Biol 1999; 9(17):967–970.
50. Oppenheim RW, Flavell RA, Vinsant S, et al. Programmed cell death of developing mammalian neurons after genetic deletion of caspases. J Neurosci 2001; 21(13): 4752–4760.
51. Yaginuma H, Shiraiwa N, Shimada T, et al. Caspase activity is involved in, but is dispensable for, early motoneuron death in the chick embryo cervical spinal cord. Mol Cell Neurosci 2001; 18(2):168–182.
52. Franklin DJ, Berges JA. Mortality in cultures of the dinoflagellate *Amphidinium carterae* during culture senescence and darkness. Proc Biol Sci 2004; 271:2099–2107.
53. Dal Canto MC, Gurney ME. Development of central nervous system pathology in a murine transgenic model of human amyotrophic lateral sclerosis. Am J Pathol 1994; 145(6):1271–1279.
54. Turmaine M, Raza A, Mahal A, et al. Nonapoptotic neurodegeneration in a transgenic mouse model of Huntington's disease. Proc Natl Acad Sci U S A 2000; 97(14): 8093–8097.
55. Regan RF, Panter SS, Witz A, et al. Ultrastructure of excitotoxic neuronal death in murine cortical culture. Brain Res 1995; 705(1–2):188–198.
56. Kang SJ, Sanchez I, Jing N, et al. Dissociation between neurodegeneration and caspase-11-mediated activation of caspase-1 and caspase-3 in a mouse model of amyotrophic lateral sclerosis. J Neurosci 2003; 23(13):5455–5460.
57. Kostic V, Jackson-Lewis V, de Bilbao F, et al. Bcl-2: prolonging life in a transgenic mouse model of familial amyotrophic lateral sclerosis. Science 1997; 277(5325): 559–562.
58. Li M, Ona VO, Guegan C, et al. Functional role of caspase-1 and caspase-3 in an ALS transgenic mouse model. Science 2000; 288(5464):335–339.
59. Sperandio S, Poksay K, de Belle I, et al. Paraptosis: mediation by MAP kinases and inhibition by AIP-1/Alix. Cell Death Differ 2004; 11:1066–1075.
60. Castro-Obregon S, Rao RV, del Rio G, et al. Alternative, nonapoptotic programmed cell death: mediation by arrestin 2, ERK2, and Nur77. J Biol Chem 2004; 279: 17543–17553.
61. Allan LA, Morrice N, Brady S, et al. Inhibition of caspase-9 through phosphorylation at Thr 125 by ERK MAPK. Nat Cell Biol 2003; 5:647–654.
62. Chang L, Karin M. Mammalian MAP kinase signalling cascades. Nature 2001; 410:37–40.
63. Kulich SM, Chu CT. Sustained extracellular signal-regulated kinase activation by 6-hydroxydopamine: implications for Parkinson's disease. J Neurochem 2001; 77(4): 1058–1066.
64. Linford NJ, Yang Y, Cook DG, et al. Neuronal apoptosis resulting from high doses of the isoflavone genistein: role for calcium and p42/44 mitogen-activated protein kinase. J Pharmacol Exp Ther 2001; 299(1):67–75.
65. Mukherjee P, Pasinetti GM. Complement anaphylatoxin C5a neuroprotects through mitogen-activated protein kinase-dependent inhibition of caspase 3. J Neurochem 2001; 77(1):43–49.
66. Regan RF, Wang Y, Ma X, et al. Activation of extracellular signal-regulated kinases potentiates hemin toxicity in astrocyte cultures. J Neurochem 2001; 79(3):545–555.

67. Runden E, Seglen PO, Haug FM, et al. Regional selective neuronal degeneration after protein phosphatase inhibition in hippocampal slice cultures: evidence for a MAP kinase-dependent mechanism. J Neurosci 1998; 18(18):7296–7305.
68. Missotten M, Nichols A, Rieger K, et al. Alix, a novel mouse protein undergoing calcium-dependent interaction with the apoptosis-linked-gene 2 (ALG-2) protein. Cell Death Differ 1999; 6(2):124–129.
69. Vito P, Pellegrini L, Guiet C, et al. Cloning of AIP1, a novel protein that associates with the apoptosis-linked gene ALG-2 in a Ca2+-dependent reaction. J Biol Chem 1999; 274(3):1533–1540.
70. Wang X., Li N, Liu B, et al. A novel human phosphatidylethanolamine-binding protein resists tumor necrosis factor alpha-induced apoptosis by inhibiting mitogen-activated protein kinase pathway activation and phosphatidylethanolamine externalization. J Biol Chem 2004; 279:45855–45864.
71. LeRoith D, Roberts CTJ. The insulin-like growth factor system and cancer. Cancer Lett 2003; 195:127–137.
72. Plymate SR, Tennant M, Birnbaum RS, et al. The effect on the insulin-like growth factor system in human prostate epithelial cells of immortalization and transformation by simian virus-40 T antigen. J Clin Endocrinol Metab 1996; 81(10):3709–3716.
73. Tennant MK, Thrasher JB, Twomey PA, et al. Protein and messenger ribonucleic acid (mRNA) for the type 1 insulin-like growth factor (IGF) receptor is decreased and IGF-II mRNA is increased in human prostate carcinoma compared to benign prostate epithelium. J Clin Endocrinol Metab 1996; 81(10):3774–3782.
74. Plymate SR, Bae VL, Maddison L, et al. Reexpression of the type 1 insulin-like growth factor receptor inhibits the malignant phenotype of simian virus 40 T antigen immortalized human prostate epithelial cells. Endocrinology 1997; 138(4): 1728–1735.
75. Plymate SS, Bae VL, Maddison L, et al. Type-1 insulin-like growth factor receptor reexpression in the malignant phenotype of SV40-T-immortalized human prostate epithelial cells enhances apoptosis. Endocrine 1997; 7(1):119–124.
76. Chott A, Sun Z, Morganstern D, et al. Tyrosine kinases expressed in vivo by human prostate cancer bone marrow metastases and loss of the type 1 insulin-like growth factor receptor. Am J Pathol 1999; 155:1271–1279.
77. Kaplan PJ, Mohan S, Cohen P, et al. The insulin-like growth factor axis and prostate cancer: lessons from the transgenic adenocarcinoma of mouse prostate (TRAMP) model. Cancer Res 1999; 59(9):2203–2209.
78. Hongo A, Yumet G, Resnicoff M, et al. Inhibition of tumorigenesis and induction of apoptosis in human tumor cells by the stable expression of a myristylated COOH terminus of the insulin-like growth factor I receptor. Cancer Res 1998; 58(11): 2477–2484.
79. Marzano C, Pellei M, Colavito D, et al. Synthesis, characterization, and in vitro antitumor properties of tris(hydroxymethyl)phosphine copper(I) complexes containing the new bis(1,2,4-triazol-1-yl)acetate ligand. J Med Chem 2006; 49: 7317–7324.
80. Tardito S, Bussolati O, Gaccioli F, et al. Non-apoptotic programmed cell death induced by a copper(II) complex in human fibrosarcoma cells. Histochem Cell Biol 2006; 126:473–482.

11

Cellular Senescence and Its Effects on Carcinogenesis

Judith Campisi
Lawrence Berkeley National Laboratory, Berkeley;
Buck Institute for Age Research, Novato, California, U.S.A.

INTRODUCTION: CELLULAR SENESCENCE

More than four decades ago, Hayflick and colleagues formally showed that normal human cells—in this case, fibroblasts from fetal and adult tissue—have only a limited ability to divide in culture (1,2). These now classic studies established that freshly explanted human fibroblasts initially proliferate robustly in culture; however, with each passage, the cultures gradually accumulate viable but nondividing cells. Eventually, the cultures become entirely postmitotic (incapable of proliferation), despite optimal culture conditions. This decline in proliferative capacity was termed cellular senescence because it was proposed to recapitulate the loss of regenerative capacity that is a hallmark of organismal aging.

Hayflick and colleagues also noted that tumor cells behaved very differently. In contrast to normal human cells, human tumor cells appeared to proliferate indefinitely in culture. This observation spawned a second hypothesis that cellular senescence also suppresses the development of cancer (2,3).

Many years after the formal description of cellular senescence, the hypothesis that this process contributes to organismal aging remains viable but still speculative (4–7). In contrast, many lines of evidence now support the idea that cellular senescence suppresses malignant tumorigenesis (8–12). Further, the apparently disparate roles that have been proposed for cellular senescence—aging and tumor suppression—are beginning to converge as our understanding of the evolutionary origins of aging phenotypes matures (13,14).

Characteristics and Causes of Cellular Senescence

The salient feature of cellular senescence is an essentially irreversible withdrawal from the cell cycle. This arrest of cell proliferation (used here interchangeably with growth) in essence converts a mitotic cell, which has the ability to proliferate, into a postmitotic cell, which has permanently lost the ability to divide. Cell proliferation is, of course, essential for tumorigenesis (12). Thus, the senescence growth arrest is the key feature of cellular senescence that suppresses the development of cancer.

Perhaps, the most intriguing aspect of cellular senescence is that the senescence growth arrest is not simply a cessation of cell division. Senescent cells develop a variety of phenotypic characteristics, collectively termed the senescent phenotype (Fig. 1). In addition to the permanent growth arrest, notable features of the senescent phenotype include resistance to certain apoptotic stimuli and widespread changes in gene expression (15–21). Altered gene expression is certainly responsible for the senescence growth arrest, and is likely responsible for other changes often associated with senescent cells: an enlarged morphology, increased lysosomal and mitochondrial biogenesis, and increased reactive oxygen species (ROS). Of particular interest, among the senescence-associated changes in

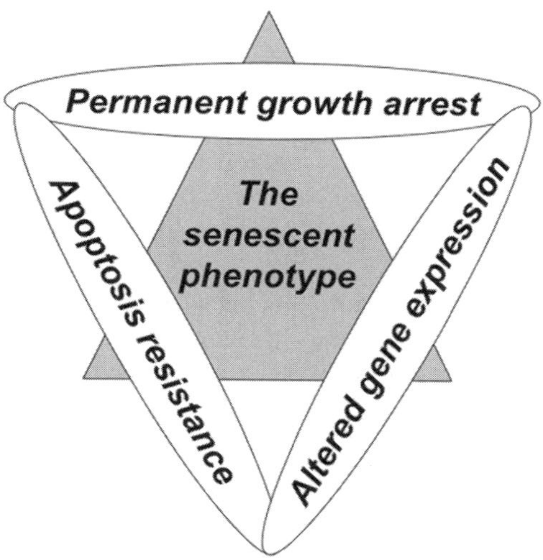

Figure 1 The senescent phenotype. Senescent cells irreversibly arrest proliferation with a complex phenotype. In addition to the permanent growth arrest, this phenotype includes resistance to apoptotic stimuli and many changes in gene expression.

gene expression, is a striking upregulation of genes encoding secreted molecules that can alter the local tissue microenvironment (14,22). As discussed later, this feature of the senescence response may be the consequence of an evolutionary trade-off between tumor suppression and longevity, and may even, ironically, fuel the development of late-life cancer.

It is now clear that the progressive shortening (and subsequent dysfunction) of telomeres that occurs during cell proliferation is a prime cause for the cellular senescence observed by Hayflick and colleagues. Telomeres are the repetitive DNA sequence (TTAGGG in vertebrates) and specialized proteins that form a protective structure, or "cap," at the ends of linear chromosomes (23). The telomeric cap prevents cellular DNA repair machineries from recognizing chromosome ends as DNA double strand breaks, and thus prevents chromosome end-to-end fusions (24). Fused chromosomes contain two centromeres, so the mitotic apparatus can literally break these chromosomes at random places during mitosis. Broken chromosomes, in turn, undergo cycles of breakage and fusion, thereby creating genomic instability, a major driving force behind the development of cancer (25–27). Functional telomeres, therefore, prevent genomic instability and hence cancer. Normal cells respond to short, dysfunctional telomeres by undergoing senescence (19,28,29).

Why do telomeres shorten during cell proliferation? Eukaryotic DNA replication is bidirectional and carried out by polymerases that are unidirectional and dependent on preexisting (normally labile RNA) primers. These biochemical constraints result in incompletely replicated 3′ ends during DNA synthesis, a phenomenon termed the end-replication problem (30). The end-replication problem causes telomeres to shorten with each round of DNA replication. Ultimately, the telomeres become critically short and incapable of forming a protective capped structure, to which most cells respond by undergoing senescence (19,28,29). The end-replication problem can be overcome by the enzyme telomerase, which can add telomeric DNA repeats to chromosome ends de novo (31). Telomerase expression is generally restricted to early embryonic cells, germ cells, and some somatic stem or progenitor cells (32,33). Most cancer cells, which proliferate indefinitely, overcome the end-replication problem by activating telomerase expression (34,35).

It is now clear that dysfunctional telomeres are but one of many stimuli that can elicit a senescence response. Other inducers of cellular senescence include severe or irreparable DNA damage, such as DNA double strand breaks and genomic damage caused by oxidative stress (36,37). Because telomeres are G-rich, they are particularly susceptible to oxidative damage, which can accelerate shortening during DNA replication (38). Short dysfunctional telomeres are now thought to induce senescence by triggering a DNA damage response similar to that caused by DNA double strand breaks (39,40). In addition, agents or genetic lesions that perturb chromatin organization can induce a senescence response, as can a variety of stresses such as suboptimal growth conditions (41–45). Furthermore, intense or unbalanced mitogenic signals can induce cellular senescence (46,47). Of especial importance, oncogenes that deliver strong mitogenic signals cause normal cells to senesce (48–50). A number of oncogenes do not cause neoplastic transformation unless the cells harbor preexisting mutations that allow them to ignore or bypass senescence-inducing signals (48–50). One commonality among all the stimuli that cause cellular senescence is the ability to promote malignant transformation.

Control of Cellular Senescence by Tumor Suppressor Pathways

Two potent tumor suppressor proteins—the p53 and retinoblastoma (pRB) proteins—are essential regulators of cellular senescence (Fig. 2) (9,51–53). These proteins lie at the heart of two tumor suppressor pathways, each comprised

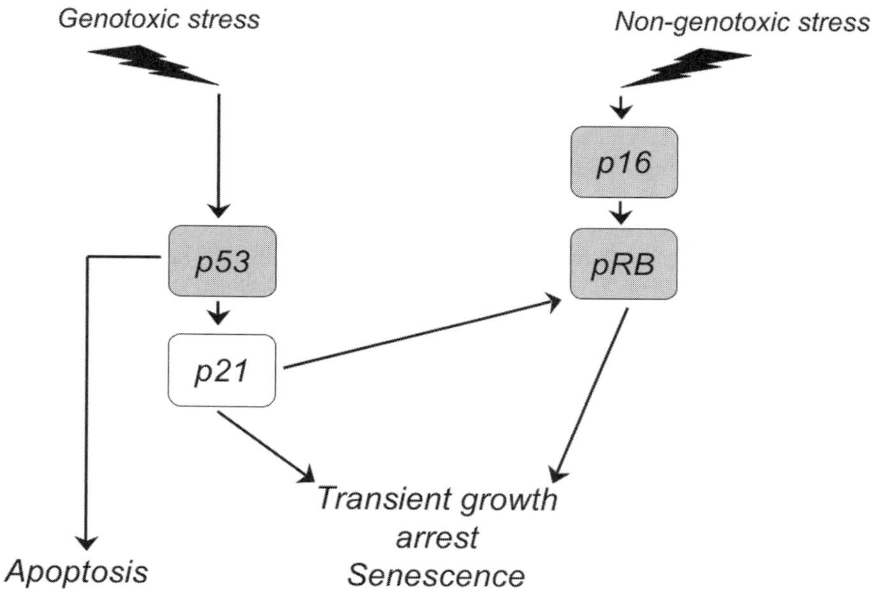

Figure 2 Cell fate controlled by the p53 and p16INK4a/pRB tumor suppressor pathways. The p53 pathway is activated primarily by genotoxic stress, which causes either a transient or permanent (senescence) growth arrest. These growth arrest states are implemented in part through the p53 target gene encoding p21, a cyclin-dependent kinase inhibitor, which, like p16INK4a, helps maintain pRB in its active unphosphorylated state. Activation of the p53 pathway can also cause apoptotic cell death. Non-genotoxic stress primarily activates the p16INK4a/pRB pathway, often by inducing expression of p16INK4a. Although p16INK4a expression is mostly associated with a senescence growth arrest, activation of pRB can also cause either a transient or a senescence arrest. The nature and strength of the stressors, cell type, and physiological context determine which cell fate is implemented by the p53 and p16INK4a/pRB pathways. *Abbreviation*: pRB, retinoblastoma protein.

of several upstream regulators and downstream effectors (54–56). Mutations in p53, pRB, or components of the pathways they govern occur in most, if not all, cancer cells.

Some components of p53 and pRB pathways are themselves tumor suppressor proteins and important regulators of the senescence response. One notable example is p16INK4a, a cyclin-dependent kinase inhibitor (CDKI). p16INK4a helps maintain pRB in its active unphosphorylated state, and thus is a positive regulator of pRB (20,56,57). In addition, the pathways governed by p53 and pRB interact (Fig. 2). Notably, p53 increases the transcription of p21 (58), another CDKI that maintains pRB in its unphosphorylated, active form. Finally, both the p53 and pRB pathways control cell fates other than senescence (Fig. 2). Activation of the p53 pathway can also lead to cell death (apoptosis), and activation of either pathway can lead to a transient, as opposed to permanent, cell cycle arrest (59,60). Very little is known about the mechanisms by which the p53 and pRB pathways trigger one cell fate over another. In all cases, both the cell type and nature of the stimulus are likely to be important variables.

The p53 and pRB pathways generally respond, at least initially, to different stimuli (Fig. 2). Genotoxic stress activates the p53 pathway, typically causing activation of posttranslational modifications to p53 (61). Genotoxic stress can be caused by direct DNA damage, or by dysfunctional telomeres or oncogenes such as mutant RAS. RAS is a GTPase that helps transduce mitogenic signals. Mitogenic oncogenes cause genotoxic stress because they deliver strong growth-promoting signals, which cause aberrant DNA replication and subsequent DNA double strand breaks (62,63). By contrast, the pRB pathway is activated by poorly defined stresses, which typically induce p16INK4a expression (56). Mitogenic signals, and oncogenes such as mutant RAS, also induce p16INK4a expression (47). Thus, the p53 pathway responds primarily to genotoxic stress, whereas the p16INK4a/pRB pathway responds primarily to non-genotoxic stress, and activation of either pathway is sufficient to cause a senescence growth arrest.

There is often cross talk between the p53 and p16INK4a/pRB pathways, and so, some senescent cells arrest growth with both pathways activated. For example, in some cells, short dysfunctional telomeres, which induce senescence by activating p53, also induce p16, albeit after a prolonged interval (64–67). In addition, both in culture (68,69) and in vivo (70,71), it is clear that the pathways cooperate to establish and maintain the tumor-suppressive senescence growth arrest. For example, some cells lose the ability to express p16INK4a, often due to promoter methylation. Such cells remain susceptible to p53-mediated senescence, but are at increased risk of neoplastic transformation. Thus, human cells that do not express p16INK4a still senesce in response to critical telomere shortening owing to activation of the p53 pathway. Subsequent inactivation of p53 in such cells causes them to proliferate, despite dysfunctional telomeres (69,72). Consequently, the cells become genomically unstable, a major risk factor for transformation. By contrast, once cells senesce owing to p16INK4a expression, the senescence growth arrest cannot be reversed by subsequent inactivation of p53, p16INK4a, or pRB (69). The mechanisms by which the p16INK4a/pRB pathway maintains the senescence growth arrest, even after p53 inactivation, may entail the establishment of heterochromatic domains that irreversibly silence genes required for cell proliferation (69,73).

Finally, it is now clear that the senescence response is not confined to normal cells. A number of human tumor cells, despite oncogenic mutations and a means to overcome the end-replication problem, retain the ability to arrest growth with senescence characteristics after treatment with DNA-damaging chemotherapeutic agents (74–76). For at least some of these tumor cells, the senescence response is mediated by the p53 pathway because the response can be altered by modulating either p21 or p53 expression (76–80). Moreover, therapy outcome was generally more favorable for tumors that retained the ability to senesce compared to tumors that failed to undergo senescence (or apoptosis). These findings suggest that the senescence response of tumor cells, like their apoptotic response, may be a prognostic indicator of the response to chemotherapy (81).

CELLULAR SENESCENCE SUPPRESSES CANCER

Tumor cells inevitably harbor mutations that derange components of the p53 and/or p16INK4a/pRB pathways, consistent with an important role of cellular senescence in suppressing cancer. Mouse models, and recent studies of premalignant and malignant lesions in humans, provide additional evidence for this idea. Some of these studies use characteristic markers, none of which are exclusive, to identify senescent cells in tissues. These markers include expression of a neutral β-galactosidase (82), expression of p16INK4a (83), and nuclear foci of heterochromatin or DNA damage proteins (84–86).

p53-deficient mice develop normally, but acquire malignant tumors within about six months of age (87). Cells from these mice are highly resistant to senescence induced by DNA damage and oncogenes such as RAS (88). However, p53-deficient cells are also defective in apoptosis (88), so the extreme cancer susceptibility of these mice is likely because of combined defects in apoptosis and senescence. p16INK4a-deficient mice, by contrast, develop tumors later in life—12 to 18 months of age—still in advance of the average age at which wild-type mice develop cancer (18–24 months, depending on the mouse strain) (89). Cells from these mice do not appear to be defective in apoptosis. Moreover, although defective in p16INK4a/pRB-mediated senescence, they retain the ability to senesce in response to genotoxic stress and p53 activation. Nonetheless, these mice are cancer-prone, suggesting that p16INK4a-mediated senescence suppresses malignant tumorigenesis in vivo.

In other mouse models, oncogenic RAS induced senescence when expressed in lymphocytes, but lymphomas formed only in mice that were deficient in Suv39h1, a histone methyltransferase that acts with pRB and mediates senescence (90). Likewise, oncogenic RAS expression caused an accumulation of senescent cells in premalignant lesions of the lung, skin, and pancreas, but not malignant tumors that developed in these tissues after a long latency (91). In a third mouse model (92), prostate-specific inactivation of the PTEN tumor suppressor, a protein tyrosine phosphatase that dampens mitogenic signals, caused premalignant prostatic lesions containing numerous senescent cells. When crossed to p53-deficient mice, the hybrid mice developed lethal invasive cancers devoid of senescent cells. Senescent cells were also found in early-stage prostate cancer in humans, but not in highly malignant later stages (92). Similarly, oncogenic mutations in BRAF (v-raf murine sarcoma viral oncogene homolog B1), a component of the RAS pathway, causes senescence in cultured human fibroblasts and melanocytes. In human skin biopsies, senescent cells were found in benign melanocytic nevi, but not malignant melanomas (50).

Together, these studies indicate that oncogenic mutations cause cellular senescence in vivo, and that the senescence response suppresses the progression of mutant premalignant cells to malignant tumors. Malignant tumors develop when cells acquire preexisting or additional mutations in the p53 or p16INK4a/pRB pathways, which allow cells with oncogenic mutations to ignore senescence signals or escape the senescence growth arrest.

CANCER AND AGING

Cancer is a major age-related disease. A vast majority of malignant tumors occur after about the midpoint of the species-specific life span (e.g., about 1.5 years in mice, 50 years in humans). From then onward, the incidence of cancer rises exponentially until well into old age (>80 years for humans) (93,94).

Cancer is a multistep process during which cells acquire increasing malignant phenotypes. What cause these phenotypic changes? First and foremost are somatic mutations. However, mutations, including oncogenic mutations, begin to accumulate very early in life and are present in apparently normal tissues (95–97) (Fig. 3). Second, malignant progression depends to a large extent on the tissue microenvironment. It is now well established that normal tissue structures and microenvironments can suppress, and in some cases even reverse, the expression of malignant phenotypes by mutant cells (98–101). These requirements for malignant tumorigenesis suggest that age-related cancers might arise owing to at least two synergistic processes—somatic mutations and a loss of normal tissue structure and function—both of which rise with age (13,94). Mutations most likely increase with age owing to imperfect repair of DNA damage from both exogenous and endogenous sources, as well as errors in DNA replication (Fig. 3). DNA replication and repair can also cause mistakes in the modifications to DNA, histones, and other chromatin-associated proteins that determine patterns of gene expression. These so-termed epi-mutations are also important causative factors for the development of cancer (102,103). The age-related decline in tissue integrity most likely has many etiologies, one of which, as discussed below, might be the accumulation of senescent cells (Fig. 3).

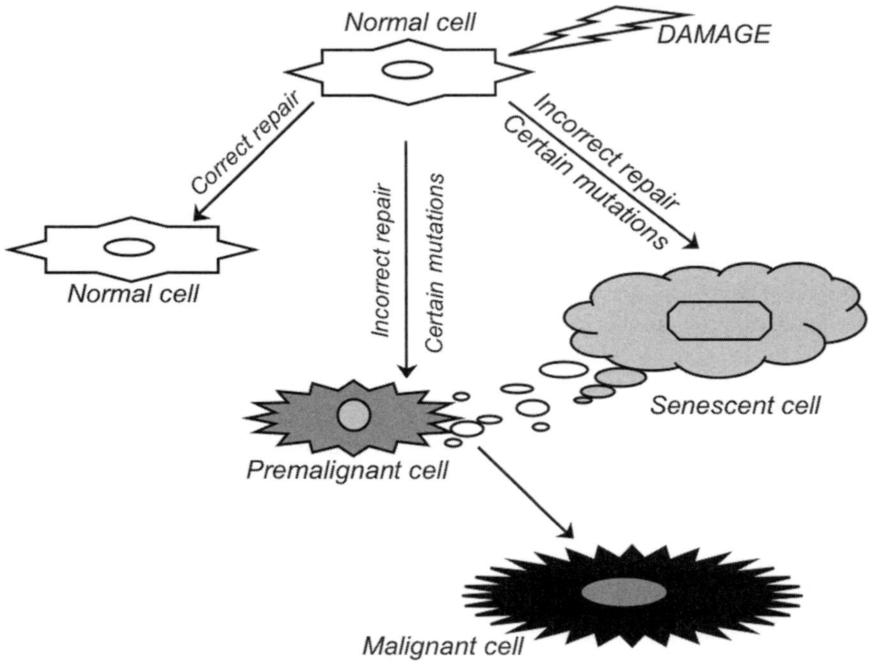

Figure 3 Mutations and a senescence microenivironment may promote malignant progression. Normal cells experience damage from endogenous and exogenous sources. If properly repaired, the cell is restored to its pre-damaged state (normal cell). If improperly repaired, a cell may harbor mutations that predispose it to malignant tumorigenesis (premalignant cells). Alternatively, irreparable damage or certain oncogenic mutations may cause a senescence response. The growth-arrested senescent cells secrete cytokines, growth factors, and extracellular matrix–remodeling proteins (small ovals) that can alter the tissue microenvironment. This altered microenvironment, in turn, can stimulate premalignant cells to progress toward more malignant phenotypes.

SENESCENT CELLS AND AGING

Senescent cells, or rather cells with one or more marker characteristic of senescent cells, increase in number with age in many mammalian tissues, particularly those that are renewable (having proliferative or regenerative capacity) (82,85,104–106). In at least some cases, these cells appear to have been induced to senesce owing to short dysfunctional telomeres (84,85). Senescent cells have also been identified at sites of age-related pathologies, particularly degenerative pathologies such as osteoarthritis, intervertebral disc degeneration, venous ulcers, and atherosclerosis (106–111). These findings support the idea that cellular senescence recapitulates some aspects of organismal aging, as initially proposed by Hayflick and colleagues. In addition, recent insights into the nature of the complex senescent phenotype raise the possibility that senescent cells might actively contribute to degenerative phenotypes and certain diseases associated with aging.

How might senescent cells contribute to aging? As discussed above, among the senescence-associated changes in gene expression is a striking increase in the expression of genes encoding secreted molecules (112–117). These secreted molecules include inflammatory cytokines, chemokines, growth factors, and matrix metalloproteinases, all of which can radically alter the local tissue microenvironment. Thus, senescent cells might contribute to the age-related decline in tissue structure and function that is a hallmark of aging organisms (14) (Fig. 3). At present, the evidence for this hypothesis is indirect. As noted above, senescent cells accumulate in aging and degenerating tissues. Moreover, in three-dimensional heterotypic cell culture models, the presence of senescent cells can, at least in principle, disrupt the normal morphological and functional differentiation of mammary and skin epithelial cells and microvascular endothelial cells (118–121).

CELLULAR SENESCENCE IN AN EVOLUTIONARY CONTEXT

Why might cellular senescence, a potent and effective tumor-suppressive mechanism, contribute to aging? This question is best phrased in a larger context: how can any fundamental process be both beneficial (tumor-suppressive) and detrimental (pro-aging) to organisms?

Cancer poses a major challenge to the longevity of organisms with renewable (regenerative) tissues. This challenge arises from intertwined nature of cell proliferation, tissue regeneration, and neoplastic growth. First, cell proliferation is essential for regeneration and repair, and hence is essential for the health of complex organisms with renewable tissues. Second, cell proliferation is also essential for tumorigenesis (12). Third, the proliferative cells that comprise renewable tissues are more prone than postmitotic cells to acquiring mutations (122), and hence, these cells are doubly at risk of neoplastic transformation. Needless to say, the risk posed by cancer was mitigated by the evolution of tumor suppressor mechanisms. Some tumor suppressor mechanisms act by suppressing the development of mutations; others act by eliminating potential cancer cells (e.g., by apoptosis) or permanently arresting their proliferation (cellular senescence). As discussed below, the evolution of at least some tumor suppressor mechanisms, cellular senescence in particular, may have entailed an evolutionary trade-off (5).

In general, organisms evolve in environments that are replete with extrinsic hazards such as predators, starvation, and infections. Therefore, throughout much of their evolutionary history, most organisms do not die of old age or age-related diseases. Rather, their life span is curtailed by death due to external catastrophes. Consequently, internal survival mechanisms such as tumor suppression must be effective only up to and during the period of peak reproduction. This period amounts to a few decades for humans and several months for mice. Should a beneficial process, such as a tumor-suppressive mechanism, be deleterious after the peak reproductive age, there would be little selective pressure to make the process less detrimental. Thus, it is at least theoretically possible for a tumor-suppressive mechanism to be both beneficial and detrimental, depending on the age of the organism. This idea that a biological process can benefit organisms early in life but have *unselected* late-life deleterious effects that have escaped the force of natural selection comprises an important evolutionary theory of aging termed antagonistic pleiotropy (123,124).

Cellular senescence might be an example of evolutionary antagonistic pleiotropy. As a beneficial tumor-suppressive mechanism, the selected phenotype was undoubtedly the arrest of cell growth, which prevents oncogenically compromised cells from proliferating. On the other hand, the changes in gene expression, particularly those that result in the secretion of molecules that alter tissue microenvironments, might be the traits that escaped the force of natural selection. The senescent secretory phenotype might have its origins in the tissue wounding response; at least for stromal fibroblasts and endothelial cells, the senescence-associated secretory phenotype resembles the response to injury (14,125). Wounding generally stimulates a transient burst of cell proliferation, followed by the secretion of molecules that attract immune cells, remodel the extracellular matrix, and stimulate the growth of neighboring cells such as adjacent epithelial cells. Senescent cells appear to be arrested in the post-replicative but activated secretory stage of wound healing. The chronic presence of senescent cells would not necessarily be beneficial because such cells can create local sites of chronic inflammation, tissue degradation, fibrosis, and hyperproliferation (of neighboring cells). Senescent cells might not affect tissue homeostasis early in life, when their numbers are relatively low, but they may well become deleterious later in life, when their numbers increase.

The idea that cellular tumor-suppressive mechanisms such as senescence (as well as apoptosis) can drive aging phenotypes is supported by two genetically engineered mouse models in which p53 was rendered chronically activated by expressing artificially generated or naturally occurring short p53 isoforms. The expression of these short p53 isoforms conferred extraordinary protection against cancer, as expected of an upregulated tumor suppressor protein. However, in both mouse models, the animals developed multiple aging phenotypes at an accelerated rate (126,127). After several months, these animals presented with gray hair, thin, wrinkled skin, osteoporosis, signs of diabetes, and other phenotypes associated with aging. When tested in culture, cells from these mice were unusually prone to both apoptosis and cellular senescence (126,128). These findings not only reinforce the idea that the cellular senescence and apoptosis suppress the development of cancer, but also suggest that these tumor-suppressive mechanisms can drive phenotypes associated with aging. In a different series of mouse models, cells that spontaneously express p16INK4a and were likely but not formally shown to be senescent, increased with age in three stem or progenitor cell compartments: the subventricular zone of the brain, the hematopoietic system, and the pancreatic islets. Concomitant with the rise in p16INK4a expression, the mice showed a decline in neurogenesis, the bone marrow showed a marked reduction in ability to reconstitute an immune system, and pancreatic regeneration and function failed, resulting in diabetes. Strikingly, these age-related decrements in tissue function were retarded in p16INK4a-deficient mice, which indicates an important role of p16INK4a in these age-associated phenotypes (129–131). Of course, p16INK4a-deficient mice are deficient in undergoing cellular senescence, and hence are cancer-prone (89).

Together, these findings support the idea that cellular senescence, and perhaps apoptosis as well, are antagonistically pleiotropic. There is little doubt that both these cellular tumor-suppressive mechanisms suppress the development of cancer, and appear to do so effectively for about half the life span in mice and humans. After about the midpoint of the life span, however, these cellular processes might promote age-related decrements in tissue structure and function and contribute to age-related diseases. As discussed in the section below, ironically, senescent cell might even contribute to the development of age-related cancer.

CELLULAR SENESCENCE AND CANCER

As discussed above, cancer incidence rises exponentially with age, and this rise is likely fueled by the dual processes of mutation accumulation and age-dependent changes in tissue integrity. Moreover, a variety of cell culture experiments show that the presence of senescent cells can, at least in principle, contribute to tissue changes commonly observed during aging (118–121). Further, the secretory phenotype of senescent fibroblasts resembles the phenotype of carcinoma- or tumor-associated fibroblasts; stromal cells that are activated by carcinomas and produce extracellular factors that promote epithelial carcinogenesis (132,133). These findings suggest that as both senescent and mutant (premalignant) cells accumulate with age, their probability of occurring in close proximity likewise increase with age. When in close proximity, molecules secreted by senescent cells might promote the growth and malignant progression of nearby premalignant cells and thus, ironically, promote the development of late-life cancer. Indeed, cell culture and mouse xenograft experiments show that senescent cells can stimulate the ability of premalignant or malignant epithelial cells to both proliferate and invade a basement membrane (115,120,134–136), two essential steps in the development of cancer in culture and in mice. In addition, senescent cells produce angiogenic factors such as VEGF (vascular endothelial growth factor), which can stimulate normal endothelial cells to invade a basement membrane in culture, and may be responsible for the increased vascularity of xenografted tumors that form in the presence of senescent fibroblasts (137). Together, these findings support the idea that cellular senescence, despite being a potent tumor-suppressive mechanism, might be antagonistically pleiotropic and therefore contribute to age-related pathology, including late-life cancer.

UNANSWERED QUESTIONS AND FUTURE DIRECTIONS

More than 40 years after the seminal description of cellular senescence by Hayflick and colleagues, the idea that this process is a potent mechanism for suppressing the development of cancer has gained substantial support. In this regard, the senescence response resembles apoptosis, the other cellular tumor-suppressive mechanism that acts at the level of cell fate. An important unanswered question is what determines whether cells undergo a transient cell cycle arrest, cellular senescence, or apoptosis in response to potentially damaging oncogenic stimuli. Further, the finding that a number of tumor cells, most of which harbor mutations in the p53 and/or p16INK4a/pRB pathways, can undergo senescence in response to DNA-damaging chemotherapeutic agents raises the possibility that there are p53- and pRB-independent pathways that can establish a senescence growth arrest. Indeed, some oncogenes, such as RAF and PRAK, appear to cause senescence by such independent pathways (138). What are the components of these pathways, and can they be harnessed to develop more effective anticancer therapies? Finally, the idea that senescent cells can create a pro-carcinogenic tissue environment is still speculative, but, if true, suggests that cancer therapies that cause apoptosis are preferable to therapies that induce senescence. On the other hand, if senescent cells are indeed deleterious and pro-carcinogenic, therapies aimed at eliminating senescent cells, or their secretory phenotype, might offer a novel preventive strategy for reducing cancer risk. A preventive strategy of this sort might be especially important for preventing cancer recurrence after exposure to DNA-damaging chemotherapy or radiation therapy.

REFERENCES

1. Hayflick L, Moorhead PS. The serial cultivation of human diploid cell strains. Exp Cell Res 1961; 25:585–621.
2. Hayflick L. The limited in vitro lifetime of human diploid cell strains. Exp Cell Res 1965; 37:614–636.
3. Sager R, Tanaka K, Lau CC, et al. Resistance of human cells to tumorigenesis induced by cloned transforming genes. Proc Natl Acad Sci U S A 1984; 80: 7601–7605.
4. Campisi J. Replicative senescence: an old lives tale? Cell 1996; 84:497–500.
5. Campisi J. Cellular senescence and apoptosis: how cellular responses might influence aging phenotypes. Exp Gerontol 2003; 38:5–11.
6. Faragher RG. Cell senescence and human aging: where's the link? Biochem Soc Trans 2000; 28:221–226.
7. Hornsby PJ. Cellular senescence and tissue aging in vivo. J Gerontol 2002; 57:251–256.

8. Braig M, Schmitt CA. Oncogene-induced senescence: putting the brakes on tumor development. Cancer Res 2006; 66:2881–2884.
9. Campisi J. Cellular senescence as a tumor-suppressor mechanism. Trends Cell Biol 2001; 11:27–31.
10. Campisi J. Suppressing cancer: the importance of being senescent. Science 2005; 309:886–887.
11. Dimri GP. What has senescence got to do with cancer? Cancer Cell 2005; 7:505–512.
12. Hanahan D, Weinberg RA. The hallmarks of cancer. Cell 2000; 100:57–70.
13. Campisi J. Cancer and ageing: rival demons? Nat Rev Cancer 2003; 3:339–349.
14. Campisi J. Senescent cells, tumor suppression and organismal aging: good citizens, bad neighbors. Cell 2005; 120:1–10.
15. Campisi J. The biology of replicative senescence. Eur J Cancer 1997; 33:703–709.
16. Goldstein S. Replicative senescence: the human fibroblast comes of age. Science 1990; 249:1129–1133.
17. Cristofalo VJ, Pignolo RJ. Replicative senescence of human fibroblast-like cells in culture. Physiol Rev 1993; 73:617–638.
18. Wang E, Lee MJ, Pandey S. Control of fibroblast senescence and activation of programmed cell death. J Cell Biochem 1994; 54:432–439.
19. Ben-Porath I, Weinberg RA. When cells get stressed: an integrative view of cellular senescence. J Clin Invest 2004; 113:8–13.
20. Campisi J, d'Adda di Fagagna F. Cellular senescence: when bad things happen to good cells. Nat Rev Mol Cell Biol 2007; 8:729–740.
21. Passos JF, Von Zglinicki T. Oxygen free radicals in cell senescence: are they signal transducers? Free Radic Res 2006; 40:1277–1283.
22. Shay JW, Roninson IB. Hallmarks of senescence in carcinogenesis and cancer therapy. Oncogene 2004; 23:2919–2933.
23. McEachern MJ, Krauskopf A, Blackburn EH. Telomeres and their control. Annu Rev Genet 2000; 34:331–358.
24. de Lange T. Protection of mammalian telomeres. Onogene 2002; 21:532–540.
25. Gray JW, Collins C. Genome changes and gene expression in human solid tumors. Carcinogenesis 2000; 21:443–452.
26. Loeb LA. Cancer cells exhibit a mutator phenotype. Adv Cancer Res 1998; 72:25–56.
27. Gisselsson D. Chromosome instability in cancer: how, when, and why? Adv Cancer Res 2003; 87:1–29.
28. Rodier F, Kim SH, Nijjar T, et al. Cancer and aging: the importance of telomeres in genome maintenance. Int J Biochem Cell Biol 2005; 37:977–990.
29. Shay JW, Wright WE. Senescence and immortalization: role of telomeres and telomerase. Carcinogenesis 2005; 26:867–874.
30. Levy MZ, Allsopp RC, Futcher AB, et al. Telomere end-replication problem and cell aging. J Mol Biol 1992; 225:951–960.
31. Blackburn EH. Telomerases. Annu Rev Biochem 1992; 61:113–129.
32. Wright WE, Piatyszek MA, Rainey WE, et al. Telomerase activity in human germline and embryonic tissues and cells. Dev Genet 1996; 18:173–179.
33. Harle-Bachor C, Boukamp P. Telomerase activity in the regenerative basal layer of the epidermis in human skin and in immortal and carcinoma-derived skin keratinocytes. Proc Natl Acad Sci U S A 1996; 93:6476–6481.

34. Meyerson M, Counter CM, Eaton EN, et al. hEST2, the putative human telomerase catalytic subunit gene, is upregulated in tumor cells and during immortalization. Cell 1997; 90:785–795.
35. Kim NW, Piatyszek MA, Prowse KR, et al. Specific association of human telomerase activity with immortal cells and cancer. Science 1994; 266:2011–2015.
36. DiLeonardo A, Linke SP, Clarkin K, et al. DNA damage triggers a prolonged p53-dependent G1 arrest and long-term induction of Cip1 in normal human fibroblasts. Genes Dev 1994; 8:2540–2551.
37. Chen Q, Fischer A, Reagan JD, et al. Oxidative DNA damage and senescence of human diploid fibroblast cells. Proc Natl Acad Sci U S A 1995; 92:4337–4341.
38. von Zglinicki T. Role of oxidative stress in telomere length regulation and replicative senescence. Ann NY Acad Sci 2000; 908:99–110.
39. Takai H, Smogorzewska A, de Lange T. DNA damage foci at dysfunctional telomeres. Curr Biol 2003; 13:1549–1556.
40. d'Adda di Fagagna F, Reaper PM, Clay-Farrace L, et al. A DNA damage checkpoint response in telomere-initiated senescence. Nature 2003; 426:194–198.
41. Ogryzko VV, Hirai TH, Russanova VR, et al. Human fibroblast commitment to a senescence-like state in response to histone deacetylase inhibitors is cell cycle dependent. Mol Cell Biol 1996; 16:5210–5218.
42. Munro J, Barr NI, Ireland H, et al. Histone deacetylase inhibitors induce a senescence-like state in human cells by a p16-dependent mechanism that is independent of a mitotic clock. Exp Cell Res 2004; 295:525–538.
43. Bandyopadhyay D, Okan NA, Bales E, et al. Down-regulation of p300/CBP histone acetyltransferase activates a senescence checkpoint in human melanocytes. Cancer Res 2002; 62:6231–6239.
44. Sherr CJ, DePinho RA. Cellular senescence: mitotic clock or culture shock? Cell 2000; 102:407–410.
45. Wright WE, Shay JW. Historical claims and current interpretations of replicative aging. Nat Biotechnol 2002; 20:682–688.
46. Blagosklonny MV. Cell senescence: hypertrophic arrest beyond the restriction point. J Cell Physiol 2006; 209:592–597.
47. Takahashi A, Ohtani N, Yamakoshi K, et al. Mitogenic signalling and the p16(INK4a)-Rb pathway cooperate to enforce irreversible cellular senescence. Nat Cell Biol 2006; 8:1291–1297.
48. Serrano M, Lin AW, McCurrach ME, et al. Oncogenic ras provokes premature cell senescence associated with accumulation of p53 and p16INK4a. Cell 1997; 88:593–602.
49. Zhu J, Woods D, McMahon M, et al. Senescence of human fibroblasts induced by oncogenic raf. Genes Dev 1998; 12:2997–3007.
50. Michaloglou C, Vredeveld LCW, Soengas MS, et al. $BRAF^{E600}$-associated senescence-like cell cycle arrest of human nevi. Nature 2005; 436:720–724.
51. Shay JW, Pereira-Smith OM, Wright WE. A role for both Rb and p53 in the regulation of human cellular senescence. Exp Cell Res 1991; 196:33–39.
52. Bringold F, Serrano M. Tumor suppressors and oncogenes in cellular senescence. Exp Gerontol 2000; 35:317–329.
53. Collins CJ, Sedivy JM. Involvement of the INK4a/Arf gene locus in senescence. Aging Cell 2003; 2:145–150.
54. Prives C, Hall PA. The p53 pathway. J Pathol 1999; 187:112–126.

55. Sherr CJ, McCormick F. The RB and p53 pathways in cancer. Cancer Cell 2002; 2:103–112.
56. Ohtani N, Yamakoshi K, Takahashi A, et al. The p16INK4a-RB pathway: molecular link between cellular senescence and tumor suppression. J Med Invest 2004; 51:146–153.
57. Gil J, Peters G. Regulation of the INK4b-ARF-INK4a tumour suppressor locus: all for one or one for all. Nat Rev Mol Cell Biol 2006; 7:667–677.
58. El-Deiry WS, Tokino T, Velculescu VE, et al. WAF1, a potential mediator of p53 tumor suppression. Cell 1993; 75:817–825.
59. Amundson SA, Myers TG, Fornace AJ. Roles for p53 in growth arrest and apoptosis: putting on the brakes after genotoxic stress. Oncogene 1998; 17:3287–3299.
60. Sherr CJ. The Pezcoller lecture: cancer cell cycles revisited. Cancer Res 2000; 60:3689–3695.
61. Caspari T. How to activate p53. Curr Biol 2000; 10:315–317.
62. Di Micco R, Fumagalli M, Cicalese A, et al. Oncogene-induced senescence is a DNA damage response triggered by DNA hyper-replication. Nature 2006; 444:638–642.
63. Mallette FA, Gaumont-Leclerc MF, Ferbeyre G. The DNA damage signaling pathway is a critical mediator of oncogene-induced senescence. Genes Dev 2007; 21:43–48.
64. Stein GH, Drullinger LF, Soulard A, et al. Differential roles for cyclin-dependent kinase inhibitors p21 and p16 in the mechanisms of senescence and differentiation in human fibroblasts. Mol Cell Biol 1999; 19:2109–2117.
65. Robles SJ, Adami GR. Agents that cause DNA double strand breaks lead to p16INK4a enrichment and the premature senescence of normal fibroblasts. Oncogene 1998; 16:1113–1123.
66. Shapiro GI, Edwards CD, Ewen ME, et al. p16INK4A participates in a G1 arrest checkpoint in response to DNA damage. Mol Cell Biol 1998; 18:378–387.
67. Jacobs JJ, de Lange T. Significant role for p16(INK4a) in p53-independent telomere-directed senescence. Curr Biol 2004; 14:2302–2308.
68. Itahana K, Zou Y, Itahana Y, et al. Control of the replicative life span of human fibroblasts by p16 and the polycomb protein Bmi-1. Mol Cell Biol 2003; 23:389–401.
69. Beausejour CM, Krtolica A, Galimi F, et al. Reversal of human cellular senescence: roles of the p53 and p16 pathways. EMBO J 2003; 22:4212–4222.
70. Crawford YG, Gauthier ML, Joubel A, et al. Histologically normal human mammary epithelia with silenced p16(INK4a) overexpress COX-2, promoting a premalignant program. Cancer Cell 2004; 5:263–273.
71. Holst CR, Nuovo GJ, Esteller M, et al. Methylation of p16(INK4a) promoters occurs in vivo in histologically normal human mammary epithelia. Cancer Res 2003; 63:1596–1601.
72. Romanov SR, Kozakiewicz BK, Holst CR, et al. Normal human mammary epithelial cells spontaneously escape senescence and acquire genomic changes. Nature 2001; 409:633–637.
73. Narita M, Nunez S, Heard E, et al. Rb-mediated heterochromatin formation and silencing of E2F target genes during cellular senescence. Cell 2003; 113:703–716.

74. Chang BD, Broude EV, Dokmanovic M, et al. A senescence-like phenotype distinguishes tumor cells that undergo terminal proliferation arrest after exposure to anticancer agents. Cancer Res 1999; 59:3761–3767.
75. te Poele RH, Okorokov AL, Jardine L, et al. DNA damage is able to induce senescence in tumor cells in vitro and in vivo. Cancer Res 2002; 62:1876–1883.
76. Schmitt CA, Fridman JS, Yang M, et al. A senescence program controlled by p53 and p16INK4a contributes to the outcome of cancer therapy. Cell 2002; 109:335–346.
77. Wang Y, Blandino G, Oren M, et al. Induced p53 expression in lung cancer cell line promotes cell senescence and differentially modifies the cytotoxicity of anti-cancer drugs. Oncogene 1998; 17:1923–1930.
78. Chang BD, Swift ME, Shen M, et al. Molecular determinants of terminal growth arrest induced in tumor cells by a chemotherapeutic agent. Proc Natl Acad Sci U S A 2002; 99:389–394.
79. Xue W, Zender L, Miething C, et al. Senescence and tumour clearance is triggered by p53 restoration in murine liver carcinomas. Nature 2007; 445:656–650.
80. Ventura A, Kirsch DG, McLaughlin ME, et al. Restoration of p53 function leads to tumour regression in vivo. Nature 2007; 445:661–665.
81. Roninson IB. Tumor cell senescence in cancer treatment. Cancer Res 2003; 63:2705–2715.
82. Dimri GP, Lee X, Basile G, et al. A novel biomarker identifies senescent human cells in culture and in aging skin in vivo. Proc Natl Acad Sci U S A 1995; 92:9363–9367.
83. Krishnamurthy J, Torrice C, Ramsey MR, et al. Ink4a/Arf expression is a biomarker of aging. J Clin Invest 2004; 114:1299–1307.
84. Herbig U, Ferreira M, Condel L, et al. Cellular senescence in aging primates. Science 2006; 311:1257.
85. Jeyapalan JC, Ferreira M, Sedivy JM, et al. Accumulation of senescent cells in mitotic tissue of aging primates. Mech Ageing Dev 2007; 128:36–44.
86. Sedelnikova OA, Horikawa I, Zimonjic DB, et al. Senescing human cells and ageing mice accumulate DNA lesions with unrepairable double-strand breaks. Nat Cell Biol 2004; 6:168–170.
87. Donehower LA, Harvey M, Slagke BL, et al. Mice deficient for p53 are developmentally normal but susceptible to spontaneous tumors. Nature 1992; 356:215–221.
88. Harvey M, Sands AT, Weiss RS, et al. In vitro growth characteristics of embryo fibroblasts isolated from p53-deficient mice. Oncogene 1993; 8:2457–2467.
89. Sharpless NE, Bardeesy N, Lee KH, et al. Loss of p16Ink4a with retention of p19Arf predisposes mice to tumorigenesis. Nature 2001; 413:86–91.
90. Braig M, Lee S, Loddenkemper C, et al. Oncogene-induced senescence as an initial barrier in lymphoma development. Nature 2005; 436:660–665.
91. Collado M, Gil J, Efeyan A, et al. Identification of senescent cells in premalignant tumours. Nature 2005; 436:642.
92. Chen Z, Trotman LC, Shaffer D, et al. Critical role of p53 dependent cellular senescence in suppression of Pten deficient tumourigenesis. Nature 2005; 436:725–730.
93. Balducci L, Ershler WB. Cancer and ageing: a nexus at several levels. Nat Rev Cancer 2005; 5:655–662.
94. DePinho RA. The age of cancer. Nature 2000; 408:248–254.

95. Jonason AS, Kunala S, Price GT, et al. Frequent clones of p53-mutated keratinocytes in normal human skin. Proc Natl Acad Sci U S A 1996; 93:14025–14029.
96. Dollé ME, Giese H, Hopkins CL, et al. Rapid accumulation of genome rearrangements in liver but not in brain of old mice. Nat Genet 1997; 17:431–434.
97. Deng G, Lu Y, Zlotnikov G, et al. Loss of heterozygosity in normal tissue adjacent to breast carcinomas. Science 1996; 274:2057–2059.
98. Park CC, Bissell MJ, Barcellos-Hoff MH. The influence of the microenvironment on the malignant phenotype. Mol Med Today 2000; 6:324–329.
99. Liotta LA, Kohn EC. The microenvironment of the tumour-host interface. Nature 2001; 411:375–379.
100. Ilmensee K. Reversion of malignancy and normalized differentiation of teratocarcinoma cells in chimeric mice. Basic Life Sci 1978; 12:3–25.
101. Bissell MJ, Radisky D. Putting tumours in context. Nat Rev Cancer 2001; 1:46–54.
102. Cairns B. Emerging roles for chromatin remodeling in cancer biology. Trends Cell Biol 2001; 11:15–21.
103. Neumeister P, Albanese C, Balent B, et al. Senescence and epigenetic dysregulation in cancer. Int J Biochem Cell Biol 2002; 34:1475–1490.
104. Paradis V, Youssef N, Dargere D, et al. Replicative senescence in normal liver, chronic hepatitis C, and hepatocellular carcinomas. Hum Pathol 2001; 32:327–332.
105. Melk A, Kittikowit W, Sandhu I, et al. Cell senescence in rat kidneys in vivo increases with growth and age despite lack of telomere shortening. Kidney Int 2003; 63:2134–2143.
106. Roberts S, Evans EH, Kletsas D, et al. Senescence in human intervertebral discs. Eur Spine J 2006; 15:312–316.
107. Vasile E, Tomita Y, Brown LF, et al. Differential expression of thymosin beta-10 by early passage and senescent vascular endothelium is modulated by VPF/VEGF: evidence for senescent endothelial cells in vivo at sites of atherosclerosis. FASEB J 2001; 15:458–466.
108. Price JS, Waters J, Darrah C, et al. The role of chondrocyte senescence in osteoarthritis. Aging Cell 2002; 1:57–65.
109. Matthews C, Gorenne I, Scott S, et al. Vascular smooth muscle cells undergo telomere-based senescence in human atherosclerosis: effects of telomerase and oxidative stress. Circ Res 2006; 99:156–164.
110. Stanley A, Osler T. Senescence and the healing rates of venous ulcers. J Vasc Surg 2001; 33:1206–1211.
111. Martin JA, Buckwalter JA. The role of chondrocyte senescence in the pathogenesis of osteoarthritis and in limiting cartilage repair. J Bone Joint Surg Am 2003; 85:106–110.
112. Shelton DN, Chang E, Whittier PS, et al. Microarray analysis of replicative senescence. Curr Biol 1999; 9:939–945.
113. Zhang H, Pan KH, Cohen SN. Senescence-specific gene expression fingerprints reveal cell-type-dependent physical clustering of up-regulated chromosomal loci. Proc Natl Acad Sci U S A 2003; 100:3251–3256.
114. Yoon IK, Kim HK, Kim YK, et al. Exploration of replicative senescence-associated genes in human dermal fibroblasts by cDNA microarray technology. Exp Gerontol 2004; 39:1369–1378.

115. Bavik C, Coleman I, Dean JP, et al. The gene expression program of prostate fibroblast senescence modulates neoplastic epithelial cell proliferation through paracrine mechanisms. Cancer Res 2006; 66:794–802.
116. Perera RJ, Koo S, Bennett CF, et al. Defining the transcriptome of accelerated and replicatively senescent keratinocytes reveals links to differentiation, interferon signaling, and Notch related pathways. J Cell Biochem 2006; 98:394–408.
117. Hampel B, Fortschegger K, Ressler S, et al. Increased expression of extracellular proteins as a hallmark of human endothelial cell in vitro senescence. Exp Gerontol 2006; 41:474–481.
118. Parrinello S, Coppe JP, Krtolica A, et al. Stromal-epithelial interactions in aging and cancer: senescent fibroblasts alter epithelial cell differentiation. J Cell Sci 2005; 118:485–496.
119. Funk WD, Wang CK, Shelton DN, et al. Telomerase expression restores dermal integrity to in vitro aged fibroblasts in a reconstituted skin model. Exp Cell Res 2000; 258:270–278.
120. Tsai KK, Chuang EY, Little JB, et al. Cellular mechanisms for low-dose ionizing radiation-induced perturbation of the breast tissue microenvironment. Cancer Res 2005; 65:6734–6744.
121. Reed MJ, Corsa AC, Kudravi SA, et al. A deficit in collagenase activity contributes to impaired migration of aged microvascular endothelial cells. J Cell Biochem 2000; 77:116–126.
122. Busuttil RA, Rubio M, Dolle ME, et al. Mutant frequencies and spectra depend on growth state and passage number in cells cultured from transgenic lacZ-plasmid reporter mice. DNA Repair 2006; 5:52–60.
123. Kirkwood TB, Austad SN. Why do we age? Nature 2000; 408:233–238.
124. Williams GC. Pleiotropy, natural selection, and the evolution of senescence. Evolution 1957; 11:398–411.
125. Kortlever RM, Bernards R. Senescence, wound healing and cancer: the PAI-1 connection. Cell Cycle 2006; 5:2697–2703.
126. Maier B, Gluba W, Bernier B, et al. Modulation of mammalian life span by the short isoform of p53. Genes Dev 2004; 18:306–319.
127. Tyner SD, Venkatachalam S, Choi J, et al. p53 mutant mice that display early aging-associated phenotypes. Nature 2002; 415:45–53.
128. Donehower LA. Does p53 affect organismal aging? J Cell Physiol 2002; 192:23–33.
129. Krishnamurthy J, Ramsey MR, Ligon KL, et al. $p16^{INK4a}$ induces an age-dependent decline in islet regenerative potential. Nature 2006; 443:453–457.
130. Janzen V, Forkert R, Fleming H, et al. Stem cell aging modified by the cyclin-dependent kinase inhibitor, $p16^{INK4a}$. Nature 2006; 443:421–426.
131. Molofsky AV, Slutsky SG, Joseph NM, et al. Declines in forebrain progenitor function and neurogenesis during aging are partially caused by increasing *Ink4a* expression. Nature 2006; 443:448–452.
132. Olumi AF, Grossfeld GD, Hayward SW, et al. Carcinoma-associated fibroblasts direct tumor progression of initiated human prostatic epithelium. Cancer Res 1999; 59:5002–5011.
133. Chang HY, Sneddon JB, Alizadeh AA, et al. Gene expression signature of fibroblast serum response predicts human cancer progression: similarities between tumors and wounds. PLoS Biol 2004; 2:E7.

134. Krtolica A, Parrinello S, Lockett S, et al. Senescent fibroblasts promote epithelial cell growth and tumorigenesis: a link between cancer and aging. Proc Natl Acad Sci U S A 2001; 98:12072–12077.
135. Dilley TK, Bowden GT, Chen QM. Novel mechanisms of sublethal oxidant toxicity: induction of premature senescence in human fibroblasts confers tumor promoter activity. Exp Cell Res 2003; 290:38–48.
136. Liu D, Hornsby PJ. Senescent human fibroblasts increase the early growth of xenograft tumors via matrix metalloproteinase secretion. Cancer Res 2007; 67:3117–3126.
137. Coppe JP, Kauser K, Campisi J, et al. Secretion of vascular endothelial growth factor by primary human fibroblasts at senescence. J Biol Chem 2006; 281:29568–29574.
138. Yaswen P, Campisi J. Oncogene-induced senescence pathways weave an intricate tapestry. Cell 2007; 128:233–234.

12

Tumor Suppressing Activities of Senescent Keratinocytes

Brian J. Nickoloff
Department of Pathology, Loyola University Medical Center, Maywood, Illinois, U.S.A.

INTRODUCTION

One of the most visible signs of chronological aging for humans involves fairly obvious clinical changes in which the aged skin appears wrinkled, saggy, discolored (1), and is more frequently associated with various types of cancer (2). Additional characteristics of the aging phenotype is skin include wound healing defects, decreased skin thickness with less subcutaneous fat tissue, and hair loss. Skin aging represents a complex process involving both intrinsic aging as well as extrinsic aging (3). For the most part in simplistic terms, intrinsic skin aging results primarily from genetically programmed molecular and cellular events in which there is progressive telomere shortening involving the ends of DNA; whereas extrinsic aging relates more to environmentally induced cellular changes in which there is more widespread DNA damage, accelerated telomere shortening, and a key role for p53 and related genes (4–6). A detailed review of the complexities of intrinsic and extrinsic skin aging is beyond the scope of this article (7), rather the focus will be on the relative biological consequences by which senescent cell types in the skin (particularly epidermal keratinocytes and dermal fibroblasts) may impact (either positively or negatively) the development of skin cancer.

There are three important cancer-related observations involving the skin that are worth pointing out at the onset: First, virtually all of the age-related skin cancers are derived from normal cells residing in the epidermis such as keratinocytes giving rise to basal cell carcinoma (BCC), or squamous cell carcinoma (SCC), or from epidermally located melanocytes that lead to malignant melanomas. These age-related malignant tendencies involving skin reflect other epithelial-based neoplasms (8,9). Only rarely do neoplasms arise in the underlying dermal compartment; perhaps the best example is the atypical fibroxanthoma, although its precise histogenesis is still unclear. Second, while there is no question that age plays a factor in developing skin cancer, it is often overlooked, or not emphasized enough, that the vast majority of the surface area of aged skin, in the vast majority of elderly individuals, never becomes involved by malignant transformation. Finally, while UV light–induced DNA mutations are undoubtedly important in the aging/carcinogenesis link, such DNA mutations alone may not be sufficient for predisposing to cancer development (10). A growing body of evidence indicates that malignant tumorigenesis involves a permissive microenvironment (such as may be provided by adjacent keratinocytes or fibroblasts) to facilitate individual tumor cells containing DNA mutations converting to invasive fully malignant cancer cells. From this context, senescence may be viewed as an example of evolutionary antagonistic pleiotropy (2). In other words, an individual cell undergoing senescence by which there is irreversible growth arrest would benefit the individual by suppressing cancer in early life, but as senescent cells accumulate with age, they may become detrimental because they secrete factors acting in a paracrine fashion that could cause local tissue dysfunction, promote inflammation or secrete growth factors or other factors that synergize with DNA mutation bearing cells to promote tumorigenesis. As will be discussed later, it appears that on one hand senescent fibroblasts can contribute to increased aging and cancer development, whereas senescent keratinocytes may have an opposite effect, thereby reducing tumor formation.

The purpose of this chapter is to review the relevant published experimental findings that provide insight into the interactions between aged skin and the development of skin cancer. While there is no question that the process of replicative senescence can serve as a tumor-suppressive mechanism for individual cells, and an underlying cause of aging, what is less clear is whether senescent cells can influence surrounding cells in a paracrine fashion, rather than solely autocrine (11). Given the accessibility of skin and obvious clinical opportunities and importance of aging and skin structure and function, there is tremendous interest in better understanding the complexities of the role of aging and cutaneous biology. Indeed, many different types of research deal with this topic ranging from cosmetic companies interested in commercializing on the next antiwrinkle cream or "fountain of youth" lotion, or injection of botulinum toxin, or other treatments currently in vogue, to those basic scientists interested in gerontology, as well as oncologists trying to find new and better ways of preventing skin cancer, and identifying nonsurgical interventions once skin cancers are detected. Due to space constraints, this chapter will focus on epithelial cells and not on melanocytic-derived benign and malignant lesions, although lessons learned from the keratinocyte and keratinocyte-derived neoplasms may pay dividends to other cell types within the skin or in other organ systems.

SENESCENCE AND THE SKIN

While it is easy to recognize aged skin with the naked eye in elderly individuals, the genetic basis of aging and the molecular pathways that regulate aging (both intrinsic and extrinsic) in the skin remain as important challenges for investigative skin biologists and gerontologists. In lower eukaryotes, distinct genetic pathways regulating aging have been identified and involve various DNA repair pathways, telomerase, growth factors, cell cycle regulators, and many other pathways (12,13). Since clinical observations may be subjective and genetic experiments in humans is not ethical, investigators have largely focused on cell culture techniques and searched for reliable biomarkers that could be used to evaluate cellular aging. Given the accessibility of human skin, it should not be surprising that several investigative teams have explored biomarkers for aging using biopsies from individuals with widely varying chronological age (14).

Before delving into the biomarker literature, a few brief key points regarding insights into aging derived from studies of cultured fibroblasts will be presented. As one considers the model systems that have contributed to our knowledge of human aging, perhaps the most widely studied cell type is the human diploid fibroblast. Indeed, by analyzing which molecular mediators regulate the life span of normal fibroblasts, as well as comparing young proliferating fibroblasts to older fibroblasts undergoing senescence, numerous candidate genes and their products have emerged as important potential contributors of aging in vivo.

In pioneering studies, Hayflick observed the limited replicative potential of cultured fibroblasts, and after a defined number of cell divisions, the fibroblasts became irreversibly growth arrested despite the replacement of the medium with various mitogenic substances (15). Since this publication, scientists have made great inroads into defining specific mechanisms that contribute to the finite or defined number of cell divisions leading to irreversible growth arrest, despite remaining viable, including differential gene expression patterns that distinguish young, proliferating versus old, growth-arrested cells (16). More recently, additional focus has complemented the original view of replicative senescence to include another response of cells that is termed accelerated senescence. In this latter response, cells exposed to oncogenic or genotoxic stresses also become irreversibly growth arrested, even though the cells may be relatively young or in an early passage number (17,18). In some ways, these aforementioned considerations of replicative senescence and accelerated senescence parallel the intrinsic versus extrinsic aging programs mentioned earlier in this text.

Perhaps the most widely used biomarker for aging is the senescence-associated beta-galactosidase (SA-β-gal) enzyme (19). The histochemically detectable activity at pH 6.0 was the specific type of β-galactosidase linked to cellular senescence (19). This marker was fortuitously identified using several different types of cultured cells including human fibroblasts and keratinocytes, and has become the gold standard by which investigators define aging on a cellular basis. In Figure 1, the SA-β-gal expression in cultured human keratinocytes (proliferating versus senescent) highlights the dramatic differences between these cells in culture. Note that even though both of the keratinocyte cultures were established from the same neonatal foreskin and maintained in the same serum free growth medium (KGM, Clonetics, San Diego, CA), proliferating, early passage keratinocytes do not express any SA-β-gal; whereas following serial passaging, when the keratinocytes undergo replicative senescence, they dramatically upregulate SA-β-gal expression.

In one of the landmark papers, Dimri et al. identified high levels of SA-β-gal in aged skin compared with young skin (19). While weak to absent SA-β-gal staining was seen in skin samples obtained from individuals under 40 years, both epidermal keratinocytes and dermal fibroblasts were strongly and diffusely SA-β-gal positive in skin obtained from individuals over the age of

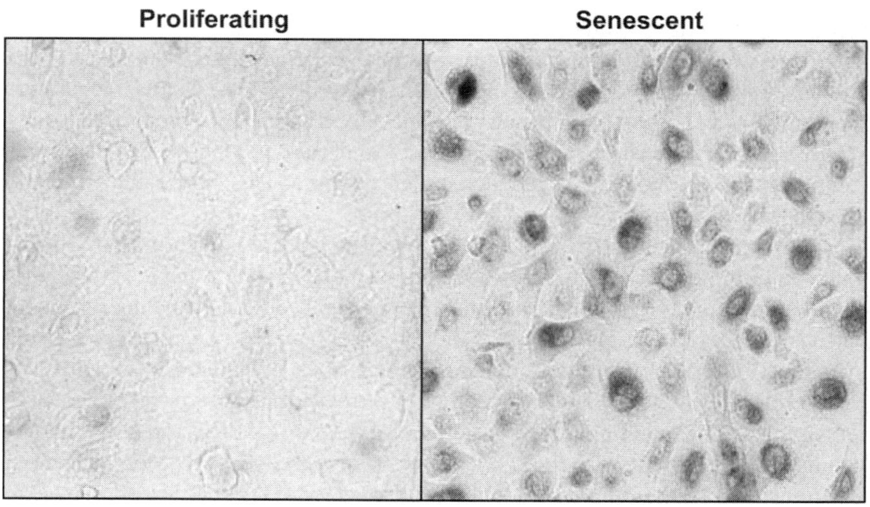

Figure 1 SA-β-gal expression in proliferating human keratinocytes versus keratinocytes undergoing replicative senescence. Note the lack of SA-β-gal expression in proliferating keratinocytes (*left side panel*), but the strong and diffuse expression of SA-β-gal in senescent keratinocytes (*right side panel*). Detection of SA-β-gal was performed at pH 6.0 as previously described (19). *Abbreviation*: SA-β-gal, senescence-associated beta-galactosidase (*The color version of this figure is provided on the DVD*).

65 years (19). More recently, it has been suggested that SA-β-gal actually represents the lysosomal variant of this enzyme, and somewhat ironically, this variant is not required for senescence by itself (20). Nonetheless, this enzyme has been an important biomarker for the past decade of research in aging, in general, and for studies involving aging of the skin. Since this enzymatic reaction requires special tissue handling, other investigators have sought to identify different biomarkers.

Using immunohistochemical staining approaches, one group followed up on an earlier study in rodents (21) indicating that the cell cycle regulatory protein INK4a/Arf could be another relevant biomarker for aging. Recently, another group has reported that the $p16^{INK4a}$ protein expression is indeed a robust biomarker for aging of human skin (22). Skin samples were obtained from three different age groups: 0–20, 21–70, and 71–95 years. Not only did these investigators find increased numbers of positively stained cells in both epidermal and dermal compartments, as a function of the age of the normal skin sampled, but there was also concomitant decrease in gene expression for BMI1. The polycomb family repressor protein BMI1 is known to regulate many genes, including being a negative upstream regulator of $p16^{INK4a}$ (23).

Our group has suggested that maspin, a protein involved in regulating angiogenesis and tumorigenesis involving various organ systems (24), may be a useful biomarker for skin aging. Interest in considering maspin as a biomarker for skin aging was derived from the observation that maspin gene expression was the most strongly upregulated amongst a panel of genes when cultured keratinocytes underwent either accelerated (e.g., induced by allowing cells to become confluent) or replicative senescence (24). In the correlative in vivo follow-up, the study of skin biopsies that had been fixed in formalin and paraffin embedded were immunostained, and a good correlation between maspin expression by epidermal keratinocytes in normal human skin and chronological aging was identified (25). Taken together, it should be clear that there is considerable interest in defining biomarkers that can be used for identifying aging in the skin. Such studies may also provide new biological insights into the relationship between senescence and skin cancers.

Senescence and Skin-Derived Cells in Culture

As mentioned earlier, many investigators have explored and defined the molecular mediators of senescence in a wide variety of cell types in vitro. For this section, a focus on keratinocytes and fibroblasts will be highlighted as regards the molecular determinants for their respective senescent programs. Perhaps no other diploid cell type has been more widely studied by investigators interested in aging than the human and murine-derived fibroblast. As it became clear that cultured fibroblasts were a good model to study senescence, my groups focused on defining the molecular determinants contributing to their irreversible growth arrest. Since a comprehensive review of this topic is beyond the scope of this review, only a few key points will be highlighted regarding cellular senescence. During every replicative cell cycle, it has become clear in many cell types that there is progressive shortening of the DNA cells, and when this telomere shortening reaches a critical point, intracellular signaling events occur involving central roles for both p53 and the Rb tumor suppressor gene (26,27). There is extensive cross talk between these DNA changes response signaling pathways and the principle downstream effector molecules that are responsible for growth arrest including two different cyclin-dependent kinase inhibitors; p21 and $p16^{INK4a}$ (28,29). Besides these pathways that are invoked by fibroblasts undergoing replicative senescence, it should also be mentioned that another aspect of cellular senescence does not involve telomere shortening. In this scenario, aptly termed accelerated senescence, stimuli that damage DNA (but not by targeting telomere ends of DNA), or strong mutagenic stimuli (such as delivered by oncogenic RAS) can also trigger the enhanced levels and activities of the two aforementioned cyclin-dependent kinase inhibitors (28,29). In patients with Werner syndrome, a rare human premature aging syndrome, the fibroblasts respond to the mutation in the gene encoding RecQ helicase WRN by enhanced telomere loss with chromosomal fusions and genomic instability thereby contributing to increased resilience of cancers in these patients (30). Another interesting set of observations related to senescence involves the cross talk between $p16^{INK4a}$ and AP-1 proteins that contribute to carcinogenesis. Two different laboratories have observed that senescence is linked to Jun proteins (31), and $p16^{INK4a}$ may prevent cell transformation by inhibiting c-Jun phosphorylation and AP-1 activity (32).

For keratinocytes, the cellular mediators mentioned above for fibroblasts are quite similar. One important distinction is that while early passage fibroblasts are not triggers to undergo senescence, when human keratinocytes become confluent they undergo accelerated senescence, and loose the capacity to further proliferate when they are trypsinized to create a single cell suspension and then reseeded in conditions with abundant growth factors. In Figure 2, the phase contrast microscopic appearance of the same batch of keratinocytes derived from an identical neonatal foreskin, are highlighted. Note that early passage proliferating human keratinocytes are relatively small and grow as individual cells forming tight clusters. However, when the keratinocytes achieve confluency and form a monolayer, the cells enlarge and become more refractile. Upon repeated passaging, the keratinocytes undergo replicative senescence and become very large, flattened with ruffled plasma membranes accompanied by prominent intracytoplasmic vacuoles. These morphological changes are accompanied by significant changes at the transcriptional and protein level that are just beginning to be defined at the molecular level. Indeed cultured human keratinocytes present interesting challenges as they often undergo a coordinated interplay of proliferation, differentiation, and senescence (33,34). Besides defining various proteins that mediate either accelerated or replicative senescence, the transcriptome has been defined for various replicative status of human keratinocytes (35).

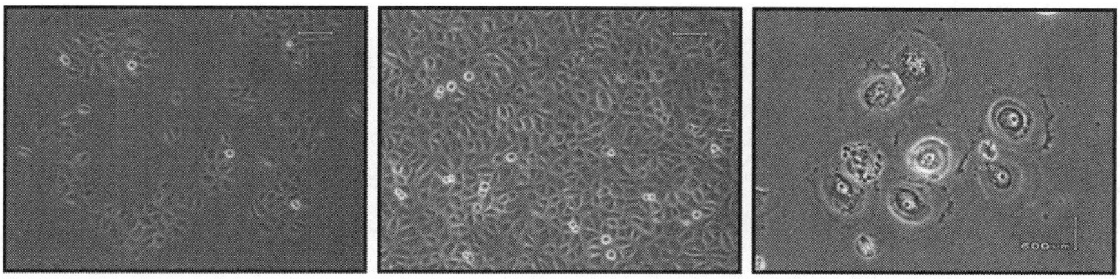

Proliferating Confluent Senescence

Figure 2 Phase contrast microscopic appearance of human keratinocytes derived from the same neonatal foreskin. Note the different appearances comparing early-passage proliferating versus early-passage confluent versus cells that underwent replicative senescence. While proliferating keratinocytes are present in small clusters with occasional dividing cells present (highly refractile and partially detached), upon confluency, the keratinocytes become more refractile with scalloped plasma membranes. Upon replicative senescence, keratinocytes enlarge and develop flattened pancake-like appearance with ruffled plasma membranes and vacuoles.

While keratinocytes undergoing senescence enlarge and loose their replicative capacity, they are far from being metabolically inert or transcriptionally silent. Rather, when whole cell RNA was extracted from either rapidly proliferating keratinocytes, keratinocytes undergoing either early confluency or late confluency–induced accelerated senescence, versus keratinocytes undergoing replicative exhaustion (replicative senescence), the subsequent mRNA array profiling revealed a large number of genes that were upregulated as well as downregulated under these different tissue culture conditions. From this study, it was also apparent that while accelerated senescence keratinocytes shared some transcriptional changes with replicative senescent keratinocytes, there were also distinct transcriptional regulation for human keratinocytes (35).

Biological Interactions Between Cultured Skin-Derived Senescent Fibroblasts and Keratinocytes with Cancer Cells

Now that we have delineated some of the molecular characteristics and signaling pathways occurring within senescent fibroblasts and keratinocytes, the next logical question is to ask how these pathways influence the behavior and phenotype of these senescent cell types. We can also ask what does senescence have to do with cancer development and/or progression (27,36). From the perspective of the individual senescent fibroblast or senescent keratinocyte, the onset of irreversible growth arrest would, at first glance, reveal a rather clear-cut conclusion by which a senescent cell would be highly resistant to transformation because of the cell cycle check point issue discussed earlier. However, there are two additional considerations to ponder. When human keratinocytes undergo senescence, they become highly resistant to apoptosis, including UV light–induced cell death (34,37). Thus, the induction of UV light-mediated apoptosis is reduced in senescent keratinocytes, and this could lead to accumulation of cells with severe genetic abnormalities not being deleted because of this apoptosis resistance. Should these genetic abnormalities or other alterations occur that would lead to a bypassing of the senescent-induced cell cycle check points, there could be a net increased risk for transformation among the population of keratinocytes?

A more widely studied cancer-related aspect of senescent cells is to explore how they may function in more of a paracrine, rather than autocrine fashion. Thus, investigators have asked what happens when either senescent fibroblasts or senescent keratinocytes are mixed with tumor cells, in other words, what are the other phenotypic characteristics of senescent cells beyond their state of being irreversibly growth arrested. The results from such experimental approaches have yielded vastly different tumor responses depending on whether senescent fibroblasts or senescent keratinocytes were used in various settings.

Beginning with senescent fibroblasts, when nontumorigenic mammary epithelial cells were mixed together, the presence of the senescent fibroblasts facilitated the production of tumors when injected into mice (38). The addition of nonsenescent fibroblasts did not facilitate tumorigenesis using this in vivo model. As a follow-up to this work, the molecular mechanism underlying this remarkable observation was linked to the ability of senescent fibroblasts to produce high levels of the metallomatrix proteinase (MMP)-3 (39). Campisi et al. have also observed that senescent fibroblasts promote epithelial cell growth and tumorigenesis by influencing the state of differentiation of the epithelial cells (40). Follow-up studies reveal that senescent fibroblasts have increased secretion of vascular endothelial cell growth function (VEGF) (41), and mediate complex stromal-epithelial interaction leading to tissue dysfunction and cancer (42,43). The increased VEGF production was able to strongly influence the behavior of normal endothelial cells, and the enhanced VEGF production occurred irrespective of whether the fibroblasts had undergone either replicative or accelerated senescence.

Turning to senescent keratinocytes, it was observed that, compared to proliferating keratinocytes, the senescent keratinocytes were able to influence the behavior of malignant tumor cells (e.g., HeLa cells). The ability of HeLa cells to form colonies in soft agar were significantly reduced in a concentration-dependent fashion (44). Interestingly, in the same system, addition of senescent fibroblasts actually enhanced colony formation of HeLa cells in soft agar (44). To explore potential mechanisms for the senescent keratinocyte-mediated inhibition of HeLa cell colony formation in soft agar, the relative levels for transforming growth factor beta (TGF-β) were examined. TGF-β was selected for further study because this cytokine was found to be able to strongly suppress HeLa cell colony formation (Fig. 3). Indeed, it was determined that senescent keratinocytes, compared to proliferating keratinocytes, significantly increase their TGF-β isoform production at both mRNA and secreted protein levels

Figure 3 Colony formation of HeLa cells in soft agar is reduced in a concentration-dependent fashion by addition of recombinant human TGF-β. Note the large number of visible colonies forming in duplicate wells representing no treatment (*upper left corner*) and minimal change when TGF-β is added at 0.1 ng/mL. However, fewer colonies and less acidification of the medium is apparent when TGF-β levels are increased at 1, 10, 50, and 100 ng/mL. The specific methods for these assays are found in Bacon et al. (44). *Abbreviation*: TGF, transforming growth factor (*The color version of this figure is provided on the DVD*).

(Fig. 4). Further studies are indicated to better define the precise mechanism by which senescent keratinocytes inhibit HeLa cell colony formation in soft agar.

SUMMARY AND FUTURE DIRECTIONS

There are many interesting aspects related to aging of the skin that go well beyond cosmetic considerations. Various cell types derived from human skin such as epidermal keratinocytes and dermal fibroblasts provide attractive cell types to further understand the complex age-related responses of human skin. At the interface between the external environment and the internal milieu, many protective mechanisms exist to maintain tissue homeostasis. The aging process has many different effects and we have touched only on a few aspects to explore potential biological consequences with a particular focus on the susceptibility and resistance to epithelial cancer formation. As reviewed in this chapter, it appears that fibroblasts undergoing senescence promote tumorigenesis, whereas keratinocytes undergoing senescence tend to inhibit tumorigenesis. Obviously, much more work remains, particularly using in vivo model systems to further explore these complex biological processes that may occur over years and even decades in human subjects. Also, the intrinsic as well as extrinsic aging processes require further study, not to mention comparing and contrasting accelerated versus replicative senescence programs for both keratinocytes and

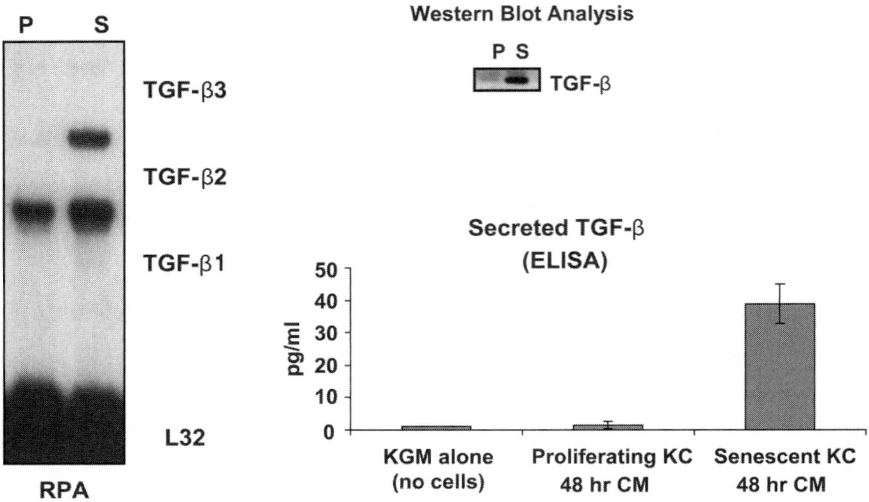

Figure 4 TGF-β at the mRNA and protein levels are increased when human keratinocytes undergo senescence. Note by RPA, mRNA levels for TGF-β1 and TGF-β2 are increased when proliferating (P) versus senescent (S) keratinocytes are compared (*left side panel*). Western blot analysis also confirms increased protein level derived from whole cell extracts comparing proliferating versus senescent keratinocytes (*right side upper panel*). Secreted levels of TGF-β measured by ELISA are also increased in senescent keratinocytes CM compared with proliferating keratinocytes. The specific methods for these assays are found in Bacon et al. (44). *Abbreviations*: RPA, RNA protection assay; TGF, transforming growth factor; ELISA, enzyme-linked immunosorbent assay; CM, conditioned media (*The color version of this figure is provided on the DVD*).

fibroblasts. Despite the complexities of this large protective coat called the skin (45), it should be clear that there are many intriguing and yet to be explored phenotypic characteristics for senescent fibroblasts and keratinocytes that go beyond simple irreversible growth arrest. The delicate balance between self-renewal of the epidermal keratinocytes, over a 100-year life span, must be seen in the context by which mutational events are occurring on a daily basis, and yet the critical barrier function of the skin must be preserved. Compared with other organ systems such as the liver or bone marrow, when genetic alterations are created in which the physiological response is apoptosis to eliminate the mutated cell because of excess stem cell–mediated renewal of the tissue, the skin must preserve the barrier function, and hence the senescent response, rather than an apoptotic response may be more physiological. Of course, much greater study of skin and senescence-related alterations are indicated, and likely to pay dividends and provide new insight into aging phenomena in other organ systems.

Finally, in Figure 5, a working model is proposed for covering the complex relationships between skin aging and epithelial-based carcinogenesis. Note that

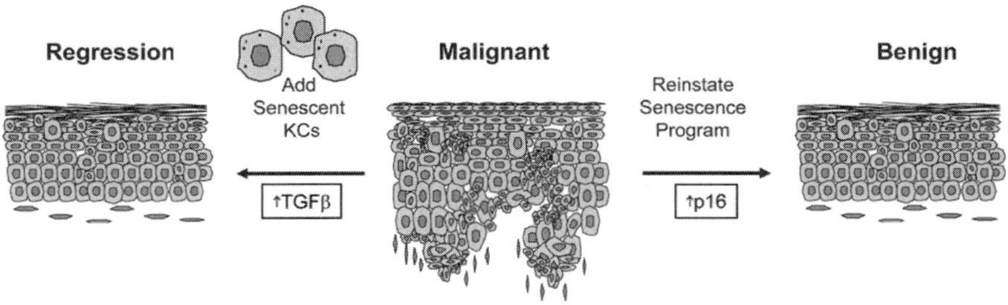

Figure 5 Working model integrating senescence and epidermal-derived skin cancers featuring a key role for senescent keratinocytes and TGF-β. Note the initial appearance of malignant keratinocytes as portrayed in the middle panel arise from bypassing the senescent program and can be potentially reversed by either coappearance of senescent keratinocytes in adjacent epidermis overexpressing TGF-β and thereby including regression (*left side panels*), or by reinstating the senescent program by which the cell cycle inhibitor $p16^{INK4a}$ is biologically active leading to a benign lesion (*right side panel*). *Abbreviation*: TGF, transforming growth factor (*The color version of this figure is provided on the DVD*).

treatment of malignant epithelial-based malignancies can be envisioned to include either the addition of senescent keratinocytes that overexpressed TGF-β triggering regression (left side panel), or reinstatement of a senescence program that may revert the malignant neoplasm. In this way, the senescence bypass mechanism that contributed to appearance of malignant cells could be reversed.

ACKNOWLEDGMENTS

The authors thank Ms. Lori Kmet for manuscript preparation. This work was partially supported by a grant from the National Institutes of Health (AR47307).

REFERENCES

1. Gupta MA, Gilchrest BA. Psychosocial aspects of aging skin. Dermatol Clin 2005; 23(4):643–648.
2. Campisi J. Cancer and ageing: rival demons? Nat Rev Cancer 2003; 3(5):339–349.
3. Yaar M, Eller MS, Gilchrest BA. Fifty years of skin aging. J Investig Dermatol Symp Proc 2002; 7(1):51–58.
4. Baumann L. Skin ageing and its treatment. J Pathol 2007; 211(2):241–251.
5. Kosmadaki MG, Gilchrest BA. The role of telomeres in skin aging/photoaging. Micron 2004; 35(3):155–159.
6. El Domyati M, Attia S, Saleh F, et al. Intrinsic aging vs. photoaging: a comparative histopathological, immunohistochemical, and ultrastructural study of skin. Exp Dermatol 2002; 11(5):398–405.
7. Uitto J. Understanding premature skin aging. N Engl J Med 1997; 337(20): 1463–1465.

8. DePinho RA. The age of cancer. Nature 2000; 408(6809):248–254.
9. Repetto L, Balducci L. A case for geriatric oncology. Lancet Oncol 2002; 3(5):289–297.
10. Bissell MJ, Radisky D. Putting tumours in context. Nat Rev Cancer 2001; 1(1):46–54.
11. Campisi J. Senescent cells, tumor suppression, and organismal aging: good citizens, bad neighbors. Cell 2005; 120(4):513–522.
12. Kirkwood TB. Genes that shape the course of ageing. Trends Endocrinol Metab 2003; 14(8):345–347.
13. Carter CS, Ramsey MM, Sonntag WE. A critical analysis of the role of growth hormone and IGF-1 in aging and lifespan. Trends Genet 2002; 18(6):295–301.
14. Dimri GP. The search for biomarkers of aging: next stop INK4a/ARF locus. Sci Aging Knowledge Environ 2004; 2004(44):e40.
15. Hayflick L. The limited in vitro lifetime of human diploid cell strains. Exp Cell Res 1965; 37:614–636.
16. Shelton DN, Chang E, Whittier PS, et al. Microarray analysis of replicative senescence. Curr Biol 1999; 9(17):939–945.
17. Serrano M, Lin AW, McCurrach ME, et al. Oncogenic ras provokes premature cell senescence associated with accumulation of p53 and p16INK4a. Cell 1997; 88(5): 593–602.
18. Yaswen P, Campisi J. Oncogene-induced senescence pathways weave an intricate tapestry. Cell 2007; 128(2):233–234.
19. Dimri GP, Lee X, Basile G, et al. A biomarker that identifies senescent human cells in culture and in aging skin in vivo. Proc Natl Acad Sci U S A 1995; 92(20): 9363–9367.
20. Lee BY, Han JA, Im JS, et al. Senescence-associated beta-galactosidase is lysosomal beta-galactosidase. Aging Cell 2006; 5(2):187–195.
21. Krishnamurthy J, Torrice C, Ramsey MR, et al. Ink4a/Arf expression is a biomarker of aging. J Clin Invest 2004; 114(9):1299–1307.
22. Ressler S, Bartkova J, Niederegger H, et al. p16INK4A is a robust in vivo biomarker of cellular aging in human skin. Aging Cell 2006; 5(5):379–389.
23. Jacobs JJ, Kieboom K, Marino S, et al. The oncogene and Polycomb-group gene bmi-1 regulates cell proliferation and senescence through the ink4a locus. Nature 1999; 397(6715):164–168.
24. Cher ML, Biliran HR Jr., Bhagat S, et al. Maspin expression inhibits osteolysis, tumor growth, and angiogenesis in a model of prostate cancer bone metastasis. Proc Natl Acad Sci U S A 2003; 100(13):7847–7852.
25. Nickoloff BJ, Lingen MW, Chang BD, et al. Tumor suppressor maspin is up-regulated during keratinocyte senescence, exerting a paracrine antiangiogenic activity. Cancer Res 2004; 64(9):2956–2961.
26. Bell JF, Sharpless NE. Telomeres, p21 and the cancer-aging hypothesis. Nat Genet 2007; 39(1):11–12.
27. Campisi J. Cellular senescence as a tumor-suppressor mechanism. Trends Cell Biol 2001; 11(11):S27–S31.
28. Schmitt CA, Fridman JS, Yang M, et al. A senescence program controlled by p53 and p16INK4a contributes to the outcome of cancer therapy. Cell 2002; 109(3): 335–346.
29. Serrano M, Lin AW, McCurrach ME, et al. Oncogenic ras provokes premature cell senescence associated with accumulation of p53 and p16INK4a. Cell 1997; 88(5): 593–602.

30. Crabbe L, Jauch A, Naeger CM, et al. Telomere dysfunction as a cause of genomic instability in Werner syndrome. Proc Natl Acad Sci U S A 2007; 104(7):2205–2210.
31. Yogev O, Anzi S, Inoue K, et al. Induction of transcriptionally active Jun proteins regulates drug-induced senescence. J Biol Chem 2006; 281(45):34475–34483.
32. Choi BY, Choi HS, Ko K, et al. The tumor suppressor p16(INK4a) prevents cell transformation through inhibition of c-Jun phosphorylation and AP-1 activity. Nat Struct Mol Biol 2005; 12(8):699–707.
33. Eckert RL, Crish JF, Robinson NA. The epidermal keratinocyte as a model for the study of gene regulation and cell differentiation. Physiol Rev 1997; 77(2):397–424.
34. Chaturvedi V, Qin JZ, Denning MF, et al. Apoptosis in proliferating, senescent, and immortalized keratinocytes. J Biol Chem 1999; 274(33):23358–23367.
35. Perera RJ, Koo S, Bennett CF, et al. Defining the transcriptome of accelerated and replicatively senescent keratinocytes reveals links to differentiation, interferon signaling, and Notch related pathways. J Cell Biochem 2006; 98(2):394–408.
36. Kahlem P, Dorken B, Schmitt CA. Cellular senescence in cancer treatment: friend or foe? J Clin Invest 2004; 113(2):169–174.
37. Chaturvedi V, Qin JZ, Stennett L, et al. Resistance to UV-induced apoptosis in human keratinocytes during accelerated senescence is associated with functional inactivation of p53. J Cell Physiol 2004; 198(1):100–109.
38. Desprez PY, Lin CQ, Thomasset N, et al. A novel pathway for mammary epithelial cell invasion induced by the helix-loop-helix protein Id-1. Mol Cell Biol 1998; 18(8):4577–4588.
39. Parrinello S, Coppe JP, Krtolica A, et al. Stromal-epithelial interactions in aging and cancer: senescent fibroblasts alter epithelial cell differentiation. J Cell Sci 2005; 118(pt 3):485–496.
40. Krtolica A, Parrinello S, Lockett S, et al. Senescent fibroblasts promote epithelial cell growth and tumorigenesis: a link between cancer and aging. Proc Natl Acad Sci U S A 2001; 98(21):12072–12077.
41. Coppe JP, Kauser K, Campisi J, et al. Secretion of vascular endothelial growth factor by primary human fibroblasts at senescence. J Biol Chem 2006; 281(40): 29568–29574.
42. Krtolica A, Campisi J. Cancer and aging: a model for the cancer promoting effects of the aging stroma. Int J Biochem Cell Biol 2002; 34(11):1401–1414.
43. Parrinello S, Coppe JP, Krtolica A, et al. Stromal-epithelial interactions in aging and cancer: senescent fibroblasts alter epithelial cell differentiation. J Cell Sci 2005; 118(pt 3):485–496.
44. Bacon P, Bodner B, Nickoloff BJ. Senescent human keratinocytes suppress colony formation of HeLa cells. J Dermatol Sci 2005; 38(1):64–66.
45. Fuchs E. Scratching the surface of skin development. Nature 2007; 445(7130): 834–842.

13

Senescence Induced by Repression of Human Papillomavirus Oncogenes in Cervical Cancer Cells

Daniel DiMaio, Kimberly Johung, Edward C. Goodwin, Stacy M. Horner, and Kristin E. Yates

Department of Genetics, Yale University School of Medicine, New Haven, Connecticut, U.S.A.

INTRODUCTION

The study of tumor viruses has provided many insights into fundamental cellular processes including cell-cycle control, signal transduction, proliferation, and apoptosis. Another important but poorly understood cellular process is replicative senescence, a permanent growth-arrested state attained by normal somatic cells after extended continuous passage in culture. Senescence is regarded as an important tumor suppressor mechanism and model of cellular aging (1). Unfortunately, replicative senescence is difficult to study because it develops in heterogenous populations of cells after several months of continuous passage. Viral oncogenes can allow cultured cells to escape replicative senescence and proliferate indefinitely, a process called cell immortalization, suggesting that studies of tumor viruses will provide insight into senescence. This chapter reviews a novel method of inducing senescence rapidly and synchronously by extinction of the expression of human papillomavirus (HPV) oncogenes in human cervical carcinoma cells. This chapter also describes some of the essential cellular and biochemical features of this model and discusses technical advantages of this system compared to replicative senescence.

THE PAPILLOMAVIRUSES AND CELL PROLIFERATION

Papillomaviruses are small non-enveloped viruses with a double-stranded circular DNA genome containing not more than ten genes. The early region of the viral genome encodes proteins primarily involved in viral DNA replication and cell transformation, and the late genes encode the structural proteins of the viral particle. A noncoding segment of DNA upstream of the early genes contains the origin of viral DNA replication and transcriptional regulatory elements including the major promoter for the early genes (2). Productive papillomavirus replication occurs exclusively in keratinocytes in stratified squamous epithelia such as the epidermis or mucous membranes, resulting in the formation of benign epithelial tumors known as papillomas, condylomas, or warts.

During the productive viral life cycle, the HPVs must establish a cellular environment conducive for high-level viral DNA replication. The HPV E7 protein binds to and destabilizes the cellular retinoblastoma (Rb) tumor suppressor protein $p105^{Rb}$ and the other members of the Rb family, p107 and p130 (3–9). The binding of the E7 protein to Rb family members and their subsequent depletion restores the activity of E2F transcription factors, which in turn, stimulate expression of cellular genes required for viral DNA synthesis (7). Similarly, the HPV E6 protein perturbs cell-cycle control by binding to the p53 tumor suppressor protein and targeting it for accelerated degradation (10–13). In combination, these E6 and E7 activities reprogram differentiating host epithelial cells to support viral DNA replication (4), which is carried out by the viral E1 and E2 proteins in concert with the cellular replication machinery. In addition, the E6 protein stimulates the expression of the hTERT catalytic subunit of telomerase, an RNA-dependent DNA polymerase that maintains the ends of chromosomes in proliferating cells (14–17). The E6 and E7 proteins also bind to other cellular proteins, but the consequences of these interactions are not clear (7,10).

HUMAN PAPILLOMAVIRUSES AND CERVICAL CARCINOMA

Persistent infection with certain types of HPV is required for the development of cervical carcinoma, a leading cause of cancer death in women (18–20). The high-risk HPV types, such as HPV16 and HPV18, are present in virtually all cervical precancerous lesions and cervical cancer, and the viral E6 and E7 oncogenes are continuously expressed in these lesions (21–24). Notably, most cervical lesions do not progress to cancers, and progression, when it occurs, usually takes many years. It is thought that the inefficient nature and long latency of carcinogenic progression may reflect in part a requirement for additional cellular mutational events in order for frank cancer to occur.

During carcinogenic progression, HPV DNA often integrates into cellular chromosomes, which usually disrupts the HPV E1 and E2 genes and prevents vegetative viral DNA replication and virus production (25–27). Prior to integration, the HPV major early promoter, which drives expression of the E6 and E7 oncogenes, is repressed by the E2 protein (28–32). Therefore, loss of E2 activity as a consequence of viral DNA integration derepresses the E6 and E7 oncogenes, allowing them to exert their growth-promoting activities.

The pathogenesis of HPV infection has also been studied in cultured cells transfected with high-risk HPV DNA. Primary human keratinocytes normally undergo a limited number of cell divisions in culture and then senesce. These cells can be efficiently immortalized by co-expression of the E6 and E7 genes from the high-risk HPV types (8,9,33–41). The ability of the E6 and E7 proteins to interfere with p53 and Rb functions and to stimulate telomerase is thought to be responsible for the immortalization activity of HPV DNA (42), because these same targets are often affected during immortalization by other viruses as well as during immortalization not associated with virus infection (43–45). Keratinocytes immortalized by the high-risk E6 and E7 proteins do not initially form tumors in experimental animals, but tumorigenic cells arise as cultures of HPV-immortalized keratinocytes are passaged, presumably because of the accumulation of additional mutations (22,34,38–41,46).

By interfering with Rb and p53 pathways, the high-risk HPV E6 and E7 proteins eliminate checkpoint control processes, allowing cells to replicate despite DNA damage and accumulate mutations (47–57). These mutations may activate cellular oncogenes, inactivate tumor suppressor genes, and disrupt other cellular growth control mechanisms, thereby stimulating further genetic instability, carcinogenic progression, and tumorigenesis (58). Strikingly, as described in the next section, although these mutations are necessary to induce the tumorigenic state of cervical cancer cells, they are not sufficient to maintain this state. Rather, expression of the HPV oncogenes is continuously required for proliferation of cervical cancer cells.

REPRESSION OF THE HPV ONCOGENES INHIBITS THE GROWTH OF CERVICAL CARCINOMA CELLS

We exploited the natural gene regulatory circuits of the papillomaviruses to test whether continuous expression of the E6 and E7 proteins is required for the survival or proliferation of cervical cancer cells. As noted above, papillomavirus E2 proteins can repress the high-risk E6/E7 promoter by binding to specific DNA sequences in the promoter (28,30,32,59,60). To determine the effect of the E2 protein on transcription of the integrated HPV genomes in cervical cancer cells, we used a replication-defective, recombinant, SV40-based viral vector to deliver the bovine papillomavirus (BPV) E2 gene into cervical cancer cell lines, including HeLa cells, which contain integrated HPV18 DNA (Fig. 1) (61). The BPV E2 protein caused dramatic repression of the HPV E6/E7 oncogenes in several different cervical carcinoma cell lines, resulting in the rapid inhibition of cell proliferation (Fig. 2) (61). In HeLa cells, expression of the resident HPV18 E6 and E7 genes was inhibited more than 20-fold by the E2 protein, and cellular DNA synthesis declined to a few percent of control levels within two days

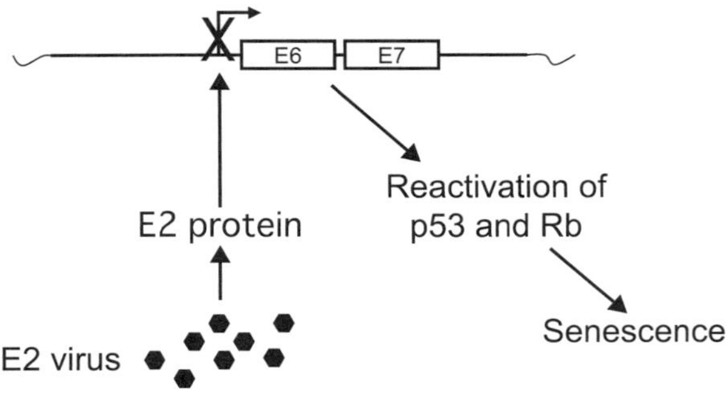

Figure 1 Schematic diagram of approach to induce senescence in cervical cancer cell lines. Expression of the BPV E2 protein represses expression of the high-risk HPV oncogenes in cervical cancer cells, resulting in activation of the p53 and Rb tumor suppressor pathways and senescence. *Abbreviations*: HPV, human papillomavirus; Rb, retinoblastoma.

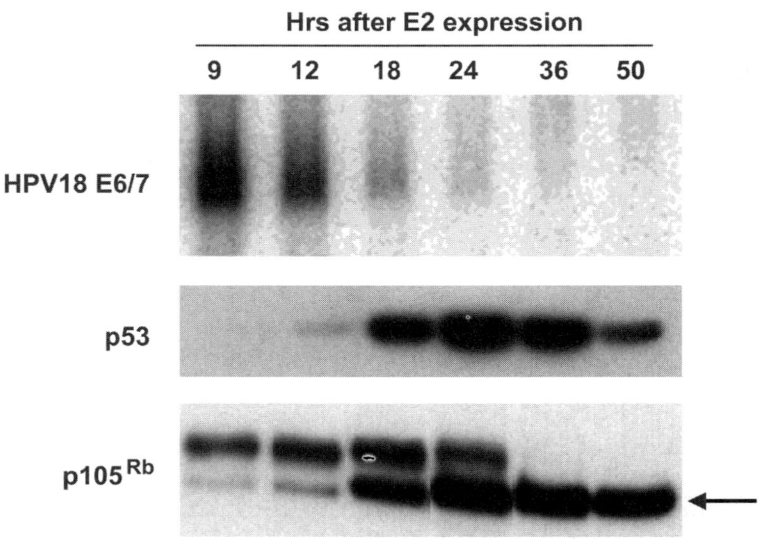

Figure 2 HPV E6/E7 repression activates p53 and Rb in HeLa cells. Extracts were prepared from HeLa cells at the indicated times (in hours) after infection with a virus expressing the BPV E2 protein. Northern blot of HPV18 E6/E7 RNA (*top*). Western blot of p53 (*middle*). Western blot of $p105^{Rb}$ (*bottom*). The arrow indicates the hypophosphorylated, growth-inhibiting form of $p105^{Rb}$. *Abbreviation*: Rb, retinoblastoma. *Source:* From Ref. 62.

(61,62). Subsequent experiments in several laboratories showed that expression of E2 genes from various papillomavirus types inhibited E6/E7 expression in cervical carcinoma cell lines containing different HPV DNA types and caused a dramatic inhibition of the ability of the transfected cells to form colonies (63–66). E2-induced growth inhibition also occurred in nontumorigenic keratinocytes immortalized by HPV16 DNA (67), demonstrating that secondary changes occurring during the late stages of carcinogenesis or long-term passage in culture are not required for this response.

E2-mediated inhibition of cell proliferation was blocked when the cells were engineered to express exogenous E6 and E7 genes from a promoter that was not repressed by the E2 protein, indicating that HPV repression is required for growth inhibition (66,68–70). However, in addition to these HPV-dependent growth-inhibitory effects, in some situations, E2 proteins from various papillomaviruses can also induce apoptosis independent of their ability to repress E6 and E7 (63,71,72). Most reports of such HPV-independent activities involve high-risk HPV E2 proteins and not the BPV E2 protein.

Expression of the BPV E2 protein also caused dramatic biochemical changes in cervical carcinoma cells. Following E6/E7 repression, p53 and the hypophosphorylated form of $p105^{Rb}$ accumulated, leading to activation of the growth-inhibitory pathways controlled by these tumor suppressor proteins (Fig. 2) (61,62,64,65,67,73–75). For example, E6/E7 repression resulted in the induction of p53-responsive genes such as p21, inhibition of cyclin-dependent kinase (cdk) activity, and repression of E2F-responsive genes required for S phase, such as cyclin A and cdc25A. An E2F site in the cdc25A promoter was required for repression, providing a molecular link to activation of the Rb pathway (75). Additional genetic experiments demonstrated that activation of the p53 or Rb pathway was required for efficient growth inhibition by the E2 protein (63,66,68,76). E6/E7 repression also inhibited telomerase activity and expression of the catalytic subunit of telomerase (15,67,77). Taken together, these results indicate that proliferation of cervical cancer cells requires continuous expression of the HPV oncogenes to maintain tumor suppressor pathways in their inactive state. Indeed, experiments in which the HPV oncogenes were repressed separately showed that continuous E7 expression was required to inactivate the Rb pathway and continuous E6 expression was required to inactive the p53 pathway (68). These experiments indicate that expression of both of the HPV oncogenes and inactivation of both of the tumor suppressor pathways are required for the optimal survival and proliferation of these cancer cells, thereby validating the viral protein and the pathways they control as therapeutic targets.

INDUCED SENESCENCE IN CERVICAL CARCINOMA CELLS

Cells induced to cease proliferation by HPV repression rapidly undergo striking phenotypic changes and resemble primary cells that have undergone replicative senescence due to serial passage (66,77–79). Specifically, following E6/E7

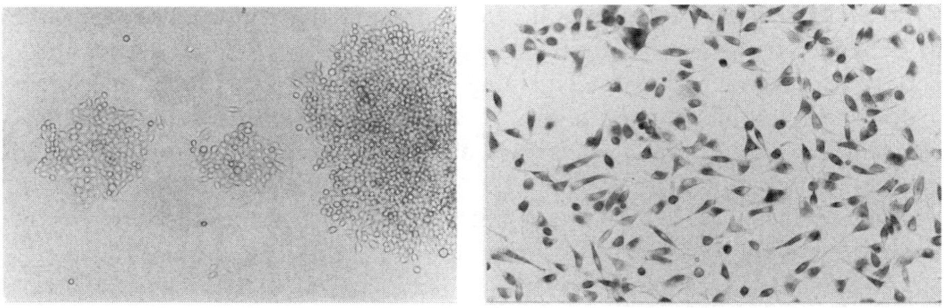

Figure 3 HPV E6/E7 repression in HeLa cells induces senescence. HeLa cells were mock-infected (*left*) or infected with a virus that expresses the BPV E2 protein (*right*). Ten days later, cells were stained for senescence-associated β-galactosidase activity at pH 6.0. *Source:* From Ref. 62 (*The color version of this figure is provided on the DVD*).

repression, HeLa cells, other cervical cancer cell lines, and HPV-immortalized, nontumorigenic cells arrest in the G1 phase of the cell cycle, display an enlarged and flattened morphology with a granular appearance, and exhibit increased autofluorescence and senescence-associated β-galactosidase activity (Fig. 3). Overexpression of the latter lysosomal enzyme is a convenient and widely used marker of senescence (79,80). As assessed by the timing and extent of growth arrest and increased autofluorescence, the senescence program begins within a day or two of E2 expression, and the cells display markedly elevated senescence-associated β-galactosidase activity by 10 days. This process is synchronous and uniform, with the vast majority of HeLa cells (>99.99%) responding to HPV repression by undergoing senescence (Fig. 3) (77). Senescence can be reversed for the first few days by extinction of E2 expression or by reexpression of E7, after which time it becomes irreversible (70,81). Following E6/E7 repression by the BPV E2 protein, the cells do not display evidence of apoptosis. Small interfering RNAs that target the HPV18 E6/E7 genes also induce senescence in HeLa cells, demonstrating that this phenotype does not require expression of the E2 protein per se (78).

Because replicative senescence in primary cells can be triggered by telomerase shortening and delayed by enforced expression of telomerase (45,82), we investigated the role of telomere length in induced senescence. Like many cancer cells, HeLa cells express abundant telomerase activity, but they have short telomeres. As noted above, repression of HPV18 E6/E7 inhibits telomerase activity (67,77). However, even though enforced expression of telomerase greatly extended the average telomere length in HeLa cells, it did not inhibit the ability of the E2 protein to induce senescence (83). Thus, senescence induced in HeLa cells by HPV repression is independent of telomere length.

Analysis of HeLa cells engineered to separately repress E6 or E7 in response to E2 expression demonstrated that E7 repression alone caused the vast majority of cells to undergo senescence in the absence of p53 activation (68).

Furthermore, following E7 repression, senescence was blocked by the expression of viral proteins that bind Rb family members or by inhibition of Rb family expression by RNA interference, indicating that activation of the Rb family was required for senescence (76,84). In contrast, although repression of E6 alone caused a significant number of cells to undergo p53-dependent senescence, a substantial fraction continued to proliferate (68,85). Many of these proliferating cells underwent p53-dependent apoptosis after about one week.

Microarray analysis provided a global view of gene expression in cervical carcinoma cells undergoing senescence after E2 expression and HPV E6/E7 repression. Several hundred genes were rapidly induced or repressed, indicating that they remain under fairly direct control of the viral oncogenes even in advanced cancer (76,81,86). When E6 and E7 were repressed together, the transcriptional response was dominated by induction of p53-responsive genes and repression of E2F-responsive genes, including those involved in mitosis and other aspects of cell-cycle progression (81,86). In cells in which only E7 was repressed, all of the transcriptional changes six days after introduction of the BPV E2 gene were due to E7 repression and not to direct effects of the E2 protein on cellular genes, and the entire transcriptional response required activation of the Rb family (76). mRNAs encoding lysosomal proteins were induced, providing a transcriptional basis for a number of phenotypic markers of senescence, including elevated senescence-associated β-galactosidase and increased autofluorescence (76). In addition, the transcriptional changes occurring during Rb-induced senescence in cervical cancer cells displayed considerable overlap with the changes occurring during replicative senescence in primary human fibroblasts, implying that these two processes involve similar biochemical pathways.

IMPLICATIONS FOR STUDIES OF SENESCENCE AND CANCER TREATMENT

Senescence was first described as the inexorable decline in proliferative potential of cultured cells during serial passage and their eventual irreversible cessation of proliferation, a process that has been studied largely as a model of cellular aging (1). Senescence has also been regarded as a tumor suppressor mechanism, and several recent studies provided direct experimental support for this notion. Precancerous lesions frequently contain cells with features of senescent cells (87,88), and animals with genetic defects that impair the ability of their cells to undergo senescence show increased tumor formation (89,90). Therefore, it appears that mobilization of a senescence program during the initial stages of carcinogenesis is an important tumor suppressor mechanism and that the development of macroscopic tumors may represent a failure of this tumor surveillance response (91). Viewed in this light, induced senescence in cultured cancer cells may be a good model for a major physiological role of senescence in vivo. Furthermore, induction of senescence in precancerous cells or even in established cancer may impose an irreversible cell-cycle block that could be put to therapeutic use.

Replicative senescence is a difficult process to study because many months of serial passage in culture are necessary to attain complete growth arrest. Furthermore, although individual cells in a passaged population undergo a fairly rapid senescence program, the time at which this program initiates in any individual cell varies widely (92). Therefore, there is significant heterogeneity in the cell population, a situation that complicates biochemical and genetic study. In contrast, induced senescence in cervical cancer cells, although it shares biochemical pathways with replicative senescence, is rapid, synchronous, and uniform. These properties of induced senescence should provide many technical advantages for studying senescence, particularly in biochemical and cell biological experiments designed to characterize the senescent phenotype. For example, the timing of biochemical changes during the senescence process is now experimentally accessible in this synchronous cell population. In addition, the rapid execution of senescence in cervical cancer cells implies that this process does not require the gradual accumulation of damage or toxic products.

An additional advantage of induced senescence is its conditional nature. Cervical carcinoma cells proliferate actively and without apparent limit. Thus, they can be readily subjected to genetic manipulation, such as specific gene knockdown or expression of exogenous genes, without expending cell generations that would bring primary cells closer to senescent growth arrest and therefore limit their use in further experiments. In contrast, cancer cells can be engineered at a deliberate pace prior to the acute activation of the senescence program.

The rapidity and reproducibility of the induced senescence response are also suitable for high-throughput screening approaches to discover inhibitors of senescence, such as small molecules or interfering RNAs. In addition, senescence induced by E6/E7 repression is very efficient, and following repeated exposure to the E2 protein, as few as one in 100,000 cells can form a colony. Therefore, induced senescence is also suitable for long-term genetic strategies to select inhibitors of senescence, that is, genes or genetic elements that allow cell proliferation and colony formation in the face of E2 expression. Such studies may reveal additional cellular components of senescence pathways and provide novel targets for agents that modulate senescence and affect cellular aging and carcinogenesis.

CONCLUSION

It is remarkable that introduction of a single viral regulatory gene can cause rapid and profound growth arrest of actively growing human cancer cell lines, including HeLa cells, which are notorious for their exuberant growth properties. It is particularly striking that HPV repression rapidly induces senescence, the first barrier to tumorigenesis. Although the acquisition of the tumorigenic phenotype in HPV-infected cells requires the accumulation of numerous mutations that are facilitated by expression of the viral oncogenes, these mutations are not sufficient to maintain the proliferative state. Furthermore, the molecules that execute the senescence response have evidently escaped mutational inactivation. The finding that sustained proliferation of these cancer cells requires the continuous expression of the viral gene products implies, of course, that other manipulations that extinguish viral activities, including small molecules that interfere with E7/Rb binding, could also induce a senescent state. Irreversible senescence triggered in primary cancer cells by HPV repression or other manipulations might provide therapeutic benefits in patients.

ACKNOWLEDGMENTS

We thank Jan Zulkeski for assistance in preparing this manuscript. Research in the DiMaio laboratory on this topic was supported by a grant to D.D. from the National Cancer Institute (CA16038).

REFERENCES

1. Campisi J. Senescent cells, tumor suppression, and organismal aging: good citizens, bad neighbors. Cell 2005; 120:513–522.
2. Chong T, Apt D, Gloss B, et al. The enhancer of human papillomavirus type 16: binding sites for the ubiquitous transcription factors oct-1, NFA, TEF-2, NF1, and AP-1 participate in epithelial cell-specific transcription. Virology 1991; 65: 5933–5943.
3. Boyer SN, Wazer DE, Band V. E7 protein of human papilloma virus-16 induces degradation of retinoblastoma protein through the ubiquitin-proteosome pathway. Cancer Res 1996; 56:4620–4624.
4. Cheng S, Schmidt-Grimminger DC, Murant T, et al. Differentiation-dependent up-regulation of the human papillomavirus E7 gene reactivates cellular DNA replication in suprabasal differentiated keratinocytes. Genes Dev 1995; 9:2335–2349.
5. Dyson N, Howley P, Munger K, et al. The human papillomavirus-16 E7 oncoprotein is able to bind to the retinoblastoma gene product. Science 1989; 243:934–936.
6. Foster SA, Galloway DA. Human papillomavirus type 16 E7 alleviates a proliferation block in early passage human mammary epithelial cells. Oncogene 1996; 12: 1773–1779.
7. Munger K, Basile JR, Duensing S, et al. Biological activities and molecular targets of the human papillomavirus E7 oncoprotein. Oncogene 2001; 20:7888–7898.
8. Munger K, Phelps WC, Bubb V, et al. The E6 and E7 genes of the human papillomavirus type 16 together are necessary and sufficient for transformation of primary human keratinocytes. J Virol 1989a; 63:4417–4421.
9. Munger K, Werness BA, Dyson N, et al. Complex formation of human papillomavirus E7 proteins with the retinoblastoma tumor suppressor gene product. EMBO J 1989b; 8:4099–4105.
10. Mantovani F, Banks L. The human papillomavirus E6 protein and its contribution to malignant progression. Oncogene 2001; 20:7874–7887.

11. Scheffner M, Huibregtse JM, Vierstra RD, et al. The HPV-16 E6 and E6-AP complex functions as a ubiquitin-protein ligase in the ubiquitination of p53. Cell 1993; 75:495–505.
12. Scheffner M, Werness BA, Huibregtse JM, et al. The E6 oncoprotein encoded by human papillomavirus type 16 and 18 promotes the degradation of p53. Cell 1990; 63:1129–1136.
13. Werness BA, Levine AJ, Howley PM. Association of human papillomavirus types 16 and 18 E6 proteins with p53. Science 1990; 248:76–79.
14. Gewin L, Galloway DA. E box-dependent activation of telomerase by human papillomavirus type 16 E6 does not require induction of c-myc. J Virol 2001; 75: 7198–7201.
15. Klingelhutz AJ, Foster SA, McDougall JK. Telomerase activation by the E6 gene product of human papillomavirus type 16. Nature 1996; 380:79–82.
16. Oh ST, Kyo S, Laimins LA. Telomerase activation by human papillomavirus type 16 E6 protein: induction of human telomerase reverse transcriptase expression through Myc and GC-rich Sp1 binding sites. J Virol 2001; 75:5559–5566.
17. Veldman T, Horikawa I, Barrett JC, et al. Transcriptional activation of the telomerase hTERT gene by human papillomavirus type 16 E6 oncoprotein. J Virol 2001; 75:4467–4472.
18. Bosch FX, Lorincz A, Munoz N, et al. The causal relation between human papillomavirus and cervical cancer. J Clin Pathol 2002; 55:244–265.
19. Schlecht NF, Kulaga S, Robitaille J, et al. Persistent human papillomavirus infection as a predictor of cervical intraepithelial neoplasia. JAMA 2001; 286:3106–3114.
20. Wallin K-L, Wiklund F, Angstrom T, et al. Type-specific persistence of human papillomavirus DNA before the development of invasive cervical cancer. N Engl J Med 1999; 341:1633–1638.
21. Boshart M, Gissmann L, Ikenberg H, et al. A new type of papillomavirus DNA, its presence in genital cancer biopsies and in cell lines derived from cervical cancer. EMBO J 1984; 3:1151–1157.
22. Durst M, Gissmann L, Ikenberg H, et al. A papillomavirus DNA from a cervical carcinoma and its prevalence in cancer biopsy samples from different geographic regions. Proc Natl Acad Sci U S A 1983; 80:3812–3815.
23. McCance DJ, Campion MJ, Clarkson PK, et al. Prevalence of human papillomavirus type 16 DNA sequences in cervical intraepithelial neoplasia and invasive carcinoma of the cervix. Br J Obstet Gynaecol 1985; 92:1101–1105.
24. Yee CL, Krishnan-Hewlett I, Baker CC, et al. Presence and expression of human papillomavirus sequences in human cervical carcinoma cell lines. Am J Pathol 1985; 119:361–366.
25. Cullen AP, Reid R, Campion M, et al. Analysis of the physical state of different human papillomavirus DNAs in intraepithelial and invasive cervical neoplasm. J Virol 1991; 65:606–612.
26. Hopman AHN, Smedts F, Dignef W, et al. Transition of high-grade cervical intraepithelial neoplasia to micro-invasive carcinoma is characterized by integration of HPV 16/18 and numerical chromosome abnormalities. J Pathol 2004; 202:23–33.
27. Schneider-Maunoury S, Croissant O, Orth G. Integration of human papillomavirus type 16 DNA sequences: a possible early event in the progression of genital tumors. J Virol 1987; 61:3295–3298.

28. Bernard BA, Bailly C, Lenoir M-C, et al. The human papillomavirus type 18 (HPV18) E2 gene product is a repressor of the HPV18 regulatory region in human keratinocytes. J Virol 1989; 63:4317–4324.
29. Bouvard V, Storey A, Pim D, et al. Characterization of the human papillomavirus E2 protein: evidence of trans-activation and trans-repression in cervical keratinocytes. EMBO J 1994; 13:5451–5459.
30. Dostatni N, Lambert PF, Sousa R, et al. The functional BPV-1 E2 trans-activating protein can act as a repressor by preventing formation of the initiation complex. Genes Dev 1991; 5:1657–1671.
31. Steger G, Corbach S. Dose-dependent regulation of the early promoter of human papillomavirus type 18 by the viral E2 protein. J Virol 7; 71:50–58.
32. Thierry F, Yaniv M. The BPV1-E2 trans-acting protein can be either an activator or a repressor of the HPV18 regulatory region. EMBO J 1987; 6:3391–3397.
33. Bedell MA, Jones KH, Grossman SR, et al. Identification of human papillomavirus type 18 transforming genes in immortalized and primary cells. J Virol 1989; 63: 1247–1255.
34. Durst M, Dzarlieva-Petrusevska RT, Boukamp P, et al. Molecular and cytogenetic analysis of immortalized human primary keratinocytes obtained after transfection with human papillomavirus type 16 DNA. Oncogene 1987; 1:251–256.
35. Halbert CL, Demers GW, Galloway DA. The E7 gene of human papillomavirus type 16 is sufficient for immortalization of human epithelial cells. J Virol 1991; 65: 473–478.
36. Hawley-Nelson P, Vousden KH, Hubbert NL, et al. HPV16 E6 and E7 proteins cooperate to immortalize human foreskin keratinocytes. EMBO J 1989; 8: 3905–3910.
37. Hudson JB, Bedell MA, McCance DJ, et al. Immortalization and altered differentiation of human keratinocytes in vitro by the E6 and E7 open reading frames of human papillomavirus type 18. J Virol 1990; 64:519–526.
38. Kaur P, McDougall JK. HPV-18 immortalization of human keratinocytes. Virology 1989; 173:302–310.
39. Pirisi L, Creek KE, Doniger J, et al. Continuous cell lines with altered growth and differentiation properties originate after transfection of human keratinocytes with human papillomavirus type 16 DNA. Carcinogenesis 1988; 9:1573–1579.
40. Pirisi L, Yasumoto S, Feller M, et al. Transformation of human fibroblasts and keratinocytes with human papillomavirus type 16 DNA. J Virol 1987; 61:1061–1066.
41. Woodworth CD, Bowden PE, Doniger J, et al. Characterization of normal human exocervical epithelial cells immortalized in vitro by papillomavirus types 16 and 18 DNA. Cancer Res 1988; 48:4620–4628.
42. Kiyono T, Foster SA, Koop JI, et al. Both Rb/p16^{INK4a} inactivation and telomerase activity are required to immortalize human epithelial cells. Nature 1998; 396:84–88.
43. Hara E, Tsuri H, Shinozaki S, et al. Cooperative effect of antisense-Rb and antisense-p53 oligomers on the extension of lifespan in human diploid fibroblasts. Biochem Biophys Res Commun 1991; 179:528–534.
44. Nevins JR. Cell transformation by viruses. In: David MK, Peter MH, eds. Field's Virology. Philadelphia: Lippincott Williams & Wilkins, 2001:245–265.
45. Vaziri H, Benchimol S. Alternative pathways for the extension of cellular life span: inactivation of p53/pRb and expression of telomerase. Oncogene 1999; 18: 7676–7680.

46. Pecoraro G, Lee M, Morgan D, et al. Evolution of in vitro transformation and tumorigenesis of HPV16 and HPV18 immortalized primary cervical epithelial cells. Am J Pathol 1991; 138:1–8.
47. Havre PA, Yuan J, Hedrick L, et al. p53 inactivation by HPV16 E6 results in increased mutagenesis in human cells. Cancer Res 1995; 55:4420–4424.
48. Hickman ES, Picksley SM, Vousden KH. Cells expressing HPV16 E7 continue cell cycle progression following DNA damage induced p53 activation. Oncogene 1994; 9:2177–2181.
49. Jones DL, Munger K. Analysis of the p53-mediated G_1 growth arrest pathway in cells expressing the human papillomavirus type 16 E7 oncoprotein. J Virol 1997; 71: 2905–2912.
50. Kessis TD, Slebos RJC, Nelson WG, et al. Human papillomavirus 16 E6 expression disrupts the p53-mediated cellular response to DNA damage. Proc Natl Acad Sci U S A 1993; 90:3988–3992.
51. Rey O, Lee S, Park NH. Impaired nucleotide excision repair in UV-irradiated human oral keratinocytes immortalized with type 16 human papillomavirus genome. Oncogene 1999; 18:6997–7001.
52. Slebos RJ, Kessis TD, Chen AW, et al. Functional consequences of directed mutations in human papillomavirus E6 proteins: abrogation of p53-mediated cell cycle arrest correlates with p53 binding and degradation in vitro. Virology 1995; 208: 111–120.
53. Slebos RJC, Lee MH, Plunkett BS, et al. p53-dependent G(1) arrest involves pRB-related proteins and is disrupted by the human papillomavirus 16 E7 oncoprotein. Proc Natl Acad Sci U S A 1994; 91:5320–5324.
54. Song S, Gulliver GA, Lambert PF. Human papillomavirus type 16 E6 and E7 oncogenes abrogate radiation-induced DNA damage responses in vivo through p53-dependent and p53-independent pathways. Proc Natl Acad Sci U S A 1998; 95: 2290–2295.
55. Thomas JT, Laimins LA. Human papillomavirus oncoproteins E6 and E7 independently abrogate the mitotic spindle checkpoint. J Virol 1998; 72:1131–1137.
56. White AE, Livanos EM, Tlsty TD. Differential disruption of genomic integrity and cell cycle regulation in normal human fibroblasts by the HPV oncoproteins. Genes Dev 1994; 8:666–677.
57. Zhang B, Spandau DF, Roman A. E5 protein of human papillomavirus type 16 protects human foreskin keratinocytes from UV B-irradiation-induced apoptosis. J Virol 2002; 76:220–231.
58. zur Hausen H. Papillomaviruses causing cancer: evasion from host-cell control in early events in carcinogenesis. J Natl Cancer Inst 2000; 92:690–698.
59. Romanczuk H, Thierry F, Howley PM. Mutational analysis of *cis* elements involved in E2 modulation of human papillomavirus type 16 P_{97} and type 18 P_{105} promoters. J Virol 1990; 64:2849–2859.
60. Thierry F, Howley PM. Functional analysis of E2-mediated repression of the HPV18 P105 promoter. New Biol 1991; 3:90–100.
61. Hwang E-S, Riese II DJ, Settleman J, et al. Inhibition of cervical carcinoma cell line proliferation by introduction of a bovine papillomavirus regulatory gene. J Virol 1993; 67:3720–3729.
62. Goodwin EC, DiMaio D. Repression of human papillomavirus oncogenes in HeLa cervical carcinoma cells causes the orderly reactivation of dormant tumor suppressor pathways. Proc Natl Acad Sci U S A 2000; 97:12513–12518.

63. Desaintes C, Demeret C, Goyat S, et al. Expression of the papillomavirus E2 protein in HeLa cells leads to apoptosis. EMBO J 1997; 16:504–514.
64. Dowhanick JJ, McBride AA, Howley PM. Suppression of cellular proliferation by the papillomavirus E2 protein. J Virol 1995; 69:7791–7799.
65. Moon MS, Lee CJ, Um SJ, et al. Effect of BPV1 E2-mediated inhibition of E6/E7 expression in HPV16-positive cervical carcinoma cells. Gynecologic Oncology 2001; 80:168–175.
66. Wells SI, Francis DA, Karpova AY, et al. Papillomavirus E2 induces senescence in HPV-positive cells via pRB- and $p21^{CIP}$-dependent pathways. EMBO J 2000; 19: 5762–5771.
67. Lee CJ, Suh EJ, Kang HT, et al. Induction of senescence-like state and suppression of telomerase activity through inhibition of HPV E6/E7 gene expression in cells immortalized by HPV16 DNA. Exp Cell Res 2002; 277:173–182.
68. DeFilippis RA, Goodwin EC, Wu L, et al. Endogenous human papillomavirus E6 and E7 proteins differentially regulate proliferation, senescence, and apoptosis in HeLa cervical carcinoma cells. J Virol 2003; 77:1551–1563.
69. Francis DA, Schmid SI, Howley PM. Repression of the integrated papillomavirus E6/E7 promoter is required for growth suppression of cervical cancer cells. J Virol 2000; 74:2679–2686.
70. Kang HT, Lee CJ, Seo EJ, et al. Transition to an irreversible state of senescence in HeLa cells arrested by repression of HPV E6 and E7 genes. Mech Ageing Dev 2004; 125:31–40.
71. Frattini MG, Hurst SD, Lim HB, et al. Abrogation of a mitotic checkpoint by E2 proteins from oncogenic human papillomaviruses correlates with increased turnover of the p53 tumor suppressor protein. EMBO J 1997; 16:318–331.
72. Webster K, Parish J, Pandya M, et al. The human papillomavirus (HPV) 16 E2 protein induces apoptosis in the absence of other HPV proteins and via a p53-dependent pathway. J Biol Chem 2000; 275:87–94.
73. Hwang E-S, Naeger LK, DiMaio D. Activation of the endogenous p53 growth inhibitory pathway in HeLa cervical carcinoma cells by expression of the bovine papillomavirus E2 gene. Oncogene 1996; 12:795–803.
74. Naeger LK, Goodwin EC, Hwang E-S, et al. Bovine papillomavirus E2 protein activates a complex growth-inhibitory program in p53-negative HT-3 cervical carcinoma cells that includes repression of cyclin A and cdc25A phosphatase genes and accumulation of hypophosphorylated retinoblastoma protein. Cell Growth Differ 1999; 10:413–422.
75. Wu L, Goodwin EC, Naeger LK, et al. E2F-Rb complexes assemble and inhibit cdc25A transcription in cervical carcinoma cells following repression of human papillomavirus oncogene expression. Mol Cell Biol 2000; 20:7059–7067.
76. Johung K, Goodwin EC, DiMaio D. Human papillomavirus E7 repression initiates a transcriptional cascade driven by the retinoblastoma family in senescing cervical carcinoma cells. J Virol 2007; 81:2102–2116.
77. Goodwin EC, Yang E, Lee C-J, et al. Rapid induction of senescence in human cervical carcinoma cells. Proc Natl Acad Sci U S A 2000; 97:10978–10983.
78. Hall AHS, Alexander KA. RNA interference of human papillomavirus type 18 E6 and E7 induces senescence in HeLa cells. J Virol 2003; 77:6066–6069.
79. Lee BY, Han JA, Im JS, et al. Senescence-associated beta-galactosidase is lysosomal beta-galactosidase. Aging Cell 2006; 5:187–195.

80. Dimri GP, Lee X, Basile G, et al. A biomarker that identifies senescent human cells in culture and in aging skin *in vivo*. Proc Natl Acad Sci U S A 1995; 92:9363–9367.
81. Wells SI, Aronow BJ, Wise TM, et al. Transcriptome signature of irreversible senescence in human papillomavirus-positive cervical cancer cells. Proc Natl Acad Sci U S A 2003; 100:7093–7098.
82. Sedivy JM. Can ends justify the means?: telomeres and the mechanisms of replicative senescence and immortalization in mammalian cells. Proc Natl Acad Sci U S A 1998; 95:9078–9081.
83. Goodwin EC, DiMaio D. Induced senescence in HeLa cervical carcinoma cells containing elevated telomerase activity and extended telomeres. Cell Growth Differ 2001; 12:525–534.
84. Psyrri A, DeFilippis RA, Edwards APB, et al. Role of the retinoblastoma pathway in senescence triggered by repression of the human papillomavirus E7 protein in cervical carcinoma cells. Cancer Res 2004; 64:3079–3086.
85. Horner SM, DeFilippis RA, Manuelidis L, et al. Repression of the human papillomavirus E6 gene initiates p53-dependent, telomerase-independent senescence and apoptosis in HeLa cervical carcinoma cells. J Virol 2004; 78:4063–4073.
86. Thierry F, Benotmane MA, Demeret C, et al. A genomic approach reveals a novel mitotic pathway in papillomavirus carcinogenesis. Cancer Res 2004; 64:895–903.
87. Collado M, Gil J, Efeyan A, et al. Tumour biology: senescence in premalignant tumours. Nature 2005; 436:642.
88. Michaloglou C, Vredeveld LC, Soengas MS, et al. BRAFE600-associated senescence-like cell cycle arrest of human naevi. Nature 2005; 436:720–724.
89. Braig M, Lee S, Loddenkemper C, et al. Oncogene-induced senescence as an initial barrier in lymphoma development. Nature 2005; 436:660–665.
90. Chen Z, Trotman LC, Shaffer D, et al. Crucial role of p53-dependent cellular senescence in suppression of Pten-deficient tumorigenesis. Nature 2005; 436: 725–730.
91. Mooi WJ, Peeper DS. Oncogene-induced cell senescence - Halting on the road to cancer. N Engl J Med 2006; 355:1037–1046.
92. Herbig U, Wei W, Dutriaux A, et al. Real-time imaging of transcriptional activation in live cells reveals rapid up-regulation of the cyclin-dependent kinase inhibitor gene CDKN1A in replicative cellular senescence. Aging Cell 2003; 2:295–304.

14

Treatment-Induced Tumor Cell Senescence and Its Consequences

Igor B. Roninson and Eugenia V. Broude
Cancer Center, Ordway Research Institute, Albany, New York, U.S.A.

SENESCENCE AS AN ANTICARCINOGENIC PROGRAM OF NORMAL CELLS

While other chapters in this book deal primarily with modes and mechanisms of cell death, this chapter addresses another antiproliferative effect of cancer therapeutics, namely, terminal cell cycle arrest through the program of senescence. Cell senescence, originally defined as proliferative arrest that occurs in normal cells after a limited number of cell divisions, is now viewed more broadly as a general biological program of terminal growth arrest. Cells that underwent senescence cannot divide even if stimulated by mitogens, but they remain metabolically and synthetically active and show characteristic changes in morphology, such as enlarged and flattened cell shape and increased granularity (1). The most widely used surrogate marker of senescent cells is the senescence-associated β-galactosidase (SA-β-gal) activity, which is detectable by X-gal staining at pH6.0 (2). SA-β-gal appears to reflect increased activity of lysosomal acid β-galactosidase (3). Senescent cells also produce many extracellular matrix (ECM) components and secreted factors that affect the growth of their

Modified with permission from Roninson, IB (2003) Tumor cell senescence in cancer treatment. Cancer Res., 63, 2705-2715.

neighboring cells as well as tissue organization. In particular, paracrine factors produced by senescent cells have major effects on the growth and survival of tumor cells in vitro and in vivo (4–6). The paracrine activities of senescent cells (a moot point in the case of cell death) are the most important aspect of the senescence response in regard to the outcome of cancer therapy. Hence, senescence should not be viewed as merely an end point in a cell's life cycle but rather as a physiological state determined by the homeostatic programs of a multicellular organism.

Several recent reviews have addressed the mechanisms of senescence in normal cells and the importance of senescence as an anticarcinogenic mechanism (1,7–9). Senescence ("growing old") was originally described in normal human cells explanted in culture; such cells undergo a finite number of divisions prior to permanent growth arrest (10). This gradual process of "replicative senescence" results from the shortening of telomeres at the ends of the chromosomes (11). Telomeric changes in cells undergoing replicative senescence show structural similarities with DNA damage or may even directly involve such damage (12,13). It is therefore not surprising that DNA damage was also found to induce rapid cell growth arrest, which is phenotypically indistinguishable from replicative senescence (14). This "accelerated senescence," which does not involve telomere shortening, is also triggered in normal cells by the expression of mutant Ras or Raf (15,16) and by some other forms of supraphysiological mitogenic signaling (1).

The key events in replicative and accelerated senescence of normal fibroblasts (the best-studied cellular system of senescence) are schematized in Figure 1. Growth arrest of senescent cells is initiated with the activation of p53. In the case of replicative senescence, p53 protein is stabilized through the involvement of $p14^{ARF}$, a tumor suppressor that sequesters the Mdm2 protein, which promotes p53 degradation. Another protein that stimulates p53 under the conditions of replicative and RAS-induced accelerated senescence is promyelocytic leukemia (PML) tumor suppressor, which regulates p53 acetylation (17,18). The activated p53 has multiple effects on gene expression, the most relevant of which in regard to senescence is transcriptional activation of $p21^{Waf1/Cip1/Sdi1}$, a pleiotropic inhibitor of different cyclin/cyclin-dependent kinase (CDK) complexes (19). p21 induction causes cell cycle arrest in senescent cells. The activation of p53 and p21 in senescent cells is only transient, and protein levels of p53 and p21 decrease after the establishment of growth arrest. While p21 expression goes down, another CDK inhibitor $p16^{Ink4A}$ becomes constitutively upregulated, suggesting that p16 may be responsible for the maintenance of growth arrest in senescent cells (20,21). Several studies have identified several positive and negative transcription-regulatory factors that play a role in transcriptional activation of p16 in senescent cells (22–24). Other CDK inhibitors $p27^{Kip1}$ and $p15^{Ink4b}$ (25,26) were also shown to play a role in fibroblast senescence. The best-known (but by no means the only) mechanism for growth arrest induced by CDK inhibitors is the blockage of CDK-mediated

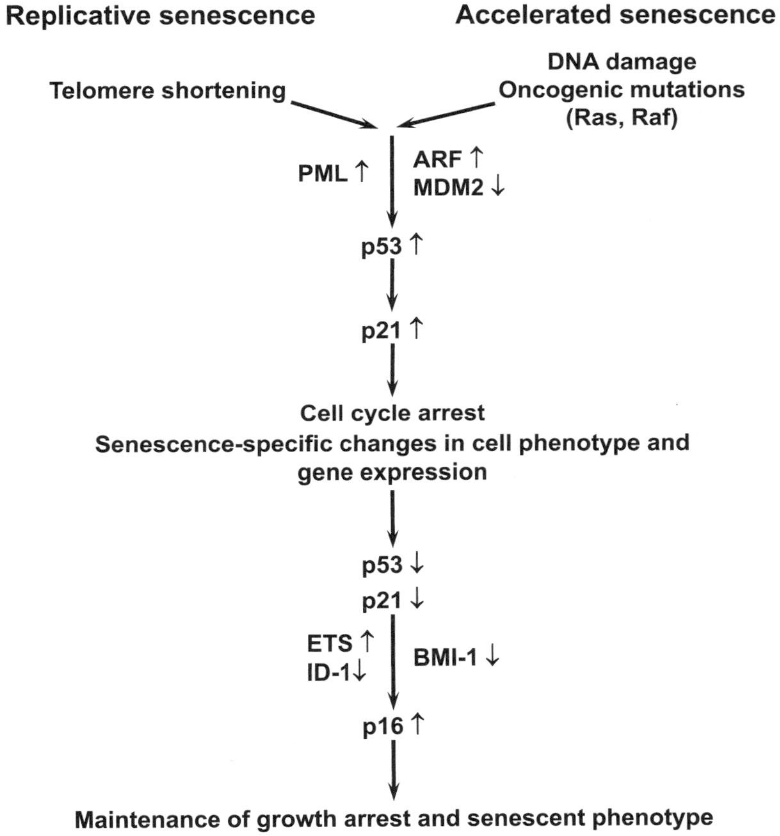

Figure 1 Key events in replicative and accelerated senescence of normal fibroblasts.

inhibitory phosphorylation of tumor suppressor protein Rb. At the onset of senescence, Rb is converted to its active hypophosphorylated form, which sequesters and inhibits E2F transcription factors necessary for cell proliferation. The cellular levels of Rb decrease, however, after the onset of senescence in at least some of the normal fibroblast cultures (21), suggesting that Rb does not always play a role in the maintenance of the senescent phenotype.

Both replicative and accelerated senescence are believed to be essential anticarcinogenic programs in normal cells. Replicative senescence imposes a limit on the total number of divisions a cell can undergo, and it should therefore be expected to interfere with tumor growth. However, studies with mice deficient in the enzyme telomerase, which counteracts the shortening of telomeres, have yielded a more complicated picture. Telomere shortening that occurs in telomerase-deficient mice after several generations was found paradoxically to promote the rate of spontaneous carcinogenesis, most probably because telomeric aberrations destabilize the genome. On the other hand, telomerase-deficient mice were more resistant to carcinogenesis under several conditions that increase the rate of tumor initiation, in agreement with the anticarcinogenic role of replicative senescence (11). The tumor-suppressive function may be more central to the program of accelerated senescence, which prevents the outgrowth of cells that have experienced oncogenic mutations (such as RAS or RAF mutations) or that underwent genome-destabilizing DNA damage.

In agreement with a role of senescence in cancer prevention, the process of carcinogenesis almost inevitably involves one or more events that inhibit senescence. Tumor cells avoid replicative senescence through the upregulation of telomerase or (less frequently) by employing alternative mechanisms of telomere maintenance (ALT) (11). Telomerase expression does not prevent accelerated senescence induced by DNA damage (27), but both replicative and accelerated forms of senescence are inhibited by the inactivation of p53 or p16, two of the most commonly disabled tumor suppressors in different types of cancer. As a result, most tumor cells have both senescence-promoting changes (short telomeres, RAS mutations) and senescence-inhibiting adaptations (activation of telomerase, inactivation of p53 and/or p16). Until late 1990s, it had been a common assumption that neoplastically transformed cells are no longer capable of senescence. Today we know, however, that tumor cells can undergo senescence and can be forced into this process by various genetic manipulations and by epigenetic factors, including conventional anticancer drugs, radiation, and differentiating agents.

INDUCING SENESCENCE IN TUMOR CELLS BY GENETIC MODIFICATIONS

The earliest means used to induce senescence in "immortal" tumor cell lines was somatic cell fusion with normal cells or with other tumor cell lines. These studies (28) have demonstrated that senescence is dominant over immortality, and they have identified four senescence-determining complementation groups. In the past years, transfection of many growth-inhibitory genes into tumor cell lines was shown to produce stable growth arrest, followed by the appearance of senescence-associated phenotypic changes and SA-β-gal expression. Many growth-inhibitory genes were shown to induce senescence-like growth arrest in tumor cells (see Table 1 in Ref. 29); most of these genes are known to play a role in the program of senescence in normal cells. These genes include RB (30), p53 (31) and two p53-related proteins (p63 and p73) (32), several CDK inhibitors (p21, p16, $p57^{Kip2}$, and $p15^{Ink4b}$) (33–36), and several other genes (29).

Is the senescence-like growth arrest induced in tumor cells by the overexpression of growth-inhibitory genes irreversible? The ability of tumor cells to recover once the expression of a growth-inhibitory gene has been turned off has been investigated for p53, p21, and p16 tumor suppressors, which were expressed in tumor cells from regulated promoters (31,33,34,37,38). In all of these cases, the ability of the cells to grow and form colonies after the promoter shutoff was inversely related to the duration of expression of the tumor suppressor, with very few cells recovering after prolonged induction (4–5 days). The failure to recover after release from p21-induced growth inhibition was also shown to depend on the level to which p21 expression was induced (38). The latter study, where p21 was expressed from an inducible promoter in HT1080 fibrosarcoma cells, has also addressed the mechanism of the failure of the cells to recover after the shutoff of p21. All the cells, despite their senescent phenotype and regardless of the duration or the magnitude of p21 induction, were found to reenter the cycle and replicate their DNAs. Upon entering mitosis, however, most of the cells that were released after prolonged arrest developed grossly abnormal mitotic figures and either died or underwent senescence-like growth arrest in a subsequent cell cycle (38). It remains to be determined if the failure to recover from senescence-like growth arrest induced by any other growth-inhibitory genes is due to a genuinely permanent cell cycle arrest that can be maintained without the inducing protein.

Another type of genetic manipulations that induce senescence in tumor cells is based on inhibiting the tumor proteins that counteract senescence. Somewhat surprisingly, inhibition of telomerase by a dominant negative mutant was found to induce primarily cell death rather than senescence in tumor cell lines (39). This result can be understood in light of the anti-apoptotic function of telomerase, which may be independent of its effect on telomeres (40) and which may also reflect the induction of mitotic catastrophe by abnormal telomeric structures. In contrast to the outcome of telomerase inhibition, senescence was readily induced in cervical carcinoma cells by inhibiting papillomavirus oncoproteins E6 and E7 that inhibit p53 and Rb tumor suppressors, respectively. Introduction of bovine papillomavirus protein E2, a negative regulator of both E6 and E7, into several human cervical carcinoma cell lines induced accelerated senescence in almost 100% of tumor cells (41,42). The effect of E2 was not accompanied by telomere shortening (41), and it was not prevented by constitutive overexpression of telomerase (43). Induction of senescence by E2 was associated with p53 stabilization and with strong induction of p21, and it was prevented by using p21-inhibiting antisense oligonucleotides or by increasing the expression of E6 or E7 (42). These results demonstrate that tumor cells are "primed" to undergo accelerated senescence once senescence-restraining mechanisms that inhibit the p53 and Rb pathways are removed. Enhancement of the extant program of accelerated senescence in tumor cells can be therefore viewed as a biologically justified approach to cancer therapy.

INDUCTION OF SENESCENCE IN TUMOR CELL LINES BY CHEMOTHERAPY, RADIATION, AND RETINOIDS

The propensity of tumor cells to undergo senescence in response to damage was demonstrated by the analysis of the effects of chemotherapeutic drugs and radiation on cell lines derived from different types of human solid tumors (44). A wide variety of anticancer agents induced senescence-like morphological changes and SA-β-gal expression in tumor cells. When equitoxic (ID_{85}) doses of different agents were applied to HT1080 fibrosarcoma cells, the strongest induction of the senescent phenotype was observed with DNA-interactive agents doxorubicin, aphidicolin, and cisplatin; a somewhat weaker response was seen with ionizing radiation, cytarabine, and etoposide; and the weakest effect was seen with microtubule-targeting drugs (taxol and vincristine). However, in contrast to taxol, another mictotubule-stabilizing agent discodermolide was found to induce senescence as strongly as doxorubicin (45). Induction of senescence by the drugs was dose dependent, and it was detectable even at the lowest drug doses that had a measurable growth-inhibitory effect. Moderate doses of doxorubicin induced the senescent phenotype in 11 of 14 cell lines derived from different types of human solid tumors (44). Other investigators have demonstrated the induction of the senescent phenotype in different tumor cell lines treated with a variety of drugs (29,46). Drug-induced senescent phenotype in tumor cells was not associated with telomere shortening and was not prevented by the overexpression of telomerase (47). Notably, in some of the cell lines, the senescent phenotype was observed in 10% to 20% of the cells even without drug treatment (44), suggesting that tumor cell senescence could develop spontaneously, possibly in response to subtle changes in the cell environment.

Drug-induced senescent phenotype was specifically associated with the tumor cells that underwent terminal growth arrest in response to treatment (37,44,48). The most conclusive evidence for this came from the analysis of growth-arrested and proliferating cells that were separated after release from the drug. This fluorescence-activated cell sorting (FACS)-based separation procedure involves labeling cells with a lipophilic fluorophore PKH2, which stably incorporates into the plasma membrane and distributes evenly between daughter cells, resulting in gradual decrease in PKH2 fluorescence with increasing number of cell divisions. In the experiment shown in Figure 2 (48), HCT116 cells were exposed to 200-μM doxorubicin for 24 hours, and then labeled with PKH2. Changes in PKH2 fluorescence were monitored on subsequent days. Drug-treated HCT116 cells remained growth arrested ($PKH2^{hi}$) for two to three days after the removal of doxorubicin, but a proliferating cell population ($PKH2^{lo}$) emerged starting from day 4. A large fraction of cells, however, remained $PKH2^{hi}$ and did not change PKH2 fluorescence throughout the experiment, indicating that these cells did not divide even once after release from the drug (Fig. 2A). Six to nine days after treatment, the cells were separated by FACS into $PKH2^{hi}$ and $PKH2^{lo}$ fractions. The $PKH2^{hi}$ cells were large, flat, and SA-β-gal+, but $PKH2^{lo}$ cells did not express the markers of senescence and were otherwise indistinguishable from the untreated cells (Fig. 2B). The $PKH2^{lo}$ but not the $PKH2^{hi}$ cells gave rise to colonies (Fig. 2C), indicating that the senescent phenotype of doxorubicin-treated HCT116 cells is associated with terminal growth arrest (48). On the other hand, when doxorubicin-treated HT1080 fibrosarcoma cells were analyzed by a similar procedure, most of the cells in the $PKH2^{hi}$ population divided once or twice after release from doxorubicin before

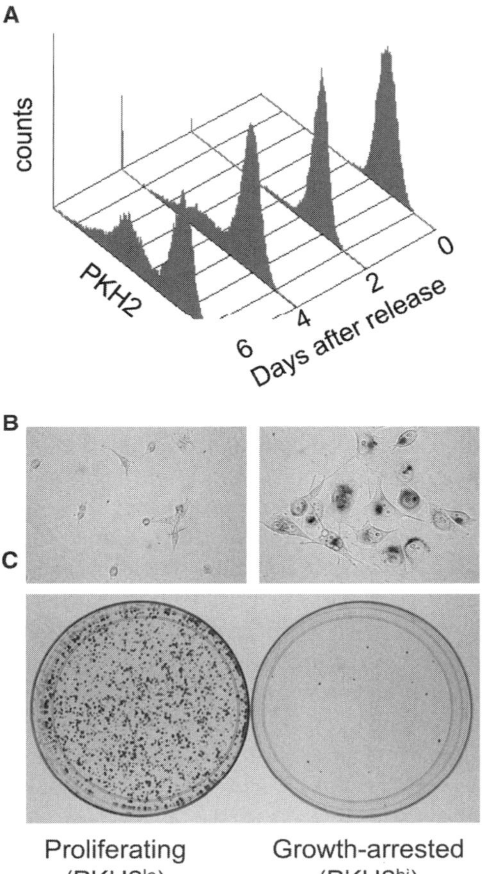

Figure 2 Doxorubicin-induced senescence in HCT116 colon carcinoma cells. (**A**) FACS profiles of PKH2 fluorescence of HCT116 cells on the indicated days after release from doxorubicin. (**B**) SA-β-gal staining of $PKH2^{hi}$ (*right*) and $PKH2^{lo}$ (*left*) populations, separated six days after release from the drug, photographed at 200×. (**C**) Colony formation by $PKH2^{hi}$ and $PKH2^{lo}$ populations, separated nine days after drug treatment and plated at 10,000 live cells per 10-cm plate. *Abbreviations*: SA-β-gal, senescence-associated β-galactosidase; $PKH2^{hi}$, growth-arrested; $PKH2^{lo}$, proliferating. *Source*: From Ref. 48 (*The color version of this figure is provided on the DVD*).

undergoing terminal growth arrest, indicating that drug-induced senescent phenotype can be associated with both an immediate and a delayed terminal growth arrest (44).

The role of senescence as a determinant of treatment response was indicated by several in vitro studies (29,46). As an example, expression of the MDR1 P-glycoprotein, which acts both as a drug efflux pump and as an inhibitor of apoptosis, protects a HeLa derivative and NIH 3T3 cells from radiation-induced apoptosis, but it does not increase their clonogenic survival. This apparent paradox was resolved by finding that a decrease in the fraction of apoptotic cells was accompanied by a commensurate increase in the fraction of cells undergoing either senescence or mitotic catastrophe indicating that the latter responses, without apoptosis, are sufficient to stop the proliferation of tumor cells (49).

Aside from cytotoxic drugs and radiation, tumor cell senescence was also found to be induced by transforming growth factor (TGF)-β (50) and by compounds that are usually referred to as "differentiating agents," including sodium butyrate (51) and retinoids. The induction of senescence has been analyzed in most detail for retinoids, natural and synthetic derivatives of vitamin A, which regulate cell growth and differentiation through their effects on gene expression. The latter effects are mediated by the binding of retinoids to retinoid receptors that act as regulators by transcription. Retinoid-induced growth arrest of tumor cells is commonly assumed to result from the induction of differentiation, and in some cases, this assumption has been corroborated by the appearance of differentiation-specific markers in retinoid-treated cells. It has now become apparent, however, that retinoid treatment can also induce senescence rather than differentiation (52). In particular, exposure of MCF-7 breast carcinoma cells to non-cytotoxic doses of *all-trans* retinoic acid (RA) and other retinoic acid receptor (RAR) ligands induced growth arrest and the senescent phenotype (44,53,54). Microarray analysis of gene expression showed that this response involves the upregulation of many growth-inhibitory genes, including those associated with senescence in other systems but not of any markers of epithelial differentiation (53,54). In another example (55), two sublines of the same neuroblastoma cell line SK-N-SH showed different morphological responses to RA. In one subline, RA induced neuronal differentiation, as defined by morphological and antigenic markers. In the other subline, RA treatment produced morphological features of senescence and SA-β-gal expression. Significantly, RA treatment increased the levels of p21 in the differentiating SK-N-SH cells, but it *decreased* p21 expression in the subline undergoing senescence, suggesting that p21 could act as a switch between differentiation and senescence in retinoid-treated cells (55). RA was also shown to decrease p21 levels in MCF-7 cells (56). In contrast to RA-induced senescence, p21 levels are increased under the conditions of senescence induced by DNA-damaging chemotherapeutic drugs (37,48). As discussed in a later section, these differences in p21 expression are paralleled by major differences in the spectra of genes that are upregulated in senescent tumor cells.

SENESCENCE AS A DETERMINANT OF IN VIVO TREATMENT RESPONSE

The induction of tumor senescence by anticancer agents is not limited to cell culture. SA-β-gal staining showed that the senescent phenotype develops in human tumor xenografts grown in nude mice and treated in vivo with retinoids or cytotoxic drugs (44,57–60). Most importantly, clinical studies on breast cancer (61) and lung cancer patients (62) revealed that the senescent phenotype, detected by staining for the SA-β-gal marker of senescence, occurs in a high fraction of tumors from patients who underwent conventional chemotherapy. te Poele et al. (61) investigated the correlation between the senescence response and the chemotherapeutic treatment in clinical breast cancer. This study used newly sectioned material from frozen archival breast tumors of patients who had or had not received neoadjuvant chemotherapy (the CAF regimen: cyclophosphamide, doxorubicin, and 5-fluorouracil). The senescent phenotype was detected by SA-β-gal activity, and the tumors were also stained for p53 and p16. Although SA-β-gal enzymatic activity is unstable even in freshly frozen tissue samples, te Poele et al. have succeeded in demonstrating SA-β-gal in 15 of 36 treated tumors (41%). Remarkably, SA-β-gal staining was confined to tumor cells but was not detected in the normal tissue, suggesting that chemotherapy-induced senescence is a specific response of tumor cells. Tumor sections of patients who had not received chemotherapy showed SA-β-gal staining of isolated tumor cells in 2 of 20 cases, suggesting spontaneous senescence. SA-β-gal staining in breast cancer was associated with low p53 staining, indicative of the lack of mutant p53, and with high staining for p16 (61).

Roberson et al. (62) used SA-β-gal staining to analyze freshly frozen sections of lung cancers and adjacent normal lung tissues from several patients who were either untreated or treated with carboplatin/taxol. Little or no staining was observed in three untreated tumors, but both tumors from patients who underwent therapy showed large areas of strong SA-β-gal positivity. While the senescence response was observed primarily in the tumor cells, normal cells in three treated patients showed detectable SA-β-gal positivity, whereas SA-β-gal was detectable in only one of three untreated normal samples.

Both senescence and apoptosis were shown to determine the response to chemotherapy in Eμ-myc lymphoma, a transgenic mouse model of B-cell lymphoma (58). When the program of apoptosis in Eμ-myc lymphoma was inhibited by transduction with a retrovirus that expresses the anti-apoptotic gene BCL-2, senescence became the principal tumor response to cyclophosphamide, as indicated by apparently complete cessation of DNA replication and mitosis and by drastic induction of the SA-β-gal marker. The senescence response (along with apoptosis) was also observed in the absence of BCL-2. Treatment-induced senescence became undetectable upon the knockout of either p53 (which also abolished the apoptotic response) or p16 (which had no effect on apoptosis). Inhibition of either apoptosis or senescence in Eμ-myc lymphoma made these tumors significantly more sensitive to chemotherapy, indicating that both of these physiological programs contribute to treatment success (58).

It should be noted, however, that senescence is not the only antiproliferative response that determines treatment response in the absence of apoptosis. Another principal effect of anticancer agents is mitotic catastrophe, abnormal mitosis that usually ends in cell death (see chap. 18 by Broude et al.). There are as yet no in vivo studies where all three responses have been analyzed at the same time. Such analysis should elucidate the relative contribution of apoptosis, senescence, and mitotic catastrophe to the overall outcome of cancer treatment.

p53, p21, AND p16 IN TUMOR CELL SENESCENCE

p53, p21, and p16, which regulate senescence in normal cells (Fig. 1), also play a role in treatment-induced senescence of tumor cells. Treatment-induced senescence in murine Eμ-myc lymphoma required wild-type p53 and p16 (58), and p16 expression correlated with SA-β-gal staining in treated breast cancers (61). As mentioned above, p53 and p16 are frequently inactivated in cancers. If these genes were required for treatment-induced senescence, one could expect that the majority of tumors, which are deficient in one or both of these genes, would be unable to undergo senescence. This, however, is not the case. Chemotherapeutic drugs readily induced senescence in p16-deficient tumor cell lines, such as HT1080 and HCT116 (44). Furthermore, moderate doses of doxorubicin induced the senescent phenotype in p53-null Saos-2 cell line, in SW480 and U251 cells carrying mutant p53, and in HeLa and Hep-2 cell lines, where p53 function has been inhibited by papillomavirus protein E6 (44). In the breast cancer study by te Poele et al. (61), 20% of the SA-β-gal+ tumors showed high p53 staining (suggestive of p53 mutations), and 13% of the SA-β-gal+ tumors did not stain for p16, indicating that wild-type p53 and p16 induction is not necessary for senescence in clinical breast cancer.

The role of p53 and p21 in treatment-induced senescence was analyzed in HT1080 fibrosarcoma cells where p53 function and p21 expression were blocked by a p53-derived genetic suppressor element and in HCT116 colon carcinoma cells with homozygous knockout of p53 or p21. In both cell lines, the inhibition or knockout of p53 or p21 strongly decreased but did not abolish drug- or radiation-induced senescence, as determined by PKH2 analysis of cell division and by SA-β-gal staining (37). Hence, p53 and p21 act as positive regulators of accelerated senescence in tumor cells, but they are not absolutely required for this response. In addition, as described above, retinoid-induced senescence involves a decrease rather than an increase in p21 expression. These observations have indicated that genes other than p53, p21, or p16 are likely to play a role in accelerated senescence of tumor cells.

INHIBITION OF CELL CYCLE PROGRESSION GENES AND INDUCTION OF INTRACELLULAR AND SECRETED GROWTH INHIBITORS IN SENESCENT TUMOR CELLS

Additional determinants of drug-induced senescence in tumor cells were identified by gene expression profiling of doxorubicin-induced senescence in HCT116 colon carcinoma cells, which are p16 deficient and wild type for p53 (48). The proliferating and senescent fractions of HCT116 cells were separated by PKH2 labeling and flow sorting six to nine days after one-day doxorubicin treatment (Fig. 2) and used for RNA extraction. cDNA microarray hybridization followed by reverse transcription–polymerase chain reaction (RT-PCR) analysis of individual genes revealed major biological clusters of genes that were either downregulated or upregulated in senescent cells. More than one half of all the genes that were strongly inhibited in senescent cells are known to play a role in cell cycle progression, with the largest groups of genes involved in mitosis and DNA replication. Inhibition of these genes became apparent one to two days after doxorubicin treatment and was likely to contribute to the maintenance of drug-induced growth arrest. Analysis of HCT116 cell lines with homozygous disruption of either p53 or p21 demonstrated that doxorubicin-induced inhibition of cell cycle progression genes was fully dependent on p21 (48). Furthermore, ectopic expression of p21 in HT1080 fibrosarcoma cells was sufficient to inhibit the transcription of the same set of genes that are downregulated in doxorubicin-treated cells (4). This effect of p21 is exerted at the level of transcription, and it is mediated at least in part by negative regulatory elements in the corresponding promoters, such as cell cycle–dependent element/cell cycle gene homology region (CDE/CHR) (63,64).

In addition to the downregulation of genes required for cell cycle progression, senescent HCT116 cells were also found to upregulate multiple genes with growth-inhibitory activities (48) (see Table 2 in Ref. 29). Concerted and sustained induction of such genes explains the growth arrest of senescent cells despite the lack of p16 and suggests that this arrest is maintained by many apparently redundant mechanisms. Several of these genes are known or putative tumor suppressors, which are silenced in the course of neoplastic transformation but become reactivated with the onset of senescence. Such genes include p21, tumor suppressor BTG1, a related gene BTG2, and candidate tumor suppressor epithelial protein lost in neoplasm (EPLIN). Of special interest, senescent HCT116 cells also overexpress several secreted growth inhibitors, including serine protease inhibitor maspin, a tumor suppressor shown to inhibit the invasion and angiogenesis of breast and prostate cancers, as well as MIC-1/GDF15 (a member of the TGF-β family) and IGF-binding protein (IGFBP)-6 (48). Expression of a secreted growth-inhibitory factor CXCL14 has also been identified as a hallmark of senescence in normal and malignant prostate epithelial cells (65). Exposure to chemotherapeutic drugs and radiation was previously shown to induce a similar set of secreted tumor-suppressing factors, and paracrine growth-inhibitory activities of the damaged cells have been documented by conditioned media and coculture assays (66). The finding that the same factors are stably overexpressed by senescent cells suggests that such cells may provide a reservoir of tumor-suppressing factors that may contribute to the long-term success of chemotherapy.

Induction of secreted tumor-suppressing factors was previously found to be mediated by p53 (66). Analysis of p53-deficient HCT116 cells revealed, however, that induction of senescence-associated growth inhibitors (intracellular or secreted) showed either no dependence on p53 (BTG1, IGFBP-6) or limited p53 dependence (BTG2, EPLIN, maspin, MIC-1, amphiregulin). The latter genes were still induced in the absence of p53, albeit their induction was delayed or diminished relative to the cells with the wild-type p53 (48). These results help to explain why p53 deficiency diminishes but does not abolish drug-induced senescence (37). p21 knockout in HCT116 cells had little or no effect on the induction of senescence-associated growth inhibitors (48), explaining why p21-deficient cells can still undergo senescence (37). On the other hand, the reduction in the senescence response of p21-deficient cells can be readily explained by a failure to inhibit the transcription of cell cycle progression genes and by the lack of p21 itself.

Concerted induction of several growth-inhibitory genes was also observed in retinoid-induced senescence of MCF-7 breast carcinoma cells (53,54). Microarray hybridization and RT-PCR analysis showed that RA-induced senescent phenotype of MCF-7 cells is associated with the induction of multiple genes with growth-inhibitory activity. These genes encode both intracellular and secreted growth-inhibitory proteins. A survey of the published genes that are inducible by retinoids in different types of tumor cells showed that retinoids induce many other intracellular and secreted growth inhibitors, most of which are also known to be upregulated in senescent cells (52). Induction of growth-inhibitory proteins was associated with paracrine growth-inhibitory activity of retinoid-arrested MCF-7 cells, which inhibited the growth of retinoid-insensitive MDA-MB231 cells in coculture, an effect that did not require the presence of retinoids in culture media (54). Many of the retinoid-induced genes encoding secreted growth inhibitors are also upregulated by doxorubicin in HCT116 cells (48), indicating that their induction can occur independently of retinoids. Production of secreted growth-inhibitory factors therefore appears to be a general phenomenon in treatment-induced senescence of tumor cells, with potentially beneficial effects for the outcome of therapy.

Testing a panel of genes upregulated in senescent tumor cells showed that genes encoding intracellular and secreted tumor-suppressive proteins are also upregulated in normal senescent keratinocytes (67). In agreement with this observation, senescent keratinocytes were found to have a paracrine growth-inhibitory activity, as determined by their effect on soft agar colony formation by HeLa cells (6). The strongest induction in senescent keratinocytes was found for an anti-invasive and antiangiogenic protein maspin. In agreement with this finding, conditioned media from senescent keratinocytes displayed antiangiogenic activity, which was abolished by a neutralizing antibody against maspin (67).

TUMOR SENESCENCE IS ALSO ASSOCIATED WITH THE INDUCTION OF TUMOR-PROMOTING SECRETED FACTORS

Inhibition of cell proliferation, however, is not the only aspect of tumor senescence with potential clinical implications. Replicative senescence of normal fibroblasts is characterized by changes in the expression of multiple proteins.

Some of the proteins that are highly expressed in senescent cells, including the precursor protein of Alzheimer's β-amyloid peptide [Alzheimer's β-amyloid precursor protein (βAPP)] (68), have long-range pathogenic effects as well as degradative enzymes, inflammatory cytokines, and growth factors, which may contribute to carcinogenesis and tumor progression (1). Indeed, coculture and conditioned media experiments showed that normal human fibroblasts undergoing either replicative or accelerated senescence stimulate the growth of transformed epithelial cells in vitro and in vivo (5). In contrast to senescent keratinocytes that inhibit soft agar colony formation by HeLa cells, senescent fibroblasts stimulated HeLa colony formation in the same assay (6). Also, in contrast to the antiangiogenic activity of senescent keratinocytes, senescent fibroblasts show proangiogenic activity (69). Senescent fibroblasts were also found to inhibit normal differentiation of the mammary gland (70). In another study, senescent prostate fibroblasts were shown to overexpress a variety of tumor-promoting factors and to stimulate the growth of preneoplastic and neoplastic prostate epithelium in vitro (71). These paracrine effects of senescent fibroblasts closely resemble the cancer-promoting activities of tumor-associated stromal fibroblasts (72).

The procarcinogenic function of normal senescent cells in vivo is also supported by the findings that SA-β-gal expression in normal human hepatocytes is strongly correlated with the presence of hepatocellular carcinoma in the surrounding liver (73), and that prostate enlargement correlates with SA-β-gal expression in prostate epithelial cells (74). In a more recent study, increased numbers of senescent stromal cells were shown to be enriched in regions juxtaposed to ovarian cancer epithelium, suggesting tumor-promoting activity of such cells (75). Interestingly, the senescent phenotype has been observed in vivo in different preneoplastic or early-neoplastic conditions, such as nevi (skin moles) with mutations of the BRAF protein (76) or early-stage prostate cancers (77), but not in advanced (untreated) malignancies of the same types. Although these findings were interpreted as indicating the role of senescence as an anticarcinogenic mechanism triggered by early carcinogenic events, it is also possible that the senescent cells appearing early in the process of carcinogenesis may provide factors that drive tumor progression.

In agreement with these observations in normal senescent cells, doxorubicin-treated senescent HCT116 carcinoma cells also showed increased expression of genes for many proteins with diverse paracrine activities (see Table 2 in Ref. 29). The secreted proteins upregulated in senescent HCT116 cells include not only the above-described growth inhibitors but also multiple proteins with mitogenic, anti-apoptotic, and angiogenic activities (48). Some of these proteins are an ECM component Cyr61 with mitogenic and angiogenic functions, an anti-apoptotic and mitogenic ECM factor prosaposin, TGFα, angiogenic factor VEGF and multiple proteases, including matrix metalloproteinases that may potentially contribute to metastatic growth. Transmembrane proteins induced in senescent cells also include βAPP, which has mitogenic activity (78). Both tumor-suppressing and tumor-promoting factors were also shown to be upregulated upon senescence in normal and transformed prostate epithelial cells (65). Thus, senescence-associated changes in gene expression involve the induction of both tumor-suppressive and tumor-promoting proteins, as well as proteins involved in age-related diseases other than cancer. Relative expression of different biological classes of senescence-associated genes is therefore likely to determine whether tumor senescence would have a mostly positive or a mostly negative effect on the outcome of treatment.

ROLE OF CDK INHIBITORS IN SENESCENCE-ASSOCIATED CHANGES IN GENE EXPRESSION: IMPLICATIONS FOR TUMOR-PROMOTING STROMAL FIBROBLASTS

About one-third of senescence-associated genes that were induced by doxorubicin in HCT116 cells showed decreased or delayed induction in a p21−/− derivative of this cell line, indicating that p21 plays a role not only in the inhibition but also in the induction of gene expression in senescent cells. Some of the genes that showed p21 dependence encode secreted mitogenic/anti-apoptotic proteins, such as prosaposin, TGFα, and βAPP (48). These findings were in accord with the results of microarray analysis of the effects of p21 on gene expression in HT1080 fibrosarcoma cells (4). p21 induction in the latter cells produces growth arrest and the senescent phenotype, inhibits transcription of multiple genes, most of which are involved in cell cycle progression, and also leads to the induction of a set of genes with important paracrine activities, including many secreted proteins and ECM components. Most of the genes induced by p21 in HT1080 cells were also induced in WI-38 normal human fibroblasts infected with a p21-expressing lentiviral vector (79), in a human melanoma cell line treated with a polyamine-depleting regimen that induces strong p21 expression (80), and in doxorubicin-induced senescence of HCT116 cells (48). Furthermore, the effects of p21 on the induction of gene expression in HT1080 cells can be largely reproduced by other senescence-associated CDK inhibitors, p16 and p27 (81).

Many p21-induced genes are known to be upregulated during replicative senescence of normal fibroblasts (4). Furthermore, products of many genes that are induced by p21 have been linked to age-related diseases, including Alzheimer's disease, amyloidosis, atherosclerosis, and arthritis. Some examples are βAPP, serum amyloid A, tissue transglutaminase, connective tissue growth factor (CTGF), and $p66^{Shc}$, a positive mediator of oxidative stress, knockout of which increases toxin resistance and the lifespan in mice (82). Another group of p21-induced genes encode secreted proteins with known mitogenic, anti-apoptotic, or angiogenic activities, such as prosaposin, epithelin/granulin, galectin-3, CTGF, or VEGF-C. The induction of such genes produces paracrine growth-promoting activities, as demonstrated by the fact that conditioned media from p21-arrested HT1080 cells has mitogenic and anti-apoptotic effects (4). These paracrine effects of p21 induction mimic the tumor-promoting activities that were demonstrated in different types of senescent fibroblasts (5) and in tumor-associated stromal fibroblasts (72). As discussed elsewhere (83), all the treatments that are known to induce the tumor-promoting functions of stromal fibroblasts also result in p21 induction, suggesting that p21 or related proteins could be responsible for the paracrine tumor-promoting functions of stromal fibroblasts.

Induction of gene expression by p21 occurs at the level of transcription, since p21 stimulated the activity of all six tested promoters of p21-inducible genes (84). Although p21 is best known as an inhibitor of cyclin/CDK complexes, it also interacts with many transcription factors and cofactors and regulators of signal transduction (19), which can account for its pleiotropic effects on gene expression. One of the effects of p21 is the augmentation of transcription factor NFκB (85). This effect was reported to be upregulated at least in part through the activation of transcription cofactors/histone deacetylases p300 and CBP, which augment not only NFκB but also many other inducible transcription factors (85).

The involvement of p21 and other CDK inhibitors in the expression of genes associated with the undesirable effects of senescence suggests that treatments that induce senescence without activating p21 or p16 may be more beneficial in the long term. In agreement with this hypothesis, very few of the genes that were found by microarray analysis to be upregulated in retinoid-induced senescence of MCF-7 breast carcinoma cells encode secreted factors with tumor-promoting activities, or other disease-promoting factors (such as amyloid proteins) (53,54). This result is likely to reflect the lack of p21 or p16 induction in retinoid-induced senescence of MCF-7 cells. These observations demonstrate that the positive effects of tumor senescence (permanent growth arrest of tumor cells and secretion of tumor-suppressing factors) can be separated from the disease-promoting activities of senescent cells.

PROGNOSTIC IMPLICATIONS OF TUMOR SENESCENCE: EXAMPLES FROM PROSTATE CANCER

Senescent cells appear within the tumor either as a consequence of treatment or spontaneously, as a result of environmental stress or sporadic inactivation of senescence-restraining mechanisms in an individual cell. Senescent cells are generally resistant to apoptosis, and senescent fibroblasts in culture are known to survive for more than a year. The persistence of senescent cells in the tumor is a double-edged sword. On one hand, the senescent tumor cells do not proliferate and, furthermore, serve as a reservoir of secreted factors that inhibit tumor growth. On the other hand, the same cells can also produce secreted factors with mitogenic, anti-apoptotic, and angiogenic activities. These tumor-promoting functions of senescent cells are determined to a large extent by the expression of p21 and p16. The presence of senescent cells in the tumor and the relative

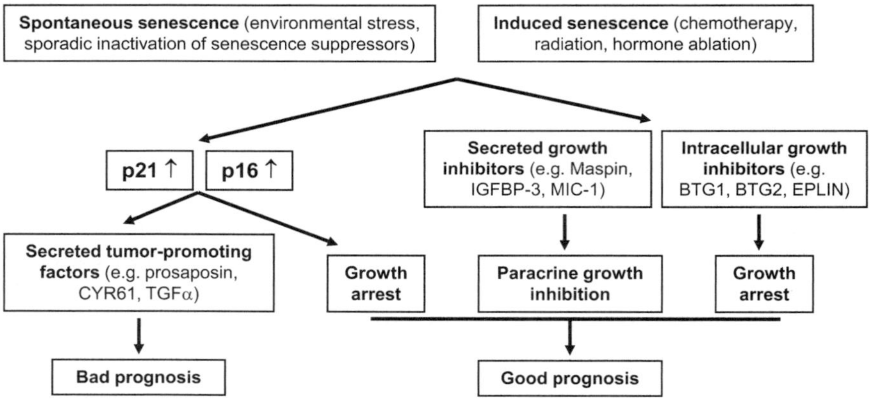

Figure 3 Tumor senescence and its consequences.

abundance of different proteins produced by the senescent cells are important biological factors that should have significant prognostic implications for the disease outcome. One would expect that more aggressive tumors may contain few or no senescent cells. Alternatively, such tumors may have a substantial fraction of senescent cells that express CDK inhibitors (p16 or p21), as well as senescence-associated tumor-promoting factors that are upregulated by CDK inhibitors. On the other hand, tumors containing senescent cells that express high levels of secreted growth inhibitors but little p16 or p21 should have a more favorable prognosis. This concept is illustrated in Figure 3.

Although only very few studies have so far directly addressed the development of tumor senescence during clinical cancer therapy, there is abundant evidence in the literature in support of various aspects of the model shown in Figure 3. This can be illustrated especially well by the results of clinical and biological studies in prostate cancer; similar examples can be found for many other tumor types.

1. A prostate carcinoma cell line LNCaP growing in culture contains 10% to 15% SA-β-gal+, suggesting that prostate cancer cells can undergo spontaneous senescence. The senescent fraction in this cell line is increased by doxorubicin (44) or by transfection with an activatable form of c-Raf (86), thus demonstrating the susceptibility of prostate cancer cells to accelerated senescence.
2. Indirect observations in radiation therapy of prostate cancer suggest that the induction of senescence may be a primary mode of treatment response. In particular, complete regression of prostate cancers was reported in some patients to take more than a year after radiation treatment (87). This slow course of tumor disappearance seems most consistent with radiation-induced

senescence. In an example from another tumor type, regression of desmoid tumors took up to two years after radiation treatment (88).

3. p16, the CDK inhibitor primarily associated with senescence, is a tumor suppressor, which is infrequently mutated in prostate cancer (89). p16 expression was not detectable by immunohistochemistry in the normal prostate (90), but it is observed in close to one half of prostate carcinomas (90). In all of these studies, p16 was found to be an unfavorable prognostic marker for prostate cancer (91,92). In particular, p16 expression was an independent indicator of early relapse after radical prostatectomy, and p16 was elevated in cancers relative to benign prostatic hyperplasia (BPH). These paradoxical adverse correlations of the tumor suppressor p16 are readily explained by the ability of p16 to upregulate secreted tumor-promoting factors. Notably, higher p16 levels in tumor cells have also been associated with the history of androgen ablation treatment (90), a result that probably reflects the induction of prostate cancer senescence by this treatment. Strong adverse correlations for p16 expression have also been reported in breast cancer (93–95).
4. Similar unfavorable prognostic correlations were found for another CDK inhibitor p21 in the majority of studies that analyzed this protein in prostate cancer (96–101). p21 expression in prostate cancer showed no correlation with p53 status, suggesting that p21 is induced in this tumor primarily by p53-independent mechanisms (96). Like p16, p21 was found to be an independent marker of early relapse after prostatectomy, and it was associated with high pathological grade and high Ki-67 index. p21 was also shown to be a highly significant marker of progression from androgen-dependent to androgen-independent cancer (101). The adverse prognostic role of p21 is disputed in some reports (102,103), but a potential explanation for such discrepancies is suggested by the findings of Sarkar et al. (99). The latter study noted a dependence of p21 correlations on the racial background (p21 is a strong independent marker of negative prognosis in Caucasians but not in African-Americans) and suggested that some as yet unknown genetic factors may affect the role of p21 in prostate cancer. Studies of p21 expression in other tumor types produced both favorable prognostic correlations (reflecting the role of p21 as a marker of wild-type p53 function) and negative correlations, similar to those in prostate cancer (83).
5. Prostate cancers also express senescence-associated growth inhibitors, with good prognostic correlations. One of the most commonly used markers in prostate cancer is IGFBP-3, a secreted protein that induces growth arrest and apoptosis. Low IGFBP-3 levels in the plasma, alone or in combination with high IGF-1, have been associated with the presence of advanced prostate cancer (104,105), and an increase in serum IGFBP-3 has been used as an indicator of treatment response (106). IGFBP-3 is induced at senescence in different types of normal and tumor cells (52,107). In particular, IGFBP-3 is consistently upregulated in senescent prostate

epithelial cells and silenced in prostate cancer cell lines and tumors (108). This senescence-associated growth inhibitor is induced in prostate cancer cells by many antiproliferative agents (109,110). Another IGFBP family member IGFBP-rP1 increases during senescence of normal prostate epithelial cells and is downregulated in prostate cancer (111). Overexpression of IGFBP-rP1 induces growth arrest and the senescent phenotype in M12 prostate cancer cell line (112).

6. The strongest correlations with good prognosis in prostate cancer have been reported so far for another senescence-associated tumor suppressor serine protease inhibitor maspin. Maspin is induced to a very high level by DNA damage in several tumor cell lines, including LNCaP prostate carcinoma (113). The absence or low levels of maspin in prostate cancer samples have been correlated to higher tumor stages, histological dedifferentiation, and early relapse (114). Conversely, maspin expression was strongly elevated in the cancers of patients treated with neoadjuvant androgen ablation therapy, and treatment-induced maspin was specifically associated with tumor cells that showed morphological effects of the treatment (115).

The above observations in prostate cancer appear to be in excellent agreement with the model depicted in Figure 3. As predicted by the model, markers of senescence are observed in untreated tumors, but their expression is elevated after treatment. Furthermore, senescence-associated growth inhibitors (maspin and IGFBP-3) correlate with good prognosis, whereas p16 and p21 correlate with bad prognosis. With the identification of multiple senescence-associated growth regulators, it should now be possible to investigate their expression and coexpression in tumor tissues by conventional immunocytochemical techniques. Such analysis should allow us to determine the relationship between the expression of these proteins, tumor staging, and treatment outcome. It should also be possible to determine whether radiation- or chemotherapy-induced tumor senescence leads to sustained secretion of senescence-associated growth inhibitors into bodily fluids. Production of such proteins could be an important factor in preventing the tumor growth. New diagnostic approaches that will arise from understanding the biology of tumor senescence may be of considerable benefit in the management of cancer patients.

POTENTIAL FOR DEVELOPING SENESCENCE-BASED ANTICANCER DRUGS

Activating the physiological program of senescence in tumor cells seems an attractive approach to cancer treatment. This response to chemotherapy is induced by a wide variety of anticancer agents, even under the conditions of minimal cytotoxicity. Even if not all of the tumor cells are rendered senescent as a result of treatment, such cells may provide a reservoir of secreted tumor-suppressing factors that will inhibit the growth of non-senescent cells. On the other hand, senescent cells can overexpress secreted tumor-promoting factors as well as proteins associated with various pathological conditions. The side effects of senescence, which are mediated at least in part by CDK inhibitors, may have potential adversarial effects in the short term (growth stimulation of non-senescent tumor cells) and in the long term (increased likelihood of de novo carcinogenesis and the development of age-related diseases). On the basis of these considerations, senescence-oriented therapeutic strategies may include two general strategies. The first direction is to develop the agents that will interfere with the induction of disease-promoting genes by CDK inhibitors. The second strategy is to develop drugs that will induce tumor cell senescence without upregulating p21 (which, unlike p16, is almost never inactivated in tumors) or p21-inducible disease-promoting genes. These approaches are schematized in Figure 4.

One of the approaches involves the use of compounds that inhibit the induction of transcription by p21 or by other CDK inhibitors. Such compounds have been identified by screening of diversified chemical libraries, based on their effect on the expression of p21-inducible promoter-reporter constructs (116). Agents that prevent the induction of gene expression by p21 are likely to interfere with the tumor-promoting paracrine activities of senescent cells that arise spontaneously or as a result of conventional chemotherapy or radiation therapy. These compounds may also block the tumor-promoting activities of stromal fibroblasts and may potentially be useful in the chemoprevention of age-related diseases, such as Alzheimer's disease or atherosclerosis (Fig. 4).

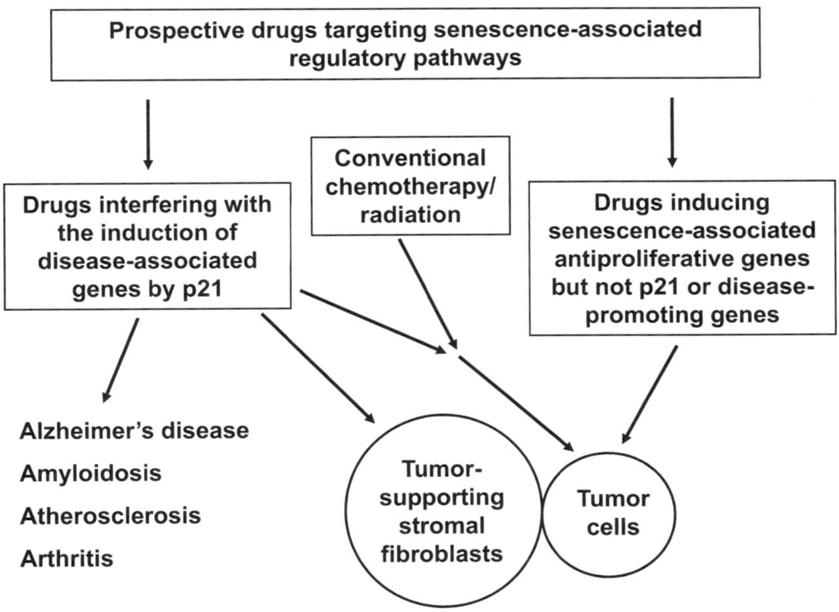

Figure 4 Strategies for senescence-based therapy in cancer treatment.

The second strategy in Figure 4 is based on the development of agents that would induce tumor cell senescence without its associated side effects. The feasibility of this strategy is suggested by the finding that retinoid-induced senescence of breast carcinoma cells involves the induction of the tumor-suppressing but not of the disease-promoting genes. The therapeutic use of retinoids is limited by the fact that these drugs act through retinoid receptors, which are readily lost in tumor cells. Senescence-associated growth-inhibitory genes, however, contain no discernible retinoid receptor–binding sites in their promoters, and they appear to be induced by retinoids through an indirect mechanism (52,54). Furthermore, the same proteins are induced by retinoids and by non-retinoid drugs, such as doxorubicin (48). It therefore seems likely that the mechanisms that produce concerted upregulation of retinoid-inducible growth inhibitors may also be stimulated by other inducers of senescence. Identification of senescence-associated growth-inhibitory genes makes it possible to develop high-throughput screening systems for agents that induce such genes. Thus, the elucidation of the biological aspects of tumor cell senescence offers plausible approaches to the development of novel therapeutic strategies to stop the growth of tumor cells.

ACKNOWLEDGMENTS

We apologize to all the colleagues whose relevant studies were not covered in this review because of space limitations. We thank former members of our laboratory who participated in the studies reviewed in this chapter, including Drs. Bey-Dih Chang, Milos Dokmanovic, Jason Poole, Mari Swift, Hongming Zhu, Keiko Watanabe, Errin Lagow, Yuhong Chen, and Donald Porter. The work from the authors' laboratory was supported by NIH grants R01 CA89636, R01 CA62099, R01 AG17921, and R01 AG028687 and grant 025-01 from Mary Kay Ash Charitable Foundation.

REFERENCES

1. Campisi J, d'Adda dF. Cellular senescence: when bad things happen to good cells. Nat Rev Mol Cell Biol 2007; 8:729–740.
2. Dimri GP, Lee X, Basile G, et al. A biomarker that identifies senescent human cells in culture and in aging skin in vivo. Proc Natl Acad Sci U S A 1995; 92:9363–9367.
3. Lee BY, Han JA, Im JS, et al. Senescence-associated beta-galactosidase is lysosomal beta-galactosidase. Aging Cell 2006; 5:187–195.
4. Chang BD, Watanabe K, Broude EV, et al. Effects of $p21^{Waf1/Cip1/Sdi1}$ on cellular gene expression: implications for carcinogenesis, senescence, and age-related diseases. Proc Natl Acad Sci U S A 2000; 97:4291–4296.
5. Krtolica A, Parrinello S, Lockett S, et al. Senescent fibroblasts promote epithelial cell growth and tumorigenesis: a link between cancer and aging. Proc Natl Acad Sci U S A 2001; 98:12072–12077.

6. Bacon P, Bodner B, Nickoloff BJ. Senescent human keratinocytes suppress colony formation of HeLa cells. J Dermatol Sci 2005; 38:64–66.
7. Collado M, Blasco MA, Serrano M. Cellular senescence in cancer and aging. Cell 2007; 130:223–233.
8. Courtois-Cox S, Jones SL, Cichowski K. Many roads lead to oncogene-induced senescence. Oncogene 2008; 27:2801–2809.
9. Prieur A, Peeper DS. Cellular senescence in vivo: a barrier to tumorigenesis. Curr Opin Cell Biol 2008; 20:150–155.
10. Hayflick L, Moorhead PS. The serial cultivation of human diploid cell strains. Exp Cell Res 1961; 37:585–621.
11. Mathon NF, Lloyd AC. Cell senescence and cancer. Nat Rev Cancer 2001; 1: 203–213.
12. Vaziri H, West MD, Allsopp RC, et al. ATM-dependent telomere loss in aging human diploid fibroblasts and DNA damage lead to the post-translational activation of p53 protein involving poly(ADP-ribose) polymerase. EMBO J 1997; 16: 6018–6033.
13. von Zglinicki T. Telomeres and replicative senescence: Is it only length that counts? Cancer Lett 2001; 168:111–116.
14. Di Leonardo A, Linke SP, Clarkin K, et al. DNA damage triggers a prolonged p53-dependent G1 arrest and long-term induction of Cip1 in normal human fibroblasts. Genes Dev 1994; 8:2540–2551.
15. Serrano M, Lin AW, McCurrach ME, et al. Oncogenic ras provokes premature cell senescence associated with accumulation of p53 and p16INK4a. Cell 1997; 88: 593–602.
16. Zhu J, Woods D, McMahon M, et al. Senescence of human fibroblasts induced by oncogenic Raf. Genes Dev 1998; 12:2997–3007.
17. Pearson M, Carbone R, Sebastiani C, et al. PML regulates p53 acetylation and premature senescence induced by oncogenic Ras. Nature 2000; 406:207–210.
18. Ferbeyre G, de Stanchina E, Querido E, et al. PML is induced by oncogenic ras and promotes premature senescence. Genes Dev 2000; 14:2015–2027.
19. Dotto GP. p21(WAF1/Cip1): more than a break to the cell cycle? Biochim Biophys Acta 2000; 1471:M43–M56.
20. Alcorta DA, Xiong Y, Phelps D, et al. Involvement of the cyclin-dependent kinase inhibitor p16 (INK4a) in replicative senescence of normal human fibroblasts. Proc Natl Acad Sci U S A 1996; 93:13742–13747.
21. Stein GH, Drullinger LF, Soulard A, et al. Differential roles for cyclin-dependent kinase inhibitors p21 and p16 in the mechanisms of senescence and differentiation in human fibroblasts. Mol Cell Biol 1999; 19:2109–2117.
22. Ohtani N, Zebedee Z, Huot TJ, et al. Opposing effects of Ets and Id proteins on p16INK4a expression during cellular senescence. Nature 2001; 409:1067–1070.
23. Itahana K, Zou Y, Itahana Y, et al. Control of the replicative life span of human fibroblasts by p16 and the polycomb protein Bmi-1. Mol Cell Biol 2003; 23: 389–401.
24. Wang W, Wu J, Zhang Z, et al. Characterization of regulatory elements on the promoter region of p16(INK4a) that contribute to overexpression of p16 in senescent fibroblasts. J Biol Chem 2001; 276:48655–48661.
25. Bringold F, Serrano M. Tumor suppressors and oncogenes in cellular senescence. Exp Gerontol 2000; 35:317–329.

26. Malumbres M, Perez DCI, Hernandez MI, et al. Cellular response to oncogenic ras involves induction of the Cdk4 and Cdk6 inhibitor p15(INK4b). Mol Cell Biol 2000; 20:2915–2925.
27. Morales CP, Holt SE, Ouellette M, et al. Absence of cancer-associated changes in human fibroblasts immortalized with telomerase. Nat Genet 1999; 21:115–118.
28. Tominaga K, Olgun A, Smith JR, et al. Genetics of cellular senescence. Mech Ageing Dev 2002; 123:927–936.
29. Roninson IB. Tumor cell senescence in cancer treatment. Cancer Res 2003; 63:2705–2715.
30. Xu HJ, Zhou Y, Ji W, et al. Reexpression of the retinoblastoma protein in tumor cells induces senescence and telomerase inhibition. Oncogene 1997; 15:2589–2596.
31. Sugrue MM, Shin DY, Lee SW, et al. Wild-type p53 triggers a rapid senescence program in human tumor cells lacking functional p53. Proc Natl Acad Sci U S A 1997; 94:9648–9653.
32. Jung MS, Yun J, Chae HD, et al. p53 and its homologues, p63 and p73, induce a replicative senescence through inactivation of NF-Y transcription factor. Oncogene 2001; 20:5818–5825.
33. Fang L, Igarashi M, Leung J, et al. $p21^{Waf1/Cip1/Sdi1}$ induces permanent growth arrest with markers of replicative senescence in human tumor cells lacking functional p53. Oncogene 1999; 18:2789–2797.
34. Dai CY, Enders GH. p16 INK4a can initiate an autonomous senescence program. Oncogene 2000; 19:1613–1622.
35. Tsugu A, Sakai K, Dirks PB, et al. Expression of p57(KIP2) potently blocks the growth of human astrocytomas and induces cell senescence. Am J Pathol 2000; 157:919–932.
36. Fuxe J, Akusjarvi G, Goike HM, et al. Adenovirus-mediated overexpression of p15INK4B inhibits human glioma cell growth, induces replicative senescence, and inhibits telomerase activity similarly to p16INK4A. Cell Growth Differ 2000; 11:373–384.
37. Chang BD, Xuan Y, Broude EV, et al. Role of p53 and $p21^{waf1/cip1}$ in senescence-like terminal proliferation arrest induced in human tumor cells by chemotherapeutic drugs. Oncogene 1999; 18:4808–4818.
38. Chang BD, Broude EV, Fang J, et al. $p21^{Waf1/Cip1/Sdi1}$-induced growth arrest is associated with depletion of mitosis-control proteins and leads to abnormal mitosis and endoreduplication in recovering cells. Oncogene 2000; 19:2165–2170.
39. Hahn WC, Stewart SA, Brooks MW, et al. Inhibition of telomerase limits the growth of human cancer cells. Nat Med 1999; 5:1164–1170.
40. Stewart SA, Hahn WC, O'Connor BF, et al. Telomerase contributes to tumorigenesis by a telomere length-independent mechanism. Proc Natl Acad Sci U S A 2002; 99:12606–12611.
41. Goodwin EC, Yang E, Lee CJ, et al. Rapid induction of senescence in human cervical carcinoma cells. Proc Natl Acad Sci U S A 2000; 97:10978–10983.
42. Wells SI, Francis DA, Karpova AY, et al. Papillomavirus E2 induces senescence in HPV-positive cells via pRB- and p21(CIP)-dependent pathways. EMBO J 2000; 19:5762–5771.
43. Goodwin EC, DiMaio D. Induced senescence in HeLa cervical carcinoma cells containing elevated telomerase activity and extended telomeres. Cell Growth Differ 2001; 12:525–534.

44. Chang BD, Broude EV, Dokmanovic M, et al. A senescence-like phenotype distinguishes tumor cells that undergo terminal proliferation arrest after exposure to anticancer agents. Cancer Res 1999; 59:3761–3767.
45. Klein LE, Freeze BS, Smith AB III, et al. The microtubule stabilizing agent discodermolide is a potent inducer of accelerated cell senescence. Cell Cycle 2005; 4:501–507.
46. Schmitt CA. Cellular senescence and cancer treatment. Biochim Biophys Acta 2007; 1775:5–20.
47. Elmore LW, Rehder CW, Di X, et al. Adriamycin-induced senescence in breast tumor cells involves functional p53 and telomere dysfunction. J Biol Chem 2002; 277:35509–35515.
48. Chang BD, Swift ME, Shen M, et al. Molecular determinants of terminal growth arrest induced in tumor cells by a chemotherapeutic drug. Proc Natl Acad Sci U S A 2002; 99:389–394.
49. Ruth AC, Roninson IB. Effects of the multidrug transporter P-glycoprotein on cellular responses to ionizing radiation. Cancer Res 2000; 60:2576–2578.
50. Katakura Y, Nakata E, Miura T, et al. Transforming growth factor beta triggers two independent-senescence programs in cancer cells. Biochem Biophys Res Commun 1999; 255:110–115.
51. Terao Y, Nishida J, Horiuchi S, et al. Sodium butyrate induces growth arrest and senescence-like phenotypes in gynecologic cancer cells. Int J Cancer 2001; 94:257–267.
52. Roninson IB, Dokmanovic M. Induction of senescence-associated growth inhibitors in the tumor-suppressive function of retinoids. J Cell Biochem 2002; 8:1–12.
53. Dokmanovic M, Chang BD, Fang J, et al. Retinoid-induced growth arrest of breast carcinoma cells involves co-induction of multiple growth-inhibitory genes. Cancer Biol Ther 2002; 1:24–27.
54. Chen Y, Dokmanovic M, Stein WD, et al. Agonist and antagonist of retinoic acid receptors cause similar changes in gene expression and induce senescence-like growth arrest in MCF-7 breast carcinoma cells. Cancer Res 2006; 66:8749–8761.
55. Wainwright LJ, Lasorella A, Iavarone A. Distinct mechanisms of cell cycle arrest control the decision between differentiation and senescence in human neuroblastoma cells. Proc Natl Acad Sci U S A 2001; 98:9396–9400.
56. Zhu WY, Jones CS, Kiss A, et al. Retinoic acid inhibition of cell cycle progression in MCF-7 human breast cancer cells. Exp Cell Res 1997; 234:293–299.
57. Roninson IB, Broude EV, Chang BD. If not apoptosis, then what? Treatment-induced senescence and mitotic catastrophe in tumor cells. Drug Resist Updat 2001; 4:303–313.
58. Schmitt CA, Fridman JS, Yang M, et al. A senescence program controlled by p53 and p16INK4a contributes to the outcome of cancer therapy. Cell 2002; 109: 335–346.
59. Christov KT, Shilkaitis AL, Kim ES, et al. Chemopreventive agents induce a senescence-like phenotype in rat mammary tumours. Eur J Cancer 2003; 39: 230–239.
60. Podtcheko A, Namba H, Saenko V, et al. Radiation-induced senescence-like terminal growth arrest in thyroid cells. Thyroid 2005; 15:306–313.
61. te Poele RH, Okorokov AL, Jardine L, et al. DNA damage is able to induce senescence in tumor cells in vitro and in vivo. Cancer Res 2002; 62:1876–1883.

62. Roberson RS, Kussick SJ, Vallieres E, et al. Escape from therapy-induced accelerated cellular senescence in p53-null lung cancer cells and in human lung cancers. Cancer Res 2005; 65:2795–2803.
63. Zhu H, Chang BD, Uchiumi T, et al. Identification of promoter elements responsible for transcriptional inhibition of Polo-like kinase 1 and Topoisomerase II genes by $p21^{WAF1/CIP1/SDI1}$. Cell Cycle 2002; 1:50–58.
64. Taylor WR, Schonthal AH, Galante J, et al. p130/E2F4 binds to and represses the cdc2 promoter in response to p53. J Biol Chem 2001; 276:1998–2006.
65. Schwarze SR, Fu VX, Desotelle JA, et al. The identification of senescence-specific genes during the induction of senescence in prostate cancer cells. Neoplasia 2005; 7:816–823.
66. Komarova EA, Diatchenko L, Rokhlin OW, et al. Stress-induced secretion of growth inhibitors: a novel tumor suppressor function of p53. Oncogene 1998; 17: 1089–1096.
67. Nickoloff BJ, Lingen MW, Chang BD, et al. Tumor suppressor maspin is upregulated during keratinocyte senescence, exerting a paracrine anti-angiogenic activity. Cancer Res 2004; 64:2956–2961.
68. Adler MJ, Coronel C, Shelton E, et al. Increased gene expression of Alzheimer disease beta-amyloid precursor protein in senescent cultured fibroblasts. Proc Natl Acad Sci U S A 1991; 88:16–20.
69. Coppe JP, Kauser K, Campisi J, et al. Secretion of vascular endothelial growth factor by primary human fibroblasts at senescence. J Biol Chem 2006; 281:29568–29574.
70. Parrinello S, Coppe JP, Krtolica A, et al. Stromal-epithelial interactions in aging and cancer: senescent fibroblasts alter epithelial cell differentiation. J Cell Sci 2005; 118:485–496.
71. Bavik C, Coleman I, Dean JP, et al. The gene expression program of prostate fibroblast senescence modulates neoplastic epithelial cell proliferation through paracrine mechanisms. Cancer Res 2006; 66:794–802.
72. Elenbaas B, Weinberg RA. Heterotypic signaling between epithelial tumor cells and fibroblasts in carcinoma formation. Exp Cell Res 2001; 264:169–184.
73. Paradis V, Youssef N, Dargere D, et al. Replicative senescence in normal liver, chronic hepatitis C, and hepatocellular carcinomas. Hum Pathol 2001; 32:327–332.
74. Choi J, Shendrik I, Peacocke M, et al. Expression of senescence-associated beta-galactosidase in enlarged prostates from men with benign prostatic hyperplasia. Urology 2000; 56:160–166.
75. Yang G, Rosen DG, Zhang Z, et al. The chemokine growth-regulated oncogene 1 (Gro-1) links RAS signaling to the senescence of stromal fibroblasts and ovarian tumorigenesis. Proc Natl Acad Sci U S A 2006; 103:16472–16477.
76. Michaloglou C, Vredeveld LC, Soengas MS, et al. BRAFE600-associated senescence-like cell cycle arrest of human naevi. Nature 2005; 436:720–724.
77. Chen Z, Trotman LC, Shaffer D, et al. Crucial role of p53-dependent cellular senescence in suppression of Pten-deficient tumorigenesis. Nature 2005; 436: 725–730.
78. Saitoh T, Sundsmo M, Roch JM, et al. Secreted form of amyloid beta protein precursor is involved in the growth regulation of fibroblasts. Cell 1989; 58:615–622.
79. E. Lagow, B. Davis, and I.B. Roninson (unpublished).
80. Kramer DL, Chang BD, Chen Y, et al. Polyamine depletion in human melanoma cells leads to G1 arrest associated with induction of $p21^{WAF1/CIP1/SDI1}$, changes in

the expression of p21-regulated genes, and a senescence-like phenotype. Cancer Res 2001; 61:7754–7762.

81. E. Lagow, B.D. Chang, and I.B. Roninson (unpublished).
82. Migliaccio E, Giorgio M, Mele S, et al. The p66shc adaptor protein controls oxidative stress response and life span in mammals (see comments). Nature 1999; 402:309–313.
83. Roninson IB. Oncogenic functions of tumour suppressor p21(Waf1/Cip1/Sdi1): association with cell senescence and tumour-promoting activities of stromal fibroblasts. Cancer Lett 2002; 179:1–14.
84. Poole JC, Thain A, Perkins ND, et al. Induction of transcription by p21Waf1/Cip1/Sdi1: role of NFkappaB and effect of non-steroidal anti-inflammatory drugs. Cell Cycle 2004; 3:931–940.
85. Perkins ND, Felzien LK, Betts JC, et al. Regulation of NF-kappaB by cyclin-dependent kinases associated with the p300 coactivator. Science 1997; 275:523–527.
86. Ravi RK, McMahon M, Yangang Z, et al. Raf-1-induced cell cycle arrest in LNCaP human prostate cancer cells. J Cell Biochem 1999; 72:458–469.
87. Cox JD, Kline RW. Do prostatic biopsies 12 months or more after external irradiation for adenocarcinoma, Stage III, predict long-term survival? Int J Radiat Oncol Biol Phys 1983; 9:299–303.
88. Bataini JP, Belloir C, Mazabraud A, et al. Desmoid tumors in adults: the role of radiotherapy in their management. Am J Surg 1988; 155:754–760.
89. Gu K, Mes-Masson AM, Gauthier J, et al. Analysis of the p16 tumor suppressor gene in early-stage prostate cancer. Mol Carcinog 1998; 21:164–170.
90. Lee CT, Capodieci P, Osman I, et al. Overexpression of the cyclin-dependent kinase inhibitor p16 is associated with tumor recurrence in human prostate cancer. Clin Cancer Res 1999; 5:977–983.
91. Halvorsen OJ, Hostmark J, Haukaas S, et al. Prognostic significance of p16 and CDK4 proteins in localized prostate carcinoma. Cancer 2000; 88:416–424.
92. Henshall SM, Quinn DI, Lee CS, et al. Overexpression of the cell cycle inhibitor p16INK4A in high-grade prostatic intraepithelial neoplasia predicts early relapse in prostate cancer patients. Clin Cancer Res 2001; 7:544–550.
93. Hui R, Macmillan RD, Kenny FS, et al. INK4a gene expression and methylation in primary breast cancer: overexpression of p16INK4a messenger RNA is a marker of poor prognosis. Clin Cancer Res 2000; 6:2777–2787.
94. Milde-Langosch K, Bamberger AM, Rieck G, et al. Overexpression of the p16 cell cycle inhibitor in breast cancer is associated with a more malignant phenotype. Breast Cancer Res Treat 2001; 67:61–70.
95. Han S, Ahn SH, Park K, et al. P16INK4a protein expression is associated with poor survival of the breast cancer patients after CMF chemotherapy. Breast Cancer Res Treat 2001; 70:205–212.
96. Osman I, Drobnjak M, Fazzari M, et al. Inactivation of the p53 pathway in prostate cancer: impact on tumor progression. Clin Cancer Res 1999; 5:2082–2088.
97. Aaltomaa S, Lipponen P, Eskelinen M, et al. Prognostic value and expression of p21 (waf1/cip1) protein in prostate cancer. Prostate 1999; 39:8–15.
98. Baretton GB, Klenk U, Diebold J, et al. Proliferation- and apoptosis-associated factors in advanced prostatic carcinomas before and after androgen deprivation therapy: prognostic significance of p21/WAF1/CIP1 expression. Br J Cancer 1999; 80:546–555.

99. Sarkar FH, Li Y, Sakr WA, et al. Relationship of p21(WAF1) expression with disease-free survival and biochemical recurrence in prostate adenocarcinomas (PCa). Prostate 1999; 40:256–260.
100. Lacombe L, Maillette A, Meyer F, et al. Expression of p21 predicts PSA failure in locally advanced prostate cancer treated by prostatectomy. Int J Cancer 2001; 95:135–139.
101. Fizazi K, Martinez LA, Sikes CR, et al. The association of p21((WAF-1/CIP1)) with progression to androgen-independent prostate cancer. Clin Cancer Res 2002; 8: 775–781.
102. Matsushima H, Sasaki T, Goto T, et al. Immunohistochemical study of p21WAF1 and p53 proteins in prostatic cancer and their prognostic significance. Hum Pathol 1998; 29:778–783.
103. Cheng L, Lloyd RV, Weaver AL, et al. The cell cycle inhibitors p21WAF1 and p27KIP1 are associated with survival in patients treated by salvage prostatectomy after radiation therapy. Clin Cancer Res 2000; 6:1896–1899.
104. Shariat SF, Lamb DJ, Kattan MW, et al. Association of preoperative plasma levels of insulin-like growth factor I and insulin-like growth factor binding proteins-2 and -3 with prostate cancer invasion, progression, and metastasis. J Clin Oncol 2002; 20:833–841.
105. Chan JM, Stampfer MJ, Ma J, et al. Insulin-like growth factor-I (IGF-I) and IGF binding protein-3 as predictors of advanced-stage prostate cancer. J Natl Cancer Inst 2002; 94:1099–1106.
106. Gupta S, Hastak K, Ahmad N, et al. Inhibition of prostate carcinogenesis in TRAMP mice by oral infusion of green tea polyphenols. Proc Natl Acad Sci U S A 2001; 98:10350–10355.
107. Suzuki T, Minagawa S, Michishita E, et al. Induction of senescence-associated genes by 5-bromodeoxyuridine in HeLa cells. Exp Gerontol 2001; 36:465–474.
108. Schwarze SR, DePrimo SE, Grabert LM, et al. Novel pathways associated with bypassing cellular senescence in human prostate epithelial cells. J Biol Chem 2002; 277:14877–14883.
109. Zi X, Zhang J, Agarwal R, et al. Silibinin up-regulates insulin-like growth factor-binding protein 3 expression and inhibits proliferation of androgen-independent prostate cancer cells. Cancer Res 2000; 60:5617–5620.
110. Tsubaki J, Hwa V, Twigg SM, et al. Differential activation of the IGF binding protein-3 promoter by butyrate in prostate cancer cells. Endocrinology 2002; 143:1778–1788.
111. Lopez-Bermejo A, Buckway CK, Devi GR, et al. Characterization of insulin-like growth factor-binding protein-related proteins (IGFBP-rPs) 1, 2, and 3 in human prostate epithelial cells: potential roles for IGFBP-rP1 and 2 in senescence of the prostatic epithelium. Endocrinology 2000; 141:4072–4080.
112. Sprenger CC, Vail ME, Evans K, et al. Over-expression of insulin-like growth factor binding protein-related protein-1(IGFBP-rP1/mac25) in the M12 prostate cancer cell line alters tumor growth by a delay in G1 and cyclin A associated apoptosis. Oncogene 2002; 21:140–147.
113. Zou Z, Gao C, Nagaich AK, et al. p53 regulates the expression of the tumor suppressor gene maspin. J Biol Chem 2000; 275:6051–6054.

114. Machtens S, Serth J, Bokemeyer C, et al. Expression of the p53 and Maspin protein in primary prostate cancer: correlation with clinical features. Int J Cancer 2001; 95:337–342.
115. Zou Z, Zhang W, Young D, et al. Maspin expression profile in human prostate cancer (CaP) and in vitro induction of maspin expression by Androgen Ablation. Clin Cancer Res 2002; 8:1172–1177.
116. B.D. Chang, D. Porter, M. Chen and I.B.R. (unpublished).

15

Senescence Regulation in Cancer Therapy

Abdelhadi Rebbaa

Children's Memorial Research Center, Department of Pediatrics, Feinberg School of Medicine, Northwestern University, Chicago, Illinois, U.S.A.

INTRODUCTION

Normal somatic cells invariably enter a state of permanent growth arrest and functional decline after a finite number of divisions (1). This phenomenon, termed cellular senescence, is believed to be a contributing factor in aging. As we age, the organisms' ability to withstand the erosive effects of internal and environmental stress diminishes, rendering us vulnerable to a variety of chronic illnesses including cancer, arthritis, diabetes, heart, and neurodegenerative diseases (2–5). Conventional wisdom holds that the body's decreased resistance to stress during aging may result from the decreased efficiency of normal maintenance and repair mechanisms. The accumulation of damage overtime may either signal for the stimulation of antistress defense, allowing somatic cells to survive longer but enhancing the risk of becoming cancerous, or to a cellular surrender leading to senescence. Accordingly, senescence induction may be considered as a barrier for cancer and thus should be exploited for the treatment of this disease. However, the respective benefits and risks of targeting senescence in cancer cells to treat tumors or in somatic cells to prevent cancer onset must be defined. For this, it is critical to define the relationship between senescence and cancer and characterize the molecular events eliciting these unique cellular stress responses.

The Senescence-Cancer Relationship

Although both cancer and cellular senescence have respective genetic causes (6,7), they share a common environmental initiator: the accumulation of cellular damage overtime. Our somatic cells (stem cells for instance) are continuously exposed to endogenous and exogenous stresses including radiation, oxidants, chemicals, hormones, and diet. The cumulative damage can cause cells to undergo either a functional decline leading to impaired proliferative ability and senescence (8–10) or a functional enhancement leading to the onset of cancer (11–13) (Fig. 1). The underlying mechanism(s) by which the same stress can cause such distinctly different cellular responses is not clear. However, since cancer is characterized by an increased ability of somatic cell to proliferate and a superior adaptation to stress (14), this disease may be viewed as an overreaction to stress. By the same logic, senescence may be considered a form of surrender to overwhelming stress (Fig. 1). In order for somatic cells to avoid either transformation to a malignant or a senescent state, a delicate balance in their antistress and proliferative ability must be maintained.

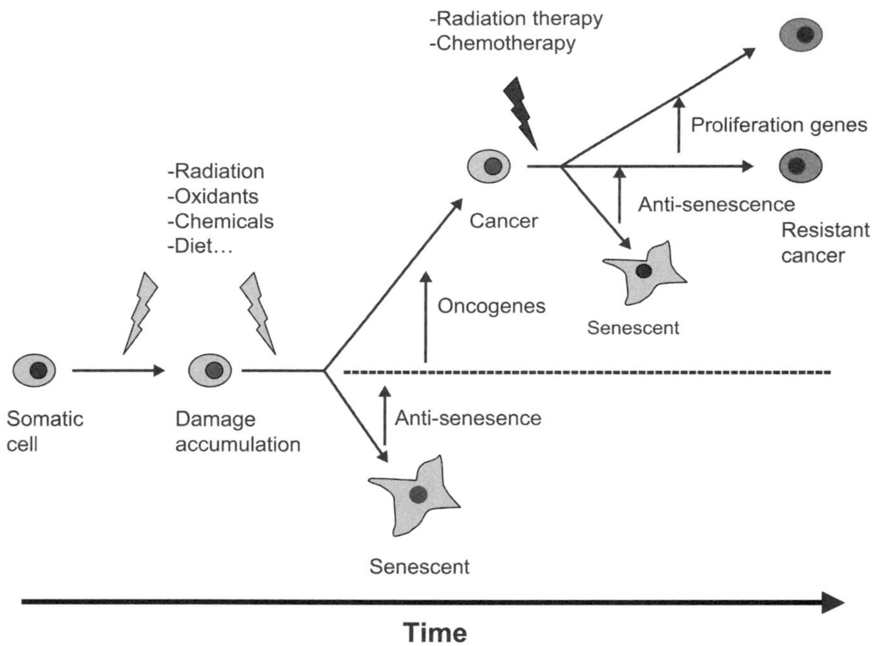

Figure 1 Stress-induced cellular damage as a cause of senescence and carcinogenesis. Continuous exposure of somatic cells (stem cells) to endogenous and exogenous stress can paradoxically lead to either induction of proliferation arrest (senescence) or increased cellular ability to proliferate (carcinogenesis). Treatment of cancer cells with chemotherapy or radiotherapy may force them into senescence, however, a subset of these cells adapt to the new toxic environment and become resistant (*The color version of this figure is provided on the DVD*).

From a molecular standpoint, the most common pathways activated in senescent cells are those mediated by tumor suppressors and cell cycle inhibitors (15,16), whereas these pathways are often inactivated (17,18) and oncogenes are overexpressed in cells destined to undergo malignant transformation. The fact that cancer cells are able to escape from environmental stress-induced senescence suggests that they must contain more effective antisenescence genes than somatic cells that stop proliferating under the same conditions. Although their response to environmental stress has diminished, it is possible to induce senescence in neoplasias particularly when they are exposed to chemotherapy or radiotherapy (15) (Fig. 1). However, not all tumor cells stop proliferating in response to these treatments and some may escapes the toxic effects of pharmacological agents giving rise to drug-resistant tumors (19). Accordingly, drug-resistant cells must express more potent antisenescence genes than their parental cells. Together, these considerations support of the notion that failure to enter senescence is not only associated with cancer onset but also with its progression.

Senescence in Somatic Cells: Role as an Inhibitor of Malignant Transformation

Evidence is accumulating that senescence act as a tumor-suppressive mechanism (20–23). A number of in vitro and in vivo studies have demonstrated that this cellular response may be considered as a safeguard program that limits the proliferative capacity of cells exposed to endogenous and exogenous stress signals (24). For instance, it has been shown in mice that senescent cells were abundant in premalignant lesions and sparse in tumors developed by these animals (25,26). Senescent cells were also found to localize mainly in benign melanocytic naevi, whereas melanomas cells displayed less staining for senescence markers (27), suggesting that this cellular response constitutes an initial barrier to neoplasia (28). On the other hand, since it has been demonstrated that senescent cells are active and able to synthesize and secrete growth factors or cytokines (29–33), it was suggested that some of these factors may act in a paracrine manner, influencing proliferation of neighboring cells and favoring cancer progression. The implications of these findings for therapy could be enormous, therefore, it is necessary to gain a clear understanding of the mechanisms by which cellular senescence contributes to cancer. For instance, it will be important to define the mechanism underlying the preferential secretion of growth-stimulatory versus growth-inhibitory factors by senescent cells. Also, assuming that growth-stimulatory factors are mostly secreted, it is critical to determine whether they act in an autocrine manner to reverse senescence in the same cells from which they were secreted, or if they act in a paracrine manner, whether they can turn noncancer cells into cancer. Considering all this, inducing senescence to suppress cancer could lead to a major dilemma in the sense that, even if this approach does not favor carinogenesis through secretion of growth factors, it will accelerate aging through the accumulation of senescent cells in the organism. Therefore, unless this paradox is solved, the clinical relevance of inducing senescence in somatic cells to prevent cancer will be severely diminished.

Senescence in Cancer Cells: Role in Preventing Development of Drug Resistance

Perhaps, the most insightful finding that helped shape our thinking about the interface between cancer and senescence was the discovery that tumor cells, when exposed to low doses of chemotherapeutic agents or radiation, display phenotypic and molecular markers similar to those found in senescent cells (15,34,35). These findings served as precursors for a new area of investigation aiming to define the relevance of senescence in cancer therapy. Pioneering studies in this area demonstrated that tumor susceptibility to senescence was associated with better prognosis (36). In addition, tissue specimens obtained from patients treated with chemotherapeutic agents expressed markers of irreversible proliferation arrest, confirming the role of senescence as an effecter program activated in response to chemotherapy (24,37). The underlying mechanisms often converged on two pathways mediated either by the tumor suppressor p53 or by the retinoblastoma gene pRB (28,30,38,39). Activation of the p53 pathway signal for enhanced expression of the cyclin kinase inhibitor p21/WAF1, which in turn participates in an orchestrated exit from the cell cycle (16). Interestingly, experimental manipulations to abolish the function of either p53 or p21/WAF1 caused a bypass of senescence and in some cases, even stimulated postmitotic senescent cells to resume growth (40,41). However, since even cells deficient in p53 or p21/WAF1 can be induced into senescence (16), alternative pathways and/or downstream targets of p21/WAF1 may be involved as well. On this point, it has been shown that inactivation of the p16/pRb pathway was required for senescence bypass in human prostate epithelial cells (42). In addition, recent findings established p16 as a bona fide senescence marker in vivo (43–45) and demonstrated that overexpression of the corresponding gene in mice have a tumor suppressing activity (46,47). Thus, the extent to which this pathway is activated must be also taken into account when evaluating the role of senescence in therapeutic response.

Considering that the stress level required for senescence induction is significantly lower that the one necessitated for apoptotic cell death (15,48), intervention to induce senescence in cancer cells may represent a nonaggressive approach to prevent tumor progression. Important reservations to this approach must be considered, since in a population of growth-arrested cells the risk of senescence reversal and initiation of proliferation is omnipresent (33), therefore, the ability of cancer cells to escape drug toxicity must be kept under control in order to achieve a lasting growth arrest. To this end, a clear understanding of the mechanisms by which cancer cells escape drug action and the participation of senescence/antisenescence pathways in this process are needed. Cancer cells have a unique ability to develop resistance to virtually any type of drug, even the newly discovered targeted therapeutics, such as Gleevec and the histone deacetylase inhibitors that are currently undergoing clinical trials (49–52). The fact that development of drug resistance can be reproduced in vitro simply by continuous exposure of cancer cells to increasing amounts of drugs over periods ranging from three to six months (19,53), has facilitated the discovery of key mechanisms. These include (*i*) activation of antistress frontline defense, such as overexpression of drug efflux pumps P-glycoprotein and multidrug resistance protein (MRP) (54,55), drug inactivating enzymes cytochrome P450 and glutathione S-transferase (GST)-pi (56), and antioxidant molecules superoxide dismutase (SOD) and catalase (57,58), (*ii*) activation of repair and/or replacement mechanisms mediated by overexpression of the DNA repair enzymes O6-methylguanine-DNA methyltransferase (MGMT), DNA-dependent protein kinase (DNA-PK), ataxia telangiectasia mutated (ATM), radiation sensitivity gene (Rad) etc. (59), or enzymes implicated in the turnover of damaged proteins, such as the proteasome and cathepsins (60,61), and (*iii*) inhibition of proliferation arrest and cell death effectors (62). This latter possibility has been investigated recently and the respective roles of apoptosis and senescence in development of drug resistance in cancer are not clearly defined at present.

The inhibition of apoptosis was introduced about a decade ago as a novel mechanism potentially able to elicit development of drug resistance in cancer (63). This was based on the notion that cancer cells that can escape drug-induced cell death will ultimately form drug-resistant tumors. Since cancer cells must not only be alive, but also able to proliferate in order for drug-resistant tumors to develop, inhibition of apoptosis alone may not be sufficient for the development of drug resistance and an intact cellular ability to proliferate may be required (48,61,64,65). Moreover, since the stress level required for induction of senescence is lower than that needed to evoke apoptosis (48), inhibition of this cell death alone may increase survival but will not guarantee cellular proliferation, a prerequisite for tumor relapse. Thus, not only apoptosis, but also senescence must be inhibited in order for drug-resistant tumors to form. Recent studies from our laboratory indicated that in the presence of subtoxic stress levels, forcing cancer cells to undergo senescence without affecting the apoptotic pathway, not only prevented, but also reversed drug resistance in cancer cells (61). Inhibition of cathepsin L, a lysosomal enzyme implicated in the turnover of damaged proteins appeared to play a key role in forcing cancer cells to undergo senescence and suppression of drug resistance. Similar findings were reported using proteasome inhibitors (23,66) suggesting that cancer cells must relay on the proper functioning of these pathways to eliminate damaged proteins and survive in a toxic environment.

An additional argument supporting the role of senescence in cancer resistance is provided by the recent finding that Sirt1, a histone deacetylase known for its ability to inhibit the p53-mediated senescence pathway (67), is overexpressed in most drug-resistant cancer cell lines and tumor specimens from patients treated with chemotherapy (68). More importantly, this gene was found to regulate expression of the multidrug resistance–gene mdr1, suggesting that it plays an active role in this phenomenon. In line with this, recent studies using chemical inhibitors of Sirt1, demonstrated that they were able to enhance cellular response to stress (69). Sirtl is in fact better known for its ability to link caloric restriction to enhanced life span and for its contribution to increased life span in organisms varying from yeast to rodents (70–72), suggesting a role of this gene in cellular resistance to stress.

Overall, these findings demonstrate that senescence may play a role in both the onset of cancer and its treatment. The possibility that alterations in senescence may affect cancer cell ability to become resistant to chemotherapy represents a strong rational for targeting this cellular response to treat cancer. Similarly, interventions to regulate cellular senescence in somatic cells may serve to suppress cancer. In light of this, a logical step will be to determine which is best to induce senescence or to inhibit it. Should senescence be targeted in cancer or in somatic cells?

Targeting Senescence to Treat or to Prevent Cancer

Induction of Senescence for the Treatment of Existing Cancers

Although the idea of forcing cancer cells into senescence to treat cancer appears to be compelling, molecules that must be targeted to accomplish this goal have yet to be precisely defined. This will require a clearer understanding of the signaling pathways initiated by different stressors and elucidation of their impact on the induction of senescence. For the purpose of simplicity, we will address only the key steps of this pathway (Fig. 2), namely those regulating (*i*) susceptibility to stress-induced damage, (*ii*) ability to repair the damage (DNA) or eliminate damaged molecules (proteins), and (*iii*) ability to execute proliferation arrest in response to damage.

Most anticancer agents induce damage to DNA, proteins or both. Cellular preparedness to prevent these alterations is often manifested by expression of a plethora of front line antistress enzymes, such as drug transporters, to reduce the intracellular amounts of damaging compounds, and drug-inactivating enzymes, and antioxidation molecules (50) to prevent reactive oxygen species (ROS) from causing damage (Fig. 2). Once damage has already occurred, DNA repair pathways (DNA-PK, ATM, ATR, etc.) and those responsible for elimination of damaged protein (proteasome and cathepsins) become implicated (Fig.2). As mentioned above, regardless of the nature of stress targets, the type of damage, or the associated repair mechanisms, common downstream senescence pathways mediated by the tumor suppressor p53 or the retinoblastoma gene pRB are invariably activated (Fig. 2). These molecules and their respective signaling partners are thus considered effectors of proliferation arrest. Based on this, three major interventions could be considered in order to force cancer cells into a senescence state (Fig. 2) (*i*) to enhance cancer cell susceptibility to stress-induced damage through inhibition of the front line antidamage enzymes, (*ii*) to

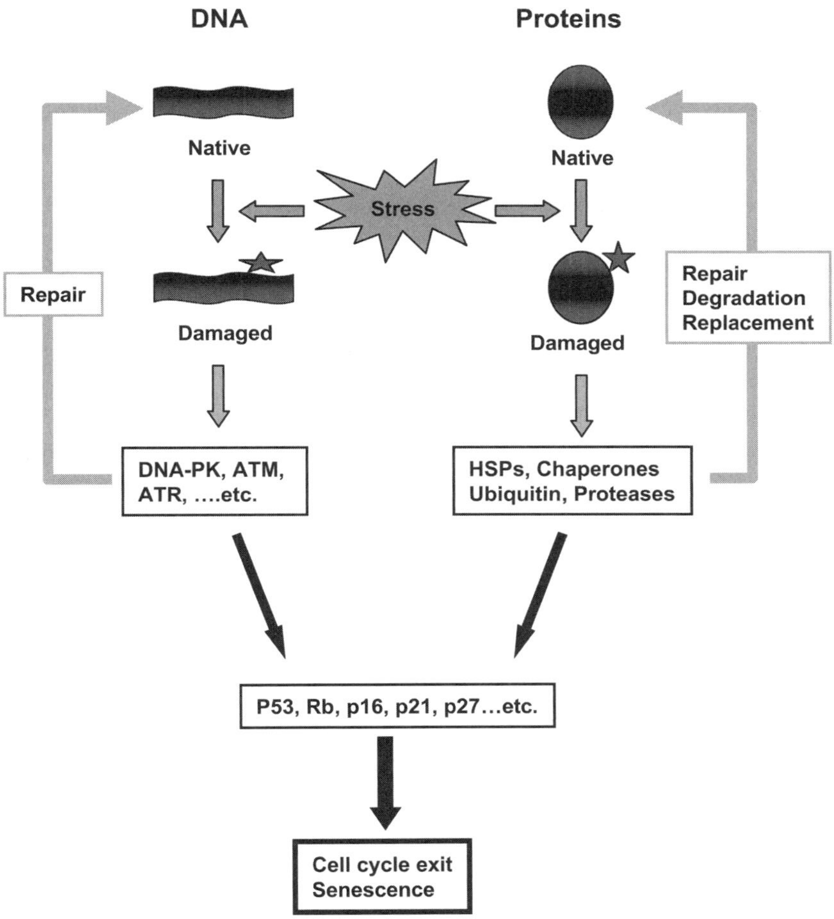

Figure 2 Stress-induced cellular senescence. Key steps in this pathway are damage initiation, repair, and execution (*The color version of this figure is provided on the DVD*).

inhibit the repair of DNA damage or the turnover of damaged proteins, and (*iii*) to enhance the availability and/or the function of proliferation arrest effectors.

Inhibition of antidamage systems. Targeting molecules that regulate drug availability and enhance the effectiveness of known chemotherapeutic agents have been used as an alternative, cost effective approach to developing new drugs. The families of molecules found to affect cancer cell sensitivity to drugs are in fact members of the antistress front line described above. Some of these enzymes (in particular those with antioxidative function) have been shown to decrease with aging (73,74) in support of a role in maintaining proper cellular function. Although the implication of these molecules in cellular senescence has not been investigated in detail, inhibiting their function in cancer cells often prevented and/or reversed drug resistance. For instance, the drug efflux inhibitors cyclosporin A, verapamil, and related analogs (some of which are currently in clinical trials) were able to reverse drug-resistance in vitro, however, their effect in vivo was not as clear due to their elevated toxicity (75). By decreasing drug efflux from the cell, these compounds may accelerate intracellular accumulation of anticancer agents and achievement of senescence-inducing stress levels. From this standpoint, drug efflux inhibitors may be considered activators of drug-induced senescence. The same principle may also apply to inhibitors of drug-metabolizing enzymes such as the cytochrome P450 family. However, this latter approach is not considered for cancer therapy because these enzymes are ubiquitously expressed in cancer as well as in normal tissues, therefore their inhibition may cause serious side effects.

Regardless of the intracellular stress level, the presence or absence of antioxidative enzymes is an important determinant in preventing the damage from occurring. In fact, numerous investigations have exploited these molecules to enhance cancer cell susceptibility to chemotherapy (76) and specific inhibitors were found to be very effective in synergizing with classical cytotoxic drugs in suppressing cancer progression. The best described among these is buthionine sulfoximine (BSO), a GST inhibitor well known for its ability to bypass drug resistance in vitro and in vivo (77–79). Although this often resulted in inhibition of cellular proliferation, the role of these inhibitors either alone or in combination with other drugs in senescence induction has not been investigated in detail.

Inhibition of damage repair pathways

Preventing repair of damaged DNA A number of anticancer agents, including doxorubicin, etoposide, cisplatin, BCNU [1,3 bis (2-chloroethyl)-1-nitrosourea], and radiation have been shown to alter the primary structure of DNA. The damage may include base substitution, cross-linking, single and/or double-strand breaks (80,81), and others. Key molecular sensors and mediators of DNA damage repair have been identified recently (82). These molecules activate checkpoints that signal for proliferation arrest to allow repair of any DNA damage before cell cycle progression is resumed, or, if the damage is too extensive, they trigger senescence and consequently cell death (Fig. 2). Among these early sensors, the DNA-PK uses damaged DNA as a cosubstrate to phosphorylate ATM and ATR (82), which in turn propagate the signal through phosphorylation and activation of additional intermediates including H2AX, NBS1, and BRCA1. Subsequently, one branch of this signaling pathway bifurcates toward proliferation arrest via p53/p21 WAF1 or p19 ING/pRB (Fig. 2). The other signaling branch leads to recruitment of additional DNA repair enzymes and restoration of native DNA structure (Fig. 2). Obviously, a fine balance between these two pathways must be maintained. If one pathway is overactivated relatively to the other, cancer cells may either enter senescence or acquire exceptional repair ability that allows them to become drug resistant regardless of their proliferation status. Accordingly, targeting this repair pathway to induce cellular senescence may hold great promise for cancer therapy, yet, not many studies have addressed the effect of inhibiting DNA damage repair pathways on cellular proliferation. The only examples are represented by studies using nonspecific chemical inhibitors known to inhibit PI3K but also found recently to affect ATM, ATR, and related kinases. Such inhibitors (i.e., wortmannin and caffeine) have been shown to be capable of enhancing cancer cell sensitivity to drugs (83,84) via induction of a senescence phenotype (85). Further studies in this area may be required to elucidate the importance of inhibiting the DNA damage repair through interference with the action of sensors and/or mediators, to accelerate senescence induction and enhance the action of classical anticancer agents.

Inhibition of damaged-protein degradation Although most cytotoxic agents do not directly target proteins, the majority induce ROS that cause damage to DNA and/or proteins (oxidation, crosslinking, aggregation, etc.). A conserved cellular response to protein damage includes expression of sensors (heat shock proteins, chaperones, and ubiquitination enzymes) as well as enzymes responsible for digesting damaged proteins (proteasome and cathepsins) (Fig. 2). It is well documented that impaired degradation and/or clearance of nonfunctional proteins leads to the accumulation of molecular aggregates that may be toxic to the cell (86–89). In fact, in most tissues of aged organisms and in most aging model systems including nematodes, fruit flies, and cultured fibroblasts, overall proteolysis decline with age (90). Since cellular homeostasis is dependent on the balance between biosynthesis and catabolism of macromolecules, perturbations of this balance are very likely to influence proliferation and survival. Two major degradation routes exist in eukaryotic cells, the ubiquitin-proteasome system and the lysosomes. Although lysosomes were discovered first, their role in cancer progression has long been underestimated. On the other hand, the entry of the ubiquitin-proteasome system into the scene has expanded the role of protein degradation from mere housekeeping to regulation of major intracellular processes, such as cell cycle and cell death (91–93).

The proteasome degrades short-lived proteins through a mechanism that implies ubiquitination of damaged substrates (94). In vitro studies demonstrated that proteosomal activity was severely impaired in senescent fibroblasts and that its inhibition in young cells induced a premature senescent-like phenotype represented by increase expression of the senescence-associated β-galactosidase (SA beta-gal) and the cell cycle inhibitor p21/WAF1 (95). The finding that proteasome inhibitors are able to suppress proliferation in a variety of cancer cell lines (96–98) suggests that these compounds may hold promise for cancer treatment. For example, PS-341 (pyrazylcarbonyl-Phe-Leu-boronate, Velcade™), directed against the 26S proteasomal subunit (99) was shown to overcome classical drug resistance in preclinical and in early clinical studies (100,101), consequently, this drug was granted the food and drug administration (FDA) fast track status for treatment of multiple myeloma in 2003. PS-341 is believed to act by inhibiting degradation of IκBα, an inhibitory protein constitutively bound to cytosolic NF-κB, thereby inhibiting its nuclear translocation, binding to DNA, and expression of target genes implicated in cell survival (102). A more general mechanism to explain the effect of proteasome on cancer proliferation has been found to be mediated through the stabilization of p53 and its downstream target p21/WAF1 (103,104), linking it directly to senescence induction. Additional in vitro and in vivo studies demonstrated that the proteasome was indeed responsible for the degradation of p21/WAF1 (105,106), further supporting the antisenescence function of this complex.

Lysosomes are the cellular reservoirs of proteases (cathepsins) responsible for degradation of long-lived proteins. Many changes can be observed in these organelles as the cells become senescent (93). These include increased lysosomal volume, decrease stability, and alterations in cathepsin activity leading to intralysosomal accumulation of nondigested material (107). Among the lysosomal proteases, cathepsin L has received particular interest in the recent few years since it was found to play key roles in carcinogenesis and cell survival (108–111). Interestingly, expression of this enzyme was also reported to decrease with age (107) suggesting a specific role in regulating cellular senescence. The implication of cathepsin L in malignant transformation was elegantly demonstrated in a mice model of multistage carcinogenesis where deficiency in its function was responsible for tumor suppression (112). Previous work from our laboratory demonstrated that chemical inhibition of cathepsin L forced cancer cells into a senescence state and reversed their resistance to chemotherapy (61). The underlying mechanism was found to be associated with increased stabilization of drug targets and increased expression of the cell cycle inhibitor p21/WAF1.

Lysosomes are also capable of eliminating damaged mitochondria by a mechanism known as autophagy (93,113). Since damaged mitochondria are source of leakage of ROS (93,114), a reduction in lysosomal proteolytic function will likely facilitate their accumulation of ROS and consequently induce cellular senescence (115,116). Interventions that facilitate ROS generation, such as cellular exposure to BSO or UV irradiation have been shown to synergize with anticancer drugs. These findings, together with those concerning proteasome targeting to induce senescence indicate that the molecular pathways responsible for eliminating unwanted (damaged) proteins can be harnessed to force cancer cells into senescence and enhance their response to chemotherapeutic agents.

Stimulation of proliferation arrest effectors. As mentioned above, regardless of the source of stress and the nature of its primary targets, induction of tumor suppressors and cell cycle inhibitors is thought to represent the ultimate event that dictates cellular fate. Therefore, in theory at least, activation of these effectors must be sufficient for the induction of senescence regardless of the status of DNA or protein susceptibility to damage or the potency of repair systems. Surprisingly, the experimental data tend to disprove this concept and suggest that overexpression of tumor suppressors and cell cycle inhibitors may not completely dictate the outcome of cellular response to drugs. For instance, although increased stabilization of p53 and enhanced expression of downstream effectors, such as p21/WAF1 and p16/ING, represent fundamental events in the signaling cascade leading to senescence, the role of these intermediates as sensitizers of cancer cell response to cytotoxic drugs is conflicting. For example, in most cases, overexpression of p53 did not enhance cellular sensitivity to antiproliferative drugs (117–119). Furthermore, overexpression of the downstream intermediates p21/WAF1, p16 ING, and p27CIP was found to synergize with cytotoxic drugs in some cases (16,61,120–123) and not in others (124–126). These findings point to the possibility that cancer cells may be capable of activating additional (perhaps unknown) antisenescence pathways to counteract the action of these genes and nullify their effect. Enhanced cellular ability to repair DNA and/or proteins damage may also account for the lack of synergy between these antiproliferative genes and anticancer agents. Once the damage has occurred and particularly if it is not sufficiently repaired, the cells are set to undergo senescence in a manner that may or may not be dependent on p53, pRB, or their direct downstream intermediates. Therefore, among the approaches described above to induce senescence in cancer cells, inhibiting the activity of antidamage enzymes may represent the most effective way to enhance cancer cell response to chemotherapy.

Targeting stress pathways in somatic cells: The potential role of antidamage systems in preventing both senescence and cancer. The possibility that senescence is a tumor suppressor mechanism creates a strong rationale for developing strategies to maintain controlled somatic cell proliferation and prevent carcinogenesis (Fig. 1). Interventions to induce senescence in somatic cells would be similar to those described above to treat cancer cells, and as mentioned earlier, although this will undeniably reduce the onset of cancer, it creates a risk of a general decline in normal cellular functions and acceleration of aging (Fig. 1). To avoid this dilemma, action must be taken to suppress early events leading to senescence and cancer. Since both may be caused by damage accumulation, particular attention should be directed toward developing strategies to prevent or to quickly repair the damage inflicted to somatic cells before it signals for the onset of either senescence or cancer. This can be achieved through one or both of the following interventions: (*i*) a reduction in the amount of stress to which somatic cells are exposed and/or (*ii*) an increase in cellular antidamage defense.

Considering the variety of endogenous and exogenous stresses to which the organism is exposed, very little can be done to reduce their sources or levels. Nevertheless, although diet represents only a minimal fraction of these stresses, reducing caloric intake has demonstrated significant capabilities in preventing cancer (13) and even in extending life span of many organisms ranging from yeast to rodents (127,128).

With regards to the second possibility, it is well recognized that as the organism ages, its antistress defense diminish (129,130) (Fig. 3A). Accordingly, the possibility exists that at a certain time point in late life, the stress/antistress balance shifts in favor of stress (Fig. 3A) and somatic cells will become overwhelmed. From this point on, cellular damage increases exponentially causing the cell either to overreact and become cancerous, or to surrender and become senescent (Fig. 3B). Activation or maintenance of antidamage defense within somatic cells should compensate for the time-dependent decline in antistress defense (Fig. 3A). Therefore, the time point at which stress overwhelms cellular defense, will be significantly postponed and consequently, the onset of both senescence and cancer may be delayed (Fig. 3C). This suggests that strategies to maintain functional antidamage systems in somatic cells or to enhance their activity (i.e., by introducing potent antistress genes) may represent far better alternatives than those relaying on induction of senescence in cancer or in somatic cells. The outstanding question, however, is where these antistress/antidamage genes would come from?

When we consider that the only cells in the organism strongly insensitive to stress are therapy-resistant cancer cells, the idea of their utilization as a source of antidamage genes becomes obvious. With the genetic advances and technology currently available, several options may be considered to identify such genes using therapy-resistant cancer cells as potential source. For example, it is possible to compare gene expression profiles between drug resistant and sensitive cells and validate the function of differentially expressed genes in protecting somatic cells from stress-induced damage. However, although many of the genes identified by this method appear to be associated with resistance to stress, they may not have a causative relationship with it. Thus, it is challenging to identify authentic antistress genes among the hundreds that are differentially expressed in drug sensitive and resistant cancer cells (131). Alternatively, a functional screen for such genes can be carried out using a cDNA library generated from drug or radiation-resistant cancer cells. With this method, the phenotype would be screened for initially, therefore, the risk of false positive results may be reduced. Overall, regardless of the strategy employed, due to the relationship between cellular susceptibility to stress-induced damage and the onset of various chronic diseases including cancer, it is necessary to develop new technologies that will allow identification of potent antistress genes so that appropriate treatment strategies can be designed to delay the onset of senescence-associated illnesses and ameliorate the life style at older age.

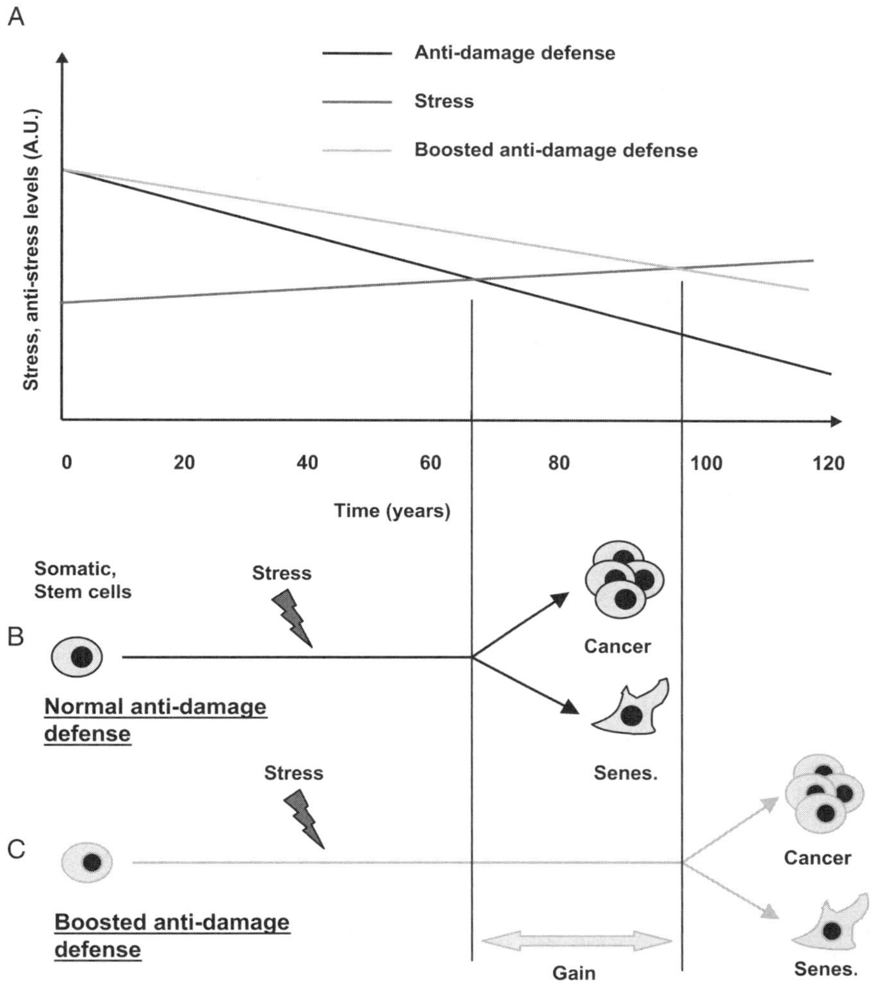

Figure 3 Protection against stress-induced cellular damage and the likelihood of delaying the onset of both cancer and senescence. Somatic cell's antistress defense declines with time (**A**). Considering that environmental stress is more likely to increase than to decrease with time, it ultimately overwhelms the antidamage defense, causing accelerated damage and the onset of cancer or senescence (**B**). If cellular antidamage system was boosted or at least maintained (**A**), the time at which it will be overwhelmed by stress will be delayed, and so do the onset of senescence or cancer (**C**). The time (in years) is arbitrary and only for the purpose of comparison (*The color version of this figure is provided on the DVD*).

CONCLUSION

The notion that cellular senescence may represent a key determinant in cancer therapy has opened new avenues for development of alternative strategies to suppress tumor progression. This research area is still however in its infancy and requires further understanding, not only of the causes and consequences of cellular senescence, but also of its relationship to carcinogenesis. In particular, the paradoxical consequences of inducing senescence to suppress cancer on the stimulation of neighboring cell proliferation or the acceleration of aging needs to be solved. A possible solution to this paradox may require suppression of cellular damages responsible for inducing senescence or cancer. From a practical standpoint, this may necessitate either reducing the source of stress or boosting the front line antidamage systems in somatic cells. Since little can be done to reduce the many sources of stress to which we are continuously exposed, identification of potent antidamage genes capable of protecting somatic cells against stress may represent a compelling approach to prevent the onset of both senescence and cancer. A proposed source for such genes would be the therapy-resistant cancer cells, often characterized by their activated antistress systems.

ACKNOWLEDGMENTS

Work in my laboratory is supported in part by the National Cancer Institute, the Anderson Foundation and the North Suburban Medical Research Junior Board. I wish to thank Dr. Bernard L. Mirkin for his continuous support and help in editing this manuscript. I also regret that the original work of several colleagues could not be cited due to space constraints.

SUMMARY

Senescence is a state of functional decline caused by continuous exposure to endogenous and exogenous stressors culminating in permanent arrest of proliferation. Carcinogenesis reflects an increase in cellular functions that may also be caused by stress, but in contrast to senescence, is characterized by enhanced proliferation and adaptation to toxic conditions. The concept that the same stress can cause somatic cells to become either senescent or cancerous suggests that these processes may be exclusive of each other. In theory at least, inducing senescence in cancer cells would be a compelling therapeutic approach, and if enforced in somatic cells, should prevent the onset of cancer. However, since it has been shown that senescent cells are metabolically active and secrete growth factors that simulate tumor proliferation, the beneficial effects of inducing senescence may be seriously compromised. Complicating this even further, is the fact that senescence induction could result in accumulation of old cells in the organism, thereby, accelerating the aging process. On the other hand, senescence inhibition in somatic cells or in tumor cells may increase the risk of cancer and development of drug resistance. A solution to this paradox must take into account the earliest and most common event in the stress response pathway leading to both senescence and cancer, namely, the accumulation of damage in somatic cells. It is anticipated that approaches to inhibit such damage would be very effective in preventing both senescence and cancer.

REFERENCES

1. Hayflick L, Moorhead PS. The serial cultivation of human diploid cell strains. Exp Cell Res 1961; 25:585–621.
2. Hamet P, Tremblay J. Genes of aging. Metabolism 2003; 52:5–9.
3. Turker MS. Somatic cell mutations: can they provide a link between aging and cancer? Mech Ageing Dev 2000; 117:1–19.
4. Migliore L, Coppede F. Genetic and environmental factors in cancer and neurodegenerative diseases. Mutat Res 2002; 512:135–153.
5. Knight JA. The biochemistry of aging. Adv Clin Chem 2000; 35:1–62.
6. Pardal R, Molofsky AV, He S, et al. Stem cell self-renewal and cancer cell proliferation are regulated by common networks that balance the activation of proto-oncogenes and tumor suppressors. Cold Spring Harb Symp Quant Biol 2005; 70:177–185.
7. Navarro CL, Cau P, Levy N. Molecular bases of progeroid syndromes. Hum Mol Genet 2006; 15(Spec No 2):R151–R161.
8. Sedelnikova OA, Horikawa I, Zimonjic DB, et al. Senescing human cells and ageing mice accumulate DNA lesions with unrepairable double-strand breaks. Nat Cell Biol 2004; 6:168–170.
9. Sohal RS, Weindruch R. Oxidative stress, caloric restriction, and aging. Science 1996; 273:59–63.
10. Dimri GP. What has senescence got to do with cancer? Cancer Cell 2005; 7: 505–512.
11. Ames BN. Mutagenesis and carcinogenesis: endogenous and exogenous factors. Environ Mol Mutagen 1989; 14(suppl 16):66–77.
12. Newell GR, Spitz MR, Sider JG. Cancer and age. Semin Oncol 1989; 16:3–9.
13. Hursting SD, Lavigne JA, Berrigan D, et al. Calorie restriction, aging, and cancer prevention: mechanisms of action and applicability to humans. Annu Rev Med 2003; 54:131–152.
14. Janssen YM, Van Houten B, Borm PJ, et al. Cell and tissue responses to oxidative damage. Lab Invest 1993; 69:261–274.
15. Chang BD, Swift ME, Shen M, et al. Molecular determinants of terminal growth arrest induced in tumor cells by a chemotherapeutic agent. Proc Natl Acad Sci U S A 2002; 99:389–394.
16. Chang BD, Xuan Y, Broude EV, et al. Role of p53 and p21waf1/cip1 in senescence-like terminal proliferation arrest induced in human tumor cells by chemotherapeutic drugs. Oncogene 1999; 18:4808–4818.
17. Laiho M, Latonen L. Cell cycle control, DNA damage checkpoints and cancer. Ann Med 2003; 35:391–397.
18. Levine AJ. The p53 tumor suppressor gene and gene product. Princess Takamatsu Symp 1989; 20:221–230.

19. Biedler JL. Drug resistance: genotype versus phenotype–thirty-second G. H. A. Clowes Memorial Award Lecture. Cancer Res 1994; 54:666–678.
20. Campisi J. Cellular senescence as a tumor-suppressor mechanism. Trends Cell Biol 2001; 11:S27–S31.
21. Campisi J. Senescent cells, tumor suppression, and organismal aging: good citizens, bad neighbors. Cell 2005; 120:513–522.
22. Ishikawa F. Cellular senescence, an unpopular yet trustworthy tumor suppressor mechanism. Cancer Sci 2003; 94:944–947.
23. Rebbaa A. Targeting senescence pathways to reverse drug resistance in cancer. Cancer Lett 2005; 219:1–13.
24. Schmitt CA. Cellular senescence and cancer treatment. Biochim Biophys Acta 2007; 1775:5–20.
25. Collado M, Serrano M. The senescent side of tumor suppression. Cell Cycle 2005; 4:1722–1724.
26. Collado M, Gil J, Efeyan A, et al. Tumour biology: senescence in premalignant tumours. Nature 2005; 436:642.
27. Gray-Schopfer VC, Cheong SC, Chong H, et al. Cellular senescence in naevi and immortalisation in melanoma: a role for p16? Br J Cancer 2006; 95:496–505.
28. Braig M, Lee S, Loddenkemper C, et al. Oncogene-induced senescence as an initial barrier in lymphoma development. Nature 2005; 436:660–665.
29. Roninson IB. Oncogenic functions of tumour suppressor p21(Waf1/Cip1/Sdi1): association with cell senescence and tumour-promoting activities of stromal fibroblasts. Cancer Lett 2002; 179:1–14.
30. Shay JW, Roninson IB. Hallmarks of senescence in carcinogenesis and cancer therapy. Oncogene 2004; 23:2919–2933.
31. Roninson IB. Tumor cell senescence in cancer treatment. Cancer Res 2003; 63:2705–2715.
32. Coppe JP, Kauser K, Campisi J, et al. Secretion of vascular endothelial growth factor by primary human fibroblasts at senescence. J Biol Chem 2006; 281: 29568–29574.
33. Kahlem P, Dorken B, Schmitt CA. Cellular senescence in cancer treatment: friend or foe? J Clin Invest 2004; 113:169–174.
34. Chang BD, Broude EV, Dokmanovic M, et al. A senescence-like phenotype distinguishes tumor cells that undergo terminal proliferation arrest after exposure to anticancer agents. Cancer Res 1999; 59:3761–3767.
35. Robles SJ, Buehler PW, Negrusz A, et al. Permanent cell cycle arrest in asynchronously proliferating normal human fibroblasts treated with doxorubicin or etoposide but not camptothecin. Biochem Pharmacol 1999; 58:675–685.
36. Schmitt CA, Fridman JS, Yang M, et al. A senescence program controlled by p53 and p16INK4a contributes to the outcome of cancer therapy. Cell 2002; 109: 335–346.
37. te Poele RH, Okorokov AL, Jardine L, et al. DNA damage is able to induce senescence in tumor cells in vitro and in vivo. Cancer Res 2002; 62:1876–1883.
38. Bringold F, Serrano M. Tumor suppressors and oncogenes in cellular senescence. Exp Gerontol 2000; 35:317–329.
39. Psyrri A, DeFilippis RA, Edwards AP, et al. Role of the retinoblastoma pathway in senescence triggered by repression of the human papillomavirus E7 protein in cervical carcinoma cells. Cancer Res 2004; 64:3079–3086.

40. Beausejour CM, Krtolica A, Galimi F, et al. Reversal of human cellular senescence: roles of the p53 and p16 pathways. EMBO J 2003; 22:4212–4222.
41. Brown JP, Wei W, Sedivy JM. Bypass of senescence after disruption of p21CIP1/WAF1 gene in normal diploid human fibroblasts. Science 1997; 277:831–834.
42. Jarrard DF, Sarkar S, Shi Y, et al. p16/pRb pathway alterations are required for bypassing senescence in human prostate epithelial cells. Cancer Res 1999; 59: 2957–2964.
43. Janzen V, Forkert R, Fleming HE, et al. Stem-cell ageing modified by the cyclin-dependent kinase inhibitor p16INK4a. Nature 2006; 443:421–426.
44. Krishnamurthy J, Ramsey MR, Ligon KL, et al. p16INK4a induces an age-dependent decline in islet regenerative potential. Nature 2006; 443:453–457.
45. Molofsky AV, Slutsky SG, Joseph NM, et al. Increasing p16INK4a expression decreases forebrain progenitors and neurogenesis during ageing. Nature 2006; 443:448–452.
46. Matheu A, Pantoja C, Efeyan A, et al. Increased gene dosage of Ink4a/Arf results in cancer resistance and normal aging. Genes Dev 2004; 18:2736–2746.
47. Kim WY, Sharpless NE. The regulation of INK4/ARF in cancer and aging. Cell 2006; 127:265–275.
48. Rebbaa A, Zheng X, Chou PM, et al. Caspase inhibition switches doxorubicin-induced apoptosis to senescence. Oncogene 2003; 22:2805–2811.
49. Shannon KM. Resistance in the land of molecular cancer therapeutics. Cancer Cell 2002; 2:99–102.
50. Moscow JA, Cowan KH. Multidrug resistance. J Natl Cancer Inst 1988; 80:14–20.
51. Roumiantsev S, Shah NP, Gorre ME, et al. Clinical resistance to the kinase inhibitor STI-571 in chronic myeloid leukemia by mutation of Tyr-253 in the Abl kinase domain P-loop. Proc Natl Acad Sci U S A 2002; 99:10700–10705.
52. von Bubnoff N, Manley PW, Mestan J, et al. Bcr-Abl resistance screening predicts a limited spectrum of point mutations to be associated with clinical resistance to the Abl kinase inhibitor nilotinib (AMN107). Blood 2006; 108(4):1328–1333.
53. Rebbaa A, Chou PM, Mirkin BL. Factors secreted by human neuroblastoma mediated doxorubicin resistance by activating STAT3 and inhibiting apoptosis. Mol Med 2001; 7:393–400.
54. Roninson IB, Chin JE, Choi KG, et al. Isolation of human mdr DNA sequences amplified in multidrug-resistant KB carcinoma cells. Proc Natl Acad Sci U S A 1986; 83:4538–4542.
55. Slovak ML, Ho JP, Bhardwaj G, et al. Localization of a novel multidrug resistance-associated gene in the HT1080/DR4 and H69AR human tumor cell lines. Cancer Res 1993; 53:3221–3225.
56. Cowan KH, Batist G, Tulpule A, et al. Similar biochemical changes associated with multidrug resistance in human breast cancer cells and carcinogen-induced resistance to xenobiotics in rats. Proc Natl Acad Sci U S A 1986; 83:9328–9332.
57. Doroshow JH, Akman S, Esworthy S, et al. Doxorubicin resistance conferred by selective enhancement of intracellular glutathione peroxidase or superoxide dismutase content in human MCF-7 breast cancer cells. Free Radic Res Commun 1991; 12–13(pt 2):779–781.
58. Akman SA, Forrest G, Chu FF, et al. Resistance to hydrogen peroxide associated with altered catalase mRNA stability in MCF7 breast cancer cells. Biochim Biophys Acta 1989; 1009:70–74.

59. Shiloh Y. ATM and ATR: networking cellular responses to DNA damage. Curr Opin Genet Dev 2001; 11:71–77.
60. Jungmann J, Reins HA, Schobert C, et al. Resistance to cadmium mediated by ubiquitin-dependent proteolysis. Nature 1993; 361:369–371.
61. Zheng X, Chou PM, Mirkin BL, et al. Senescence-initiated reversal of drug resistance: specific role of cathepsin L. Cancer Res 2004; 64:1773–1780.
62. Bunz F, Hwang PM, Torrance C, et al. Disruption of p53 in human cancer cells alters the responses to therapeutic agents. J Clin Invest 1999; 104:263–269.
63. Pommier Y, Sordet O, Antony S, et al. Apoptosis defects and chemotherapy resistance: molecular interaction maps and networks. Oncogene 2004; 23:2934–2949.
64. Roninson IB, Broude EV, Chang BD. If not apoptosis, then what? Treatment-induced senescence and mitotic catastrophe in tumor cells. Drug Resist Updat 2001; 4:303–313.
65. Borst P, Borst J, Smets LA. Does resistance to apoptosis affect clinical response to antitumor drugs? Drug Resist Updat 2001; 4:129–131.
66. Voorhees PM, Orlowski RZ. The proteasome and proteasome inhibitors in cancer therapy. Annu Rev Pharmacol Toxicol 2006; 46:189–213.
67. Vaziri H, Dessain SK, Ng Eaton E, et al. hSIR2(SIRT1) functions as an NAD-dependent p53 deacetylase. Cell 2001; 107:149–159.
68. Chu F, Chou PM, Zheng X, et al. Control of multidrug resistance gene mdr1 and cancer resistance to chemotherapy by the longevity gene sirt1. Cancer Res 2005; 65:10183–10187.
69. Heltweg B, Gatbonton T, Schuler AD, et al. Antitumor activity of a small-molecule inhibitor of human silent information regulator 2 enzymes. Cancer Res 2006; 66:4368–4377.
70. Kaeberlein M, McVey M, Guarente L. The SIR2/3/4 complex and SIR2 alone promote longevity in Saccharomyces cerevisiae by two different mechanisms. Genes Dev 1999; 13:2570–2580.
71. Hekimi S, Guarente L. Genetics and the specificity of the aging process. Science 2003; 299:1351–1354.
72. Howitz KT, Bitterman KJ, Cohen HY, et al. Small molecule activators of sirtuins extend Saccharomyces cerevisiae lifespan. Nature 2003; 425:191–196.
73. Landis GN, Tower J. Superoxide dismutase evolution and life span regulation. Mech Ageing Dev 2005; 126:365–379.
74. Blander G, de Oliveira RM, Conboy CM, et al. Superoxide dismutase 1 knock-down induces senescence in human fibroblasts. J Biol Chem 2003; 278: 38966–38969.
75. Thomas H, Coley HM. Overcoming multidrug resistance in cancer: an update on the clinical strategy of inhibiting p-glycoprotein. Cancer Control 2003; 10:159–165.
76. Lau DH, Lewis AD, Ehsan MN, et al. Multifactorial mechanisms associated with broad cross-resistance of ovarian carcinoma cells selected by cyanomorpholino doxorubicin. Cancer Res 1991; 51:5181–5187.
77. Ford JM, Yang JM, Hait WN. Effect of buthionine sulfoximine on toxicity of verapamil and doxorubicin to multidrug resistant cells and to mice. Cancer Res 1991; 51:67–72.
78. Vanhoefer U, Cao S, Minderman H, et al. d,l-buthionine-(S,R)-sulfoximine potentiates in vivo the therapeutic efficacy of doxorubicin against multidrug resistance protein-expressing tumors. Clin Cancer Res 1996; 2:1961–1968.

79. O'Dwyer PJ, Hamilton TC, LaCreta FP, et al. Phase I trial of buthionine sulfoximine in combination with melphalan in patients with cancer. J Clin Oncol 1996; 14:249–256.
80. Binaschi M, Capranico G, Dal Bo L, et al. Relationship between lethal effects and topoisomerase II-mediated double-stranded DNA breaks produced by anthracyclines with different sequence specificity. Mol Pharmacol 1997; 51:1053–1059.
81. Reed E, Kohn EC, Sarosy G, et al. Paclitaxel, cisplatin, and cyclophosphamide in human ovarian cancer: molecular rationale and early clinical results. Semin Oncol 1995; 22:90–96.
82. Durocher D, Jackson SP. DNA-PK, ATM and ATR as sensors of DNA damage: variations on a theme? Curr Opin Cell Biol 2001; 13:225–231.
83. West KA, Castillo SS, Dennis PA. Activation of the PI3K/Akt pathway and chemotherapeutic resistance. Drug Resist Updat 2002; 5:234–248.
84. Kyriazis AP, Kyriazis AA, Yagoda A. Enhanced therapeutic effect of cis-diamminedichloroplatinum(II) against nude mouse grown human pancreatic adenocarcinoma when combined with 1-beta-D-arabinofuranosylcytosine and caffeine. Cancer Res 1985; 45:6083–6087.
85. Mirzayans R, Pollock S, Scott A, et al. Relationship between the radiosensitizing effect of wortmannin, DNA double-strand break rejoining, and p21WAF1 induction in human normal and tumor-derived cells. Mol Carcinog 2004; 39:164–172.
86. Merker K, Stolzing A, Grune T. Proteolysis, caloric restriction and aging. Mech Ageing Dev 2001; 122:595–615.
87. Lynch G, Bi X. Lysosomes and brain aging in mammals. Neurochem Res 2003; 28:1725–1734.
88. Szweda PA, Camouse M, Lundberg KC, et al. Aging, lipofuscin formation, and free radical-mediated inhibition of cellular proteolytic systems. Ageing Res Rev 2003; 2:383–405.
89. Ciechanover A. Intracellular protein degradation: from a vague idea, through the lysosome and the ubiquitin-proteasome system, and onto human diseases and drug targeting (Nobel lecture). Angew Chem Int Ed Engl 2005; 44:5944–5967.
90. Cuervo AM, Dice JF. When lysosomes get old. Exp Gerontol 2000; 35:119–131.
91. Duffy MJ. Proteases as prognostic markers in cancer. Clin Cancer Res 1996; 2: 613–618.
92. Ciechanover A. Intracellular protein degradation: from a vague idea thru the lysosome and the ubiquitin-proteasome system and onto human diseases and drug targeting. Cell Death Differ 2005; 12:1178–1190.
93. Cuervo AM, Bergamini E, Brunk UT, et al. Autophagy and aging: the importance of maintaining "clean" cells. Autophagy 2005; 1:131–140.
94. Pickart CM. Targeting of substrates to the 26S proteasome. FASEB J 1997; 11:1055–1066.
95. Torres C, Lewis L, Cristofalo VJ. Proteasome inhibitors shorten replicative life span and induce a senescent-like phenotype of human fibroblasts. J Cell Physiol 2006; 207:845–853.
96. Ogiso Y, Tomida A, Lei S, et al. Proteasome inhibition circumvents solid tumor resistance to topoisomerase II-directed drugs. Cancer Res 2000; 60:2429–2434.
97. Cusack JC Jr., Liu R, Houston M, et al. Enhanced chemosensitivity to CPT-11 with proteasome inhibitor PS-341: implications for systemic nuclear factor-kappaB inhibition. Cancer Res 2001; 61:3535–3540.

98. Ogiso Y, Tomida A, Tsuruo T. Nuclear localization of proteasomes participates in stress-inducible resistance of solid tumor cells to topoisomerase II-directed drugs. Cancer Res 2002; 62:5008–5012.
99. Lara PN Jr., Davies AM, Mack PC, et al. Proteasome inhibition with PS-341 (bortezomib) in lung cancer therapy. Semin Oncol 2004; 31:40–46.
100. Adams J, Kauffman M. Development of the proteasome inhibitor Velcade (Bortezomib). Cancer Invest 2004; 22:304–311.
101. Richardson PG, Barlogie B, Berenson J, et al. A phase 2 study of bortezomib in relapsed, refractory myeloma. N Engl J Med 2003; 348:2609–2617.
102. Cusack JC. Rationale for the treatment of solid tumors with the proteasome inhibitor bortezomib. Cancer Treat Rev 2003; 29(suppl 1):21–31.
103. An WG, Hwang SG, Trepel JB, et al. Protease inhibitor-induced apoptosis: accumulation of wt p53, p21WAF1/CIP1, and induction of apoptosis are independent markers of proteasome inhibition. Leukemia 2000; 14:1276–1283.
104. Sheaff RJ, Singer JD, Swanger J, et al. Proteasomal turnover of p21Cip1 does not require p21Cip1 ubiquitination. Mol Cell 2000; 5:403–410.
105. Bendjennat M, Boulaire J, Jascur T, et al. UV irradiation triggers ubiquitin-dependent degradation of p21(WAF1) to promote DNA repair. Cell 2003; 114: 599–610.
106. Coleman ML, Marshall CJ, Olson MF. Ras promotes p21(Waf1/Cip1) protein stability via a cyclin D1-imposed block in proteasome-mediated degradation. EMBO J 2003; 22:2036–2046.
107. Amano T, Nakanishi H, Kondo T, et al. Age-related changes in cellular localization and enzymatic activities of cathepsins B, L and D in the rat trigeminal ganglion neuron. Mech Ageing Dev 1995; 83:133–141.
108. Joyce JA, Baruch A, Chehade K, et al. Cathepsin cysteine proteases are effectors of invasive growth and angiogenesis during multistage tumorigenesis. Cancer Cell 2004; 5:443–453.
109. Hashmi S, Britton C, Liu J, et al. Cathepsin L is essential for embryogenesis and development of Caenorhabditis elegans. J Biol Chem 2002; 277:3477–3486.
110. Kane SE, Gottesman MM. The role of cathepsin L in malignant transformation. Semin Cancer Biol 1990; 1:127–136.
111. Rousselet N, Mills L, Jean D, et al. Inhibition of tumorigenicity and metastasis of human melanoma cells by anti-cathepsin L single chain variable fragment. Cancer Res 2004; 64:146–151.
112. Gocheva V, Zeng W, Ke D, et al. Distinct roles for cysteine cathepsin genes in multistage tumorigenesis. Genes Dev 2006; 20:543–556.
113. Ogier-Denis E, Codogno P. Autophagy: a barrier or an adaptive response to cancer. Biochim Biophys Acta 2003; 1603:113–128.
114. Huang H, Manton KG. The role of oxidative damage in mitochondria during aging: a review. Front Biosci 2004; 9:1100–1117.
115. Hyland P, Barnett C, Pawelec G, et al. Age-related accumulation of oxidative DNA damage and alterations in levels of p16(INK4a/CDKN2a), p21(WAF1/CIP1/SDI1) and p27(KIP1) in human CD4+ T cell clones in vitro. Mech Ageing Dev 2001; 122:1151–1167.
116. Macip S, Igarashi M, Fang L, et al. Inhibition of p21-mediated ROS accumulation can rescue p21-induced senescence. EMBO J 2002; 21:2180–2188.

117. Trepel M, Groscurth P, Malipiero U, et al. Chemosensitivity of human malignant glioma: modulation by p53 gene transfer. J Neurooncol 1998; 39:19–32.
118. Maeda T, Matsubara H, Koide Y, et al. Radiosensitivity of human breast cancer cells transduced with wild-type p53 gene is influenced by the p53 status of parental cells. Anticancer Res 2000; 20:869–874.
119. Pestell KE, Hobbs SM, Titley JC, et al. Effect of p53 status on sensitivity to platinum complexes in a human ovarian cancer cell line. Mol Pharmacol 2000; 57:503–511.
120. Ruan S, Okcu MF, Ren JP, et al. Overexpressed WAF1/Cip1 renders glioblastoma cells resistant to chemotherapy agents 1,3-bis(2-chloroethyl)-1-nitrosourea and cisplatin. Cancer Res 1998; 58:1538–1543.
121. St Croix B, Florenes VA, Rak JW, et al. Impact of the cyclin-dependent kinase inhibitor p27Kip1 on resistance of tumor cells to anticancer agents. Nat Med 1996; 2:1204–1210.
122. Zhang X, Xu LS, Wang ZQ, et al. ING4 induces G2/M cell cycle arrest and enhances the chemosensitivity to DNA-damage agents in HepG2 cells. FEBS Lett 2004; 570:7–12.
123. Lincet H, Poulain L, Remy JS, et al. The p21(cip1/waf1) cyclin-dependent kinase inhibitor enhances the cytotoxic effect of cisplatin in human ovarian carcinoma cells. Cancer Lett 2000; 161:17–26.
124. Wang Y, Blandino G, Givol D. Induced p21waf expression in H1299 cell line promotes cell senescence and protects against cytotoxic effect of radiation and doxorubicin. Oncogene 1999; 18:2643–2649.
125. Naumann U, Weit S, Rieger L, et al. p27 modulates cell cycle progression and chemosensitivity in human malignant glioma. Biochem Biophys Res Commun 1999; 261:890–896.
126. Hama S, Sadatomo T, Yoshioka H, et al. Transformation of human glioma cell lines with the p16 gene inhibits cell proliferation. Anticancer Res 1997; 17:1933–1938.
127. Cohen HY, Miller C, Bitterman KJ, et al. Calorie restriction promotes mammalian cell survival by inducing the SIRT1 deacetylase. Science 2004; 305:390–392.
128. Frame LT, Hart RW, Leakey JE. Caloric restriction as a mechanism mediating resistance to environmental disease. Environ Health Perspect 1998; 106(suppl 1): 313–324.
129. Mattson MP, Magnus T. Ageing and neuronal vulnerability. Nat Rev Neurosci 2006; 7:278–294.
130. Vijg J, Suh Y. Genetics of longevity and aging. Annu Rev Med 2005; 56:193–212.
131. Kang HC, Kim IJ, Park JH, et al. Identification of genes with differential expression in acquired drug-resistant gastric cancer cells using high-density oligonucleotide microarrays. Clin Cancer Res 2004; 10:272–284.

16

Exploiting Drug-Induced Senescence in Transgenic Mouse Models

Mehtap Kilic

Department of Hematology/Oncology, Charité - Universitätsmedizin Berlin, Berlin, Germany

Clemens A. Schmitt

Department of Hematology/Oncology, Charité - Universitätsmedizin Berlin, and Max-Delbrück-Center for Molecular Medicine Berlin-Buch, Berlin, Germany

INTRODUCTION

During the past 20 years, experimental strategies addressing tumor cell responses to—by and large empirically developed—anticancer drugs mostly aimed to measure short-term viability in vitro. However, cancer cell lines used in these studies were typically expanded for many passages after their establishment from primary tumor material under non-physiological culture conditions, thereby being prone to select for genetic defects that may contribute to a drug-resistant phenotype. For example, mutations at the *INK4a/ARF* locus that encodes the cell cycle regulators $p16^{INK4a}$ and ARF have been found in culture-established cell lines at a much higher frequency when compared with primary tumors (1), and have been linked to chemoresistance in various clinical scenarios (2,3). Indeed, primary tumor cells propagated in culture for several weeks were shown to be much less susceptible to anticancer agents than their matched primary, untreated counterparts, suggesting that genetic components of treatment responses acquired inactivating mutations (4). Even when performed with primary tumor cells freshly isolated from a cancer patient and thus possibly reflecting a better approximation of the patient response to therapy, in vitro cytotoxicity assays often failed to anticipate drug action in vivo possibly because of the fact that local drug concentrations, exposure times, and kinetics of the biological effects experienced in vivo were insufficiently recapitulated in vitro. Importantly, cell culture systems cannot even come close to model non-cell autonomous factors of the tumor microenvironment, whose importance is increasingly acknowledged. Accordingly, it is not too much of a surprise that many studies failed to identify a direct correlation between in vitro chemosensitivity and long-term patient outcome (5,6).

Undoubtedly, it is difficult to conduct functional analyses in biopsied patient material. Outcome stratification by biological parameters in clinical studies must remain correlative in nature, and genetic or pharmacological manipulations of the tumor site are often restricted by technical limitations, and of course, ethical concerns.

Therefore, more and more sophisticated genetically engineered mouse models in which malignancies form based on gain- or loss-of-function mutations are generated and serve as experimental platforms to assess tumor development, biology, and treatment responses under more physiological conditions than cell culture models ever could. Despite species-related differences that certainly need to be taken into consideration, mouse models are increasingly recognized to bridge the gap between culture-based experimental settings and human cancer specimens.

Activated oncogenes act as driving forces of malignant growth but trigger, in turn, cellular fail-safe mechanisms such as apoptosis and senescence. Given the critical overlap between tumor-suppressive safeguard mechanisms on one hand and cellular effector programs executing anticancer drug action on the other, transgenic mouse models of cancer can be utilized to provide key insights into the principles of both tumor formation and treatment responses.

Apoptosis and Cellular Senescence Are Oncogene-Provoked Cellular Safeguard Responses In Vitro and In Vivo

The mitogenic properties of activated oncogenes are sensed by cell cycle checkpoints that trigger fail-safe responses, namely, apoptosis or cellular senescence. Whether apoptosis or senescence is chosen as the primary response depends, at least in part, on the specific properties of the respective prototypic oncogene (7–10). Apoptotic cell death reflects a programmed cascade of mediators ultimately leading to the self-destruction of potentially damaged or tumorigenic cells (11). Activation of oncogenes such as Myc or the viral oncoprotein E1A drives proliferation on one hand but promotes an apoptotic response that involves the activation of the ARF-p53 pathway on the other hand

(8,9). Consequently, successful Myc-initiated tumor formation requires suppression or disruption of the cell death machinery, for example, by acquiring genetic defects in the apoptotic ARF-p53 signaling cascade. In fact, mutations inactivating ARF or p53 have been shown to accelerate Myc-driven tumor development in transgenic mouse models by disabling apoptosis without significantly affecting proliferation (12). Likewise, Myc activation collaborates with antiapoptotic mediators, for example, Bcl2 and Bcl-X_L, or with the disruption of proapoptotic Bax signaling in vivo (13,14). While p53 alleles are typically lost in Myc-driven lymphomas forming in $p53^{+/-}$ mice, expression of Bcl2 in lymphoma progenitor cells was shown to alleviate the pressure to inactivate p53 in upcoming tumors (15). This demonstrates that apoptosis is the pivotal p53-governed fail-safe mechanism that counteracts Myc action.

Alternatively, oncogenic activation of the mitogenic Ras/Raf cascade provokes a terminal G1 arrest in vitro with features of cellular senescence including increased senescence-associated β-galactosidase (SA-β-gal) activity, elevated amounts of $p16^{INK4a}$ protein, and a retinoblastoma protein (Rb) held predominantly in the active, hypophosphorylated condition (16,17). Cells entering senescence become insensitive to mitogenic stimuli, display a gene expression profile that distinguishes them from quiescent cells, and acquire typical morphological changes: they enlarge, their granularity increases because of an accumulation of cytoplasmic vacuoles, and their borders tend to vanish (18–21). Biochemical characteristics include changes in cell metabolism and changes in gene expression levels of factors such as $p16^{INK4a}$, $p21^{CIP1}$, ARF, p53, PML (the product of the promyelocytic leukemia gene), and the plasminogen activator inhibitor-1, although not all of these indicators have to be simultaneously present or must be detectable in all senescent cells of different origins (21–23). Work by Scott Lowe and colleagues demonstrated that Ras-induced senescence is associated with alterations in the chromatin structure (24). In such cells, Rb-mediated senescence-associated heterochromatic foci (SAHFs)—sites of inactive transcription—are assembled via chromatin regulators such as TUP1-like enhancer of split protein 1(HIRA) and anti-silencing function protein 1 homolog A (ASF1α), display characteristic histone modifications, and contain a variety of chromatin proteins such as heterochromatin protein 1 (HP1), macroH2A (histone variant), and high-mobility group A proteins (HMGA, nonhistone architectural chromatin proteins) (25,26). Mechanistically, histone H3 methylated at its lysine 9 residue (H3K9me)—the repressive histone modification that SAHFs are enriched for is mediated via an Rb-bound histone methyltransferase activity. Since active Rb also physically interacts with E2F transcription factors, Rb serves as a scaffold for the histone methyltransferase moiety to mediate transcriptional silencing restricted to the vicinity of E2F-responsive S-phase genes (24). Hence, these observations suggest that epigenetic regulation is involved in the mitogen-refractory G1 arrest known as the hallmark phenotype of oncogene-induced senescence.

Initial studies claiming the finding of oncogene-induced senescence in cell culture–based experiments were questioned because of the possibility that cells with enforced Ras expression might simply arrest due to a "culture shock" artifact (27). However, recent reports were able to confirm the relevance of oncogene-induced senescence as a bona fide tumor suppressor mechanism in a variety of tumor models in vivo (28–32). Studies from our group and collaborators demonstrated that primary mouse lymphocytes entered a Ras-induced SA-β-gal-positive growth arrest depending on the presence of the histone methyltransferase Suv39h1 as the critical activity that catalyzes trimethylation of the lysine 9 residue at histone H3 (29). Importantly, when a Ras-transgenic mouse model, that is the *Eμ-N-Ras* transgenic mouse, was employed to probe the role of Suv39h1 function in executing oncogene-provoked cellular senescence as a tumor suppressor program relevant in vivo, the absence of Suv39h1 permitted the manifestation of early-onset T-cell lymphomas in virtually all animals, while Suv39h1-proficient mice developed this entity only in rare, sporadic cases and succumbed significantly later to a different tumor type of macrophage origin. Thus, the Ras-provoked and Suv39h1-dependent growth arrest precludes cells from progressing to a full-blown malignant phenotype, thereby instating an early barrier against tumorigenesis.

Importantly, additional evidence underscored the relevance of oncogene-induced senescence throughout different cell types and over species barriers: senescent cells were identified in early or premalignant tumor lesions in human naevi, a benign tumor type arising from melanocytes driven by oncogenic V-RAF murine sarcoma viral oncogene homolog B1(BRAF) (32), in murine K-Ras-initiated lung adenomas and pancreatic intraductal neoplasia (31), in phosphatase and tensin homolog (PTEN)-deficient (and thus Akt-activated) prostatic intraepithelial neoplastic lesions (30), and in the pituitary gland hyperplasia that forms in a mouse model with deregulated E2F3 activity (28). Thus, oncogene-induced senescence has been identified as another fail-safe program that limits tumor development at an early stage.

At present, it remains controversial whether autophagy acts as another fail-safe mechanism in response to oncogenic stress. Autophagy, at least in its initial phase, is a caspase-independent survival program characterized by the appearance of double- or multi-membrane cytoplasmic vacuoles that engulf cellular organelles such as parts of the endoplasmic reticulum or mitochondria (33). Autophagy can be activated as a cellular safeguard principle to remove damaged and therefore potentially hazardous organelles or to simply "recycle" them as a source of energy and cellular metabolites under conditions of nutrient shortage. Indeed, autophagy can be induced by nutrient deprivation in human cancer cells: withdrawal of interleukin-3 in the growth factor–dependent and apoptosis-incompetent $Bax^{-/-}$/$Bak^{-/-}$ hematopoietic cell line induced autophagy (34). Moreover, disruption of autophagy by RNA interference (RNAi)–mediated suppression of autophagy genes such as *Atg5* led to rapid cell death, suggesting that autophagy is essential to maintain cell survival following growth factor deprivation (35). In contrast, the mammalian autophagy regulator gene *beclin 1* was shown to act as a haploinsufficient tumor suppressor gene (36). Mice lacking one copy of the *beclin1* gene displayed an increased incidence in lymphomas, lung and liver carcinomas, indicating that autophagic degradation might contribute to prevention of tumor formation. Hence, apoptosis, cellular senescence, and autophagic death may serve as bona fide tumor suppressor mechanisms in vivo, whereby the primary response chosen seems to depend on functional properties of the driving oncogenic force as well as cell type–specific and environmental conditions.

Oncogenes Provoke a DNA-Damage Response in Premalignant Lesions

Understanding the consequences of DNA-damaging cancer therapies is directly linked to the availability and functionality of a cellular network that senses DNA damage, which may have been selected against as a result of oncogene-induced intermediates during tumor formation. How inappropriate growth signals elicited by oncogenic action are mechanistically sensed and converted into cellular responses remains largely unclear. Activation of oncogenic *H-ras* acutely produces DNA damage and genomic instability (37). Likewise, Myc may damage DNA via induction of reactive oxygen species (ROS) and DNA hyper-replication stress as a consequence of inappropriate S-phase entry (38,39). Emerging data now suggest that oncogene-induced DNA damage responses (DDRs) may ultimately lead to apoptosis or cellular senescence, which, in turn, cancel early tumorigenesis. Primary keratinocytes derived from a Myc-induced mouse skin cancer model displayed increased DNA damage associated with Atm activation and accumulation of p53. Importantly, inactivation of ataxia-telangiectasia-mutated (Atm) significantly reduced the level of p53-dependent apoptosis and promoted tumorigenesis in this model (40). Moreover, recent studies suggested that oncogene-deregulated cell cycle progression is associated with DNA-replicational stress resulting in an ataxia-telangiectasia and Rad3-related protein (Atr)-governed cellular DDR (41). While the initiating types of DNA damage, that is, oxidative DNA lesions versus stalled DNA replication forks, remain an issue of debate, premalignant stages of various types of human cancers originating from bladder, breast, colon, or lung were indeed found to display activated components of the DDR machinery, and oncogene-induced senescence became detectable as a DDR-associated cellular effector mechanism (42–45). Given the tumor-suppressive potential of the oncogene-provoked DDR via executing apoptosis or senescence, full-blown malignancies must have successfully selected against these principles by inactivating either components of the DDR or the respective downstream effector programs. In fact, more advanced tumors tend not to display activated DDR components anymore (42,43). As a consequence, oncogene-provoked selection against the functional DDR machinery should have important ramifications for the subsequent responsiveness to chemotherapy since most anticancer agents exert their therapeutic potential via causing DNA damage and initiating a DDR as well.

Oncogene-Provoked and Chemotherapy-Induced Responses Are Intimately Linked

DNA-damaging agents and DNA-damaging γ-irradiation reflect the major nonsurgical modalities to treat cancer. Beyond their individual pharmacological properties, most anticancer agents produce DNA damage via direct interaction with the DNA itself or in an indirect manner on targeting enzymatic activities that participate in DNA replication (e.g., topoisomerases or DNA polymerase). The DNA damage, in turn, executes a rather uniform DDR cascade, leading to a temporary growth arrest or ultimately to apoptosis or cellular senescence (46–48). Hence, DNA-damaging anticancer therapies trigger a cellular response that shares signaling components with the DDR seen on activation of mitogenic oncogenes.

Compared to drug-inducible programmed cell death, cellular senescence is much less intensely studied drug-responsive effector mechanism, although it may become particularly important when an intact apoptotic machinery is no longer available. Induction of senescence by DNA-damaging anticancer agents has been demonstrated in vitro in different types of cultured human tumor cells by assessment of SA-β-gal activity, lack of DNA synthesis, and accumulation of p53, $p21^{CIP1}$, and $p16^{INK4A}$ proteins (49–51). Meanwhile, numerous anticancer drugs including the topoisomerase inhibitors adriamycin and campthothecin or the cross-linking agent cisplatin have been shown to induce senescence. In contrast, antimicrotubule agents such as paclitaxel or vincristine that do not primarily damage DNA fail to produce a significant senescence response (49). Taken together, activated oncogenes and anticancer agents may initiate overlapping responses such as apoptosis or cellular senescence because both trigger DNA damage. Conversely, oncogenes that cause DNA damage may not only select against components of the DDR machinery to attenuate tumor-suppressive fail-safe responses but might also simultaneously predetermine drug resistance as a consequence of impaired sensing of drug-induced genotoxicity.

Assessing Drug Responses In Vitro

Multi-passage cancer cell lines serve as the most widely used experimental models to study drug responses and genetic mechanisms of chemoresistance. They are typically well characterized and available in unlimited amounts and should give rise to reproducible results in a reasonable period of time. Furthermore, cancer cell lines can be utilized for matched pair analyses based on homologous gene inactivation, small-interfering RNA(siRNA)-based gene knockdown, or overexpression strategies. However, tumor-derived cell lines may have become substantially altered during their establishment in cell culture, thereby masking effects that might be relevant in primary tumors (4). In fact, it has not been demonstrated in larger comparisons that established cell lines reflect the treatment sensitivity of their corresponding primary tumors, which is, at least in part, due to the fact that cell culture systems fail to properly recapitulate components of the tumor microenvironment that impact on both tumor growth and treatment sensitivity. Hence, the critical components contributing to drug action in vivo might be missed by studying cellular responses in vitro.

Assessing Drug Responses in Patient Material In Vivo

Given the limitations of cell culture–based assays for studying chemoresistance, assessing drug responses directly in human tumors in vivo appears as the obvious strategy. Accordingly, patient data on clinical courses or outcome might be stratified by certain markers related to cancer biology, and cancer material can be used for flanking mutational or functional analyses. However, clinical studies are correlative at best. For example, while some of the clinical studies found a correlation between inactivated p53 and poor clinical outcome, others remained inconclusive or did not find any correlation, possibly because of unknown extragenic mutations in the p53 pathway or heterogeneity of the material (52). Moreover, studies that focused on specific mechanisms such as apoptosis might have missed other components of treatment outcome. Finally, further limitations apply to patient material, such as impracticality of using fresh tumor material *ex vivo* as well as ethical concerns particularly for experimental in vivo settings. Thus, mouse models developing primary tumors with defined genetic lesions may bridge the gap between well-characterized multi-passage tumor cell lines and the heterogeneity of human tumors studied at their natural sites in vivo. Certainly, mouse models of cancer have been used to address genetics of tumorigenesis for years, but they have been understudied as experimental platforms to investigate short- and long-term consequences of drug-inducible effector programs.

Transgenic Mouse Models to Link Genetic Defects with Treatment Outcome

Most of the mouse models that were established to study treatment responses are still based on xenograft models. These immunocompromised mice harbor transplant tumors that formed on propagation of human tumor cell lines at a site unrelated to the actual organ of origin (*ectopic implantation*) with or without additional genetic lesions (Fig. 1) (53). Ectopic implantation has important implications such that tumor dissemination cannot be controlled by local barriers, for instance, by a basal membrane or an organ capsule. Moreover, angiogenic signals at sites not necessarily serving as prime locations for metastases of this particular type of cancer may force the host to generate a neovasculature, which forms together with other cellular components of the ectopic site, a species-mixed microenvironment that might be quite different from the local conditions these tumor cells were exposed to in the originating

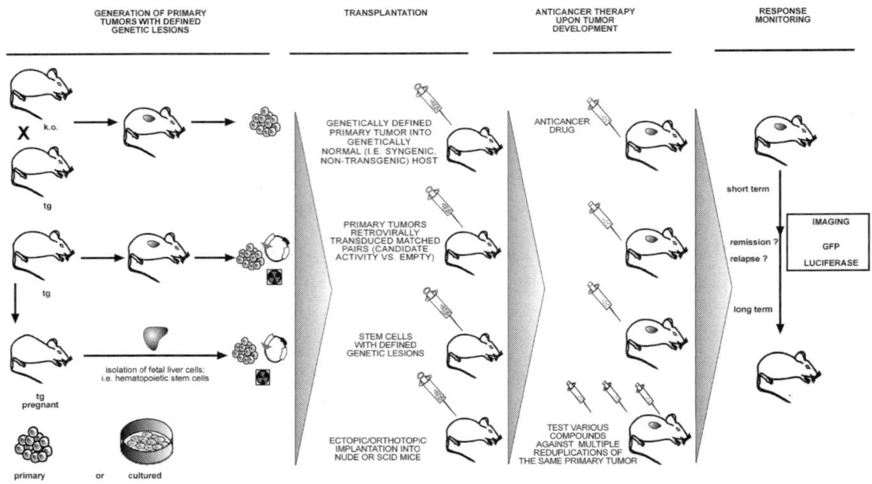

Figure 1 Mouse models can be used to study genetic mechanisms of tumor development and drug responses in malignancies with defined genetic lesions in vivo. Generation of primary tumors. By intercrossing transgenic (tg) mice to other genetically engineered mice such as knockout (k.o.) mice, transgenic offspring with defined additional genetic lesions can be obtained. Primary tumor cells or their progenitors [i.e., oncogene transgenic fetal liver cells (as a source of hematopoietic stem cells)] may be isolated from preneoplastic or tumor-bearing animals, and can be transduced *ex vivo* with retroviral vectors to stably transfer a gene-of-interest and/or a marker gene (i.e., GFP). Transplantation of primary tumors. Genetically modified primary tumor cells or stem cells can be propagated in (multiple, if required) syngeneic (but non-transgenic) immunocompetent or immunocompromised recipient mice. Drug treatment. Numerous animals harboring a series of primary or transplant tumors can be treated either with the same compound to test for treatment responses based on a defined genetic defect, or animals bearing biological reduplications of the same primary tumor may undergo treatments with different compounds to test the superiority of a certain drug against a specific tumor scenario. Response monitoring. Tumor latencies and drug responses can be easily studied and monitored using in vivo imaging technologies. Tumor cells from relapse tumors can be isolated and further used for functional analyses.

donor patient. In contrast, orthotopic models based on the surgical implantation of an intact tissue block or an *ex vivo* reengineered tissue equivalent at the homing site certainly come closer to the architecture of a genuine tumor However, such approaches are technically demanding and laborious because of the requirement of anesthesia, surgical skills, and the inherent risk of lethality, and therefore, cannot be easily multiplied to produce higher numbers of biologically comparable tumor-bearing mice for subsequent treatment studies.

Unlike xenograft models, which still rely on culture-adapted cell lines, transgenic cancer models develop primary malignancies, and their tumors are typically transplantable into syngeneic immunocompetent normal recipient mice (Fig. 1). Of note, although the underlying mechanisms of tumor surveillance are far from being fully understood yet, studying treatment responses in an immunocompetent model without a species barrier between tumor and host might indeed reflect an improved approximation to the biological complexity of human tumors growing in their natural organismic context. Transgenic mice, often utilized as a basic approach to study oncogenic action, are generated by enforced expression of an additional gene either under control of a universal promoter throughout the organism or in predefined organs driven by a tissue-specific promoter (53,54). In contrast, knockout mice are generated by the targeted deletion of a gene in every cell and mostly utilized to globally investigate gene function, but also to more specifically test whether a candidate gene may possess tumor suppressor activity. Conditional knockouts are generated by flanking the to-be-deleted genomic sequence of the targeted gene by lox-P sites that are sensitive to the Cre recombinase activity (55). On provision of Cre, expressed from a tissue-specific promoter, the floxed exon will be deleted in this particular cellular compartment. This approach has been taken to the next level in mouse tumor models carrying alleles of activated oncogenes with an expression-blocking floxed stop sequence knocked into the endogenous locus. Based on rare stochastic recombination events, the stop cassette is spontaneously lost and thus mimics the sporadic mutational activation of an oncogene to initiate tumorigenesis then, as demonstrated for K-Ras-driven lung cancer mouse models (56,57). Conditional knockout or knockin technology have paved the road for the generation of models in which gene inactivation or activation is executed in a way that resembles patient reality where genetic events occur sporadically in the tissue of origin as a seed for the subsequent process of multistep tumor formation.

Of course, multiple genetic modifications can be introduced into single individuals by various ways. For example, mice with compound genetic backgrounds are generated by intercrossing the transgenic models to mice expressing a secondary oncogene (i.e., *Eμ-myc* to *Eμ-bcl2*) as well as to mice harboring defects in tumor suppression (i.e., *Eμ-myc* to *p53*-knockout mice) (Fig. 1). As an alternative, stable retroviral infection, particularly of cell populations with stem cell characteristics and their subsequent reimplantation at the site of origin in recipient mice, has emerged as a novel approach in cancer research (Fig. 1). Models based on this approach are now widely used to study the role of a candidate gene in leukemia and lymphoma development since hematopoietic stem cells are particularly accessible and readily infectible (15,58). Retrovirally transduced stem cells obtained from bone marrow or fetal liver reconstitute the hematopoietic compartment of recipient mice that underwent myeloablative total-body γ-irradiation prior to stem cell transplantation (Fig. 1). Moreover, retroviral integration of an RNA polymerase promoter and a gene-specific small hairpin RNA can be used for long-term silencing of target genes in somatic cells (59,60) or even in the germ line compartment to generate inheritable "knockdown" mice as a more rapidly available alternative to knockout mice—similar to retroviral germ line strategies to avoid the need for blastocyst injection in the process of conventional mouse transgenesis (61,62).

Of note, emerging novel microscopic and macroscopic in vivo imaging technologies define a rapidly growing field and may contribute to the progress of cancer research made by the genetic engineering of mice. For example, *positron emission tomography* (PET) imaging is capable of distinguishing between a metabolically active, that is, viable tumor lesion, and a scar-like residual mass of connective tissue as the remainder after chemotherapy or spontaneous regression (63), and *magnetic resonance imaging* (MRI) can be utilized to visualize tumor sites in their anatomical context in viable mice (64). *Bioluminescence* and *fluorescence imaging* are based on endogenous or exogenous provision of the respective marker in the animal prior to scanning (15,65–67). For example, in a model for pituitary gland tumorigenesis, bioluminescence imaging was used for temporospatial noninvasive monitoring of the tumor size by site-specific luciferase expression, eventually allowing in vivo quantification of tumor onset, progression, and response to therapy (68). As an alternative to luciferase, tumor cells tagged with red or green fluorescent proteins can also be tracked throughout the body to measure their expansion and dissemination (67). Thus, improved imaging possibilities and the development of complex reporter systems integrated into genetically engineered mice or tumors allow the examination of certain biological processes via external detection.

Exploiting Treatment Responses in Transgenic Mice

Genetically tractable, immunocompetent transgenic mice prone to spontaneous tumor development serve as a versatile tool to identify genes important for cancer development and progression, to exploit the role of drug responsive programs in vivo, and to dissect the signaling pathways involved in drug action and resistance. For example, mouse mammary tumor virus (*MMTV*)-*Ras* and *MMTV-myc* transgenic mice that develop mammary and salivary gland tumors were intercrossed to *p53*-knockout mice to examine the impact of p53 defects on the responsiveness to the anticancer agents adriamycin and paclitaxel (69). Interestingly, in these solid tumor models, responses observed were mostly mediated by changes in cell cycle distribution rather then via apoptosis, although the levels of spontaneous apoptosis found in the untreated tumors were predictive for outcome to either agent. Moreover, *myc*- and *Ras*-transgenic mice have been used to test compounds that specifically interfere with oncogenic activation (70,71). Pharmacological inhibition of farnesyltransferase, a key enzyme mediating Ras activation via posttranslational modification, induced tumor regression in *MMTV-Ras* mice, and in vivo administration of antisense oligonucleotides targeting c-*myc* delayed lymphoma onset in *Eμ-myc* transgenic mice. In a doxycycline-inducible H-Ras$^{(V12G)}$ mouse melanoma model, melanoma development and maintenance were found to be strictly dependent on the persistent expression of the activated *Ras* oncogene, because withdrawal of doxycycline to cancel oncogenic Ras expression resulted in clinical and histological regression of the primary lesion and explanted tumors in this model (72). Likewise, it was shown that mice expressing a regulatable *myc* transgene in their hematopoietic compartment developed malignant lymphoid and myeloid tumors, which also regressed on inactivation of the transgene (73). Despite the appreciation of tumorigenesis as a multistep process, these studies support the notion that mitogenic oncogenes as the often initiating and driving forces of cancer formation and progression remain essential throughout even more advanced stages of the malignant condition, termed the "oncogene addiction."

The *Eμ-myc* transgenic mouse has emerged as probably one of the most versatile models for studying treatment responses of genetically defined malignancies (53,74,75). In this model, the c-*myc* oncogene is constitutively expressed in the B-cell lineage under the control of the *Eμ* immunoglobulin enhancer. The histopathological and genetic features of *Eμ-myc* lymphomas closely resemble human Non-Hodgkin's lymphomas. By intercrossing to other genetically engineered mice, *Eμ-myc* transgenic offspring that carries additional defined genetic lesions can be obtained. *Eμ-myc* lymphoma cells may also be cultured and transduced with retroviral vectors to overexpress a gene-of-interest, and these genetically modified primary lymphoma cells can be propagated in numerous non-transgenic but otherwise syngeneic and hence entirely wild-type recipient mice to conduct, for example, a direct matched-pair comparison of individual lymphomas with and without stable expression of a candidate moiety (Fig. 1). The non-transgenic recipients develop lymphomas with an associated leukemia within two to three weeks, and the malignancy is histopathologically indistinguishable from tumor patterns found in primary *Eμ-myc* transgenic mice. Tumor latencies (as the time-to-onset) and drug responses (e.g., the time-to-relapse) can easily be monitored by palpation of enlarged peripheral lymph nodes or by assessing the leukemic burden in white blood cell counts and peripheral blood smears (76).

In the *Eμ-myc* model, p53 and its upstream regulator ARF, the alternate reading frame product of the *INK4a/ARF* locus, limit lymphomagenesis (12,77). *Eμ-myc* lymphomas arising in a $p53^{+/-}$ or in an $INK4a/ARF^{+/-}$ background not only invariably lose the remaining wild-type allele and manifest at a much faster rate compared to mice harboring two intact alleles of the respective locus but also display resistance to therapy (12). Although the *INK4a* component has been shown not to influence *Eμ-myc*-driven lymphoma development, it critically improves treatment outcome in the absence of ARF, since mice harboring ARF-deficient lymphomas that retain $p16^{INK4a}$ expression achieve an excellent long-term outcome to therapy despite the fact that drug-inducible apoptosis is impaired in these lymphomas, presumably because of critically lower p53 protein levels available for DDR-mediated posttranslational activation (78).

Like loss of p53, overexpression of the antiapoptotic protein Bcl2 leads to a particularly aggressive phenotype of *Eμ-myc* lymphomas (4,12,13,79,80). However, a remarkable difference between p53-deficient and Bcl2-overexpressing lymphomas was noticed when tumor-bearing mice were treated with the alkylating agent cyclophosphamide (CTX) (78). Mice harboring *Eμ-myc* lymphomas with no additional defined genetic defect (hereafter referred to as control lymphomas) responded well to therapy and achieved an excellent long-term outcome. In contrast, loss of p53 compromised drug sensitivity, and lymphomas, after responding in a delayed fashion, invariably relapsed. Notably, Bcl2-overexpressing *Eμ-myc* lymphomas—although they failed to regress following CTX treatment—showed no signs of progression for extended periods of time. Thus, mice bearing tumors either lacking *p53* alleles or overexpressing Bcl2 equally failed to achieve an immediate remission as a consequence of impaired apoptotic tumor cell lysis, but p53 loss, unlike Bcl2 overexpression, apparently disabled an additional principle of long-term disease control that must be different from classic forms of programmed cell death. Indeed, further analyses unveiled that the type of growth arrest observed in Bcl2-protected control lymphomas reflects a drug-inducible form of premature cellular senescence (78). When the complete cessation of growth became detectable a few days after treatment, Bcl2-expressing control lymphomas were characterized by lack of bromodeoxyuridine (BrdU) incorporation, lack of Ki67 immunoreactivity, and increased SA-β-gal activity. Notably, the drug-inducible terminal cell cycle arrest and the positive SA-β-gal staining depend on the presence of p53. Since mice harboring senescence-competent Bcl2-expressing lymphomas lived much longer when compared to mice bearing p53-deficient lymphomas and exposed to the same type and dose of treatment, it is suggestive of that drug-inducible senescence improves outcome to anticancer therapy.

Not only the *p53* but also the *INK4A/ARF* locus plays a key role in the genetic control of drug-inducible senescence (78). ARF-deficient lymphomas that still express $p16^{INK4a}$ responded much better to therapy than lymphomas deficient for both ARF and $p16^{INK4a}$, and the superior outcome is reflected by an SA-β-gal-positive growth arrest associated with increased protein levels of $p16^{INK4a}$. In line, the poorly responding *INK4a/ARF*-null tumors lack any detectable SA-β-gal staining following CTX treatment. Thus, $p16^{INK4a}$ seems to contribute to treatment outcome in vivo through its pro-senescence activity at least in the absence of ARF, and this capability depends on p53, since p53-deficient lymphomas expressing high levels of $p16^{INK4a}$ can still proliferate. Taken together, senescence has been identified as a p53- and $p16^{INK4a}$-controlled drug-responsive program that might be executed as a "backup" in the absence of an available apoptotic response. Conversely, drug-inducible senescence extends host survival at least in tumor scenarios where apoptosis became unavailable as a drug effector program.

Complementing this model of drug-inducible senescence in the absence of apoptosis, another cancer model addressed the role of apoptosis on genetic ablation of a critical cell cycle arrest gene (81). Two isogenic human tumor cell lines differing in their status of the cyclin-dependent kinase (CDK) inhibitor and growth arrest gene *CIP1* showed no differences in sensitivity to γ-irradiation in clonogenic survival assays, whose readout proliferation cannot distinguish between antiproliferative responses such as apoptotic death and any kind of growth arrest. However, when tested in vivo, the cells deficient for p21^{CIP1} appeared to be more radiosensitive, suggesting that a genetic defect to properly halt in cycle increased the cellular sensitivity to radiation-induced apoptosis. These data indicate that apoptosis and forms of long-term arrests including senescence can be induced by anticancer therapy in tumors based on their availability, substitute for each other under certain circumstances, and both may contribute to treatment outcome.

The therapeutic potential of senescence raises concerns whether there might be any additional biological side effects, particularly since long-term fate and functional properties of senescent cells remain poorly understood. Indeed, it has been proposed that senescent cells may promote growth of non-senescent tumor cells in their vicinity, because they are capable of secreting mitogenic and other growth-supportive factors including certain matrix metalloproteinases and inflammatory cytokines (82). Indeed, at least in some settings, senescent bystanders may promote tumor progression as shown in a study by Campisi and coworkers in which senescent fibroblasts stimulated proliferation of premalignant epithelial cells and facilitated tumorigenesis (83). Thus, the overall contribution of drug-induced senescence to outcome is not only determined by the fact that it locks cancer cells into a terminally growth-arrested state but also depends on additional non-cell autonomous capabilities by which the senescent tumor cell population modulates functional properties of its neighbors.

Mechanisms Overcoming Cellular Senescence

If drug-inducible senescence improves the outcome to cancer therapy, mechanisms that counter or overcome this terminal growth arrest should result in or add to chemoresistance. For example, any cellular switch that could permit senescent cells to divide again might also produce treatment failure. In fact, in vitro experiments demonstrated that senescent cells might reenter the cell cycle on acquisition of additional senescence-compromising mutations (84). For instance, acute inactivation of the floxed *Rb* locus via recombinase-mediated gene deletion in a senescent cell population allowed the cells to reemerge from the growth-arrested state. Of note, senescent and therefore DNA-non-replicating tumor cells residing in their natural environment are presumably much less prone to acquire additional mutations that would eventually permit to cell cycle reentry. While potential reversal of the senescent condition remains a formal possibility of resistance, it appears more likely that senescent cells, if they ever manage to reduplicate their genomic information, will ultimately enter secondary forms of cell death such as mitotic catastrophe (49).

Nevertheless, a number of the recently discovered histone lysine demethylases indeed possess the capability to remove lysine 9's critical methyl modifications of histone H3 and are thus candidate moieties to potentially revert the H3K9me-mediated senescent cell cycle arrest (24,29,85–87). For instance, the nuclear amine oxidase homolog and H3K4me demethylase LSD1 reportedly gained H3K9me demethylation activity in prostate cells in the presence of the androgen receptor (88–90), and a rapidly growing list of histone demethylases with varying activities on tri-, di-, and monomethylation marks of different histone marks, namely numerous members of the jumonji family, are putative senescence revertases (91,92). However, it remains to be determined whether any of the H3K9me demethylases can actually cancel or prevent oncogene- or drug-inducible senescence (93–95).

Histone methylation is often reciprocally linked to histone acetylation, that is, acetyl modifications preclude a residue from getting methylated. As a consequence, histone deacetylase inhibitors—despite their broad action throughout the chromatin—that should also block removal of the transcriptionally activating acetyl modification at H3K9 could prevent pro-senescent methylation at this site, thereby mimicking loss of Suv39h1 as demonstrated in a *ras* transgenic mouse model (29). Moreover, a matrix of transplantable *Eμ-myc* transgenic lymphomas with defined genetic lesions in numerous cell cycle– and apoptosis-controlling genes has been exploited to test the therapeutic activities of a variety of histone deacetylase inhibitors, based on the specific genetic defect, with particular emphasis on apoptotic responses and the overall survival of the tumor-bearing recipient animals exposed to different compounds (96). Certainly, future studies based on this "low-throughput lymphoma in vivo array" will also allow to address the impact of cellular senescence as a response to therapies that target histone modifications such as acetylation, or possibly in the future, methylation as well.

In fact, how tumor cells, if ever, overcome a senescent arrest in their natural surroundings is not known, but growth after senescence-inducing stresses may occur for different reasons: tumor cells that resume to grow after a terminally arresting chemotherapy may have escaped from the senescent state, or alternatively, might have bypassed the arrest a priori. For instance, all mice harboring drug-induced senescent *Eμ-myc* lymphomas ultimately succumbed to their neoplastic disease when lymphomas regained the potential to expand after several weeks (78). Moreover, genetic approaches directed to identify critical players of senescence also provide evidence that a heterozygous status of loci such as *p53* or *INK4a/ARF* facilitated disruption of a lasting drug-induced senescent phenotype by genetic inactivation of the respective remaining wild-type allele in the relapse tumor. However, genetic alterations that enable bypassing of oncogene-induced fail-safe mechanisms do not always shut off the drug-inducible responses (29). For example, while Ras-driven lymphomas deficient for Suv39h1 or p53 develop on disruption of oncogene-induced senescence and co-exhibit refractoriness to drug-inducible senescence in vitro, sporadic Ras-lymphomas with no defined p53 or Suv39h1 defect—that sufficiently overcame oncogene-provoked senescence via unknown genetic co-events—still underwent senescence in response to adriamycin. Hence, in some Ras/Raf-driven scenarios—although the oncogene-induced senescence program must have been bypassed to allow tumor formation—the "pharmacological dose of DNA damage" provided by conventional anticancer agents may suffice to induce a senescence response.

Eμ-Myc Mice and More

The *Eμ-myc* model has also been utilized to exploit the impact of novel targeted therapies such as the mammalian target of rapamycin (mTOR) inhibitor rapamycin to treat tumors that were engineered to harbor a constitutively active PI3 kinase/Akt survival pathway (97). Introduction of a myristoylated Akt allele into *Eμ-myc* transgenic hematopoietic stem cells gave rise to accelerated lymphoma formation in recipient mice, and these lymphomas appeared to be chemoresistant because of an apoptotic defect. Disruption of Akt downstream signaling by using the mTOR inhibitor rapamycin restored apoptotic sensitivity to conventional chemotherapeutic agents such as adriamycin, which, therefore, synergizes with rapamycin in its anti-lymphoma efficacy. Overexpression of the translational regulator eIF4E, which acts downstream of Akt and mTOR, recapitulates Akt action in tumorigenesis and chemoresistance, but eIF4E-overexpressing lymphomas were refractory to rapamycin as a chemo-resensitization strategy. Hence, this work exemplifies how targeting therapies that restore apoptotic sensitivity can reestablish susceptibility to chemotherapeutic drugs when precisely interfering with the relevant signaling component of the deregulated pathway.

A very recent study introduced another Myc-induced lymphoma model to address the question whether anticancer therapy may induce autophagy and to what extent this program will interfere with treatment outcome (98). This mouse model is based on a regulatable *p53* gene, that is, a p53-estrogen receptor fusion gene knocked into the *p53* locus ($p53ER^{TAM}$), which allows the in vivo dissection of the effects of p53 restoration on exposure to the estrogen derivative tamoxifen. In brief, bone marrow cells from $p53ER^{TAM}/p53ER^{TAM}$ mice were infected with a Myc-expressing retrovirus. Utilizing this model, the authors showed that induction of p53 in *Myc-driven* $p53ER^{TAM}$ tumors resulted in apoptosis, whereby surviving tumor cells displayed signs of autophagy. Importantly, genetic ablation of the autophagy mediator Atg5 enhanced the extent of cell death produced by either p53 restoration or anticancer therapy with alkylating agents. Hence, at least the Atg5-dependent phase of autophagy serves as a survival mechanism that antagonizes drug-inducible and p53-mediated cell death in vivo, suggesting that autophagy might be an adaptive mechanism that contributes to tumor cell survival and resistance to therapy-induced apoptosis.

Taken together, transgenic mouse models can be exploited in many ways for studying the role of apoptosis and cellular senescence as drug effector programs in vivo, dissecting the signaling pathways important for tumor development, progression, drug action and resistance, and for identifying genetic lesions that may be pharmacologically targeted by novel compounds alone or in addition with conventional chemotherapy. Mouse models have revolutionized our insights into cancer genetics and mechanisms of treatment responses; in particular, as they provide experimental platforms where primary malignancies with defined genetic lesions can be examined, tumor biology can be recapitulated at natural sites, and tumor burden and extent of dissemination can be monitored by non-invasive imaging techniques. Thus, mouse models increasingly serve as complex but controlled experimental tools and may complement, flank, and sometimes outperform cell culture–based assays and the difficult-to-examine materials obtained from clinical trials for studying responses to anticancer therapies.

CONCLUSION AND PERSPECTIVES

Senescence not only reflects a physiological response of "aging" cells that exhausted their proliferative capacity in culture but is also considered as an important acute stress response initiated by oncogenic signals and DNA-damaging therapies. Senescence contributes to tumor suppression and most likely improves the outcome to cancer therapy as well. In this regard, mouse models will serve as the critical experimental platform to elucidate the long-term impact of cellular senescence on treatment outcome.

Frequently, malignant tumors are dependent on persistent oncogenic signals, dubbed as "oncogene addiction" (99), rendering those cells vulnerable to inhibitors of the driving oncogenic force, for example, to genetic inactivation of the Myc or the Ras oncogene, or to pharmacological inhibition as clinically pioneered by the abl kinase inhibitor imatinib used to treat bcr/abl-driven leukemias and related malignancies. Currently, an analogous idea has emerged for the concept of "tumor suppressor addiction," that is, the hypothesis that the persistent inactivation of a tumor suppressor protein may similarly be required for tumor maintenance, thereby opening a new window for therapeutic approaches against cancer (60,100,101). Indeed, various mouse models of epithelial and lymphatic tumors driven by Myc or Ras unveiled that brief restoration of p53 expression was sufficient to cause profound tumor regression, indicating that no independence of p53 signaling was achieved during tumor formation and that the mere reactivation of p53 in the presence of an oncogenic activity was instantly converted into a cellular fail-safe response. Interestingly, whether apoptosis or senescence occurred as the primary response appeared to be cell type- and oncogene-dependent. Importantly, in a Ras-driven liver cancer model, p53 restoration produced, expectedly, cellular senescence, but tumor-bearing animals achieved remissions based on a secondary mechanism of tumor cell clearance out of senescence (60). The observation that senescent and not non-senescent tumor cells secreted pro-inflammatory cytokines and other immune modulators (e.g., Csf1, Mcp1, Cxcl1, and interleukin-15) that subsequently attracted macrophages, neutrophils and natural killer cells, underscores the possibility that senescent cells may become "immunogenic" by eliciting an innate immune response. Besides, this finding also points toward an undervalued aspect of tumor dynamics: if senescent cells do not accumulate but rather disappear via immune-mediated engulfment-like or cytotoxic mechanisms, their overall impact on growth kinetics of the tumor based on steady state assessments of the fraction of senescent cells would be very different. However, it still remains to be determined whether this mode of clearance applies to senescent cells in precancerous or aging tissues as well. Clearly, these first and still preliminary data support the view that mouse models are needed to address persistence, immunological implications, and other functional properties of senescent tumor cells in their tumor microenvironment.

REFERENCES

1. Spruck CH 3rd, Gonzalez-Zulueta M, Shibata A, et al. p16 gene in uncultured tumours. Nature 1994; 370(6486):183–184.
2. Carter TL, Reaman GH, Kees UR. INK4A/ARF deletions are acquired at relapse in childhood acute lymphoblastic leukaemia: a paired study on 25 patients using real-time polymerase chain reaction. Br J Haematol 2001; 113(2):323–328.
3. Calero Moreno TM, Gustafsson G, Garwicz S, et al. Deletion of the Ink4-locus (the p16ink4a, p14ARF and p15ink4b genes) predicts relapse in children with ALL treated according to the Nordic protocols NOPHO-86 and NOPHO-92. Leukemia 2002; 16(10):2037–2045.
4. Schmitt CA, Rosenthal CT, Lowe SW. Genetic analysis of chemoresistance in primary murine lymphomas. Nat Med 2000; 6(9):1029–1035.
5. Phillips RM, Bibby MC, Double JA. A critical appraisal of the predictive value of in vitro chemosensitivity assays. J Natl Cancer Inst 1990; 82(18):1457–1468.
6. Nara N, Suzuki T, Nagata K, et al. Relationship between the in vitro sensitivity to cytosine arabinoside of blast progenitors and the outcome of treatment in acute myeloblastic leukaemia patients. Br J Haematol 1988; 70(2):187–191.
7. Evan GI, Wyllie AH, Gilbert CS, et al. Induction of apoptosis in fibroblasts by c-myc protein. Cell 1992; 69(1):119–128.
8. de Stanchina E, McCurrach ME, Zindy F, et al. E1A signaling to p53 involves the p19(ARF) tumor suppressor. Genes Dev 1998; 12(15):2434–2442.
9. Zindy F, Eischen CM, Randle DH, et al. Myc signaling via the ARF tumor suppressor regulates p53-dependent apoptosis and immortalization. Genes Dev 1998; 12(15):2424–2433.
10. Serrano M. The tumor suppressor protein p16INK4a. Exp Cell Res 1997; 237(1): 7–13.
11. Kerr JF, Winterford CM, Harmon BV. Apoptosis. Its significance in cancer and cancer therapy. Cancer 1994; 73(8):2013–2026 (published erratum appears in Cancer 1994; 73(12):3108).
12. Schmitt CA, McCurrach ME, de Stanchina E, et al. INK4a/ARF mutations accelerate lymphomagenesis and promote chemoresistance by disabling p53. Genes Dev 1999; 13(20):2670–2677.
13. Strasser A, Harris AW, Bath ML, et al. Novel primitive lymphoid tumours induced in transgenic mice by cooperation between myc and bcl-2. Nature 1990; 348(6299): 331–333.

14. Eischen CM, Roussel MF, Korsmeyer SJ, et al. Bax loss impairs Myc-Induced apoptosis and circumvents the selection of p53 mutations during Myc-mediated lymphomagenesis. Mol Cell Biol 2001; 21(22):7653–7662.
15. Schmitt CA, Yang M, Fridman JS, et al. Dissecting p53 tumor suppressor functions in vivo. Cancer Cell 2002; 1(3):289–298.
16. Serrano M, Lin AW, McCurrach ME, et al. Oncogenic ras provokes premature cell senescence associated with accumulation of p53 and p16INK4a. Cell 1997; 88(5): 593–602.
17. Zhu J, Woods D, McMahon M, et al. Senescence of human fibroblasts induced by oncogenic Raf. Genes Dev 1998; 12(19):2997–3007.
18. Mason DX, Jackson TJ, Lin AW. Molecular signature of oncogenic ras-induced senescence. Oncogene 2004; 23(57):9238–9246.
19. Shelton DN, Chang E, Whittier PS, et al. Microarray analysis of replicative senescence. Curr Biol 1999; 9(17):939–945.
20. Lundberg AS, Hahn WC, Gupta P, et al. Genes involved in senescence and immortalization. Curr Opin Cell Biol 2000; 12(6):705–709.
21. Collado M, Serrano M. The power and the promise of oncogene-induced senescence markers. Nat Rev Cancer 2006; 6(6):472–476.
22. Kurz DJ, Decary S, Hong Y, et al. Senescence-associated (beta)-galactosidase reflects an increase in lysosomal mass during replicative ageing of human endothelial cells. J Cell Sci 2000; 113(pt 20):3613–3622.
23. Beausejour CM, Krtolica A, Galimi F, et al. Reversal of human cellular senescence: roles of the p53 and p16 pathways. EMBO J 2003; 22(16):4212–4222.
24. Narita M, Nunez S, Heard E, et al. Rb-mediated heterochromatin formation and silencing of E2F target genes during cellular senescence. Cell 2003; 113(6): 703–716.
25. Zhang R, Poustovoitov MV, Ye X, et al. Formation of MacroH2A-containing senescence-associated heterochromatin foci and senescence driven by ASF1a and HIRA. Dev Cell 2005; 8(1):19–30.
26. Narita M, Narita M, Krizhanovsky V, et al. A novel role for high-mobility group a proteins in cellular senescence and heterochromatin formation. Cell 2006; 126(3): 503–514.
27. Sherr CJ, DePinho RA. Cellular senescence: mitotic clock or culture shock? Cell 2000; 102(4):407–410.
28. Lazzerini Denchi E, Attwooll C, Pasini D, et al. Deregulated E2F activity induces hyperplasia and senescence-like features in the mouse pituitary gland. Mol Cell Biol 2005; 25(7):2660–2672.
29. Braig M, Lee S, Loddenkemper C, et al. Oncogene-induced senescence as an initial barrier in lymphoma development. Nature 2005; 436(7051):660–665.
30. Chen Z, Trotman LC, Shaffer D, et al. Crucial role of p53-dependent cellular senescence in suppression of Pten-deficient tumorigenesis. Nature 2005; 436(7051): 725–730.
31. Collado M, Gil J, Efeyan A, et al. Tumour biology: senescence in premalignant tumours. Nature 2005; 436(7051):642.
32. Michaloglou C, Vredeveld LC, Soengas MS, et al. BRAFE600-associated senescence-like cell cycle arrest of human naevi. Nature 2005; 436(7051):720–724.
33. Liang XH, Jackson S, Seaman M, et al. Induction of autophagy and inhibition of tumorigenesis by beclin 1. Nature 1999; 402(6762):672–676.

34. Lum JJ, Bauer DE, Kong M, et al. Growth factor regulation of autophagy and cell survival in the absence of apoptosis. Cell 2005; 120(2):237–248.
35. Boya P, Gonzalez-Polo RA, Casares N, et al. Inhibition of macroautophagy triggers apoptosis. Mol Cell Biol 2005; 25(3):1025–1040.
36. Qu X, Yu J, Bhagat G, et al. Promotion of tumorigenesis by heterozygous disruption of the beclin 1 autophagy gene. J Clin Invest 2003; 112(12):1809–1820.
37. Denko NC, Giaccia AJ, Stringer JR, et al. The human Ha-ras oncogene induces genomic instability in murine fibroblasts within one cell cycle. Proc Natl Acad Sci U S A 1994; 91(11):5124–5128.
38. Tanaka S, Diffley JF. Deregulated G1-cyclin expression induces genomic instability by preventing efficient pre-RC formation. Genes Dev 2002; 16(20):2639–2649.
39. Vafa O, Wade M, Kern S, et al. c-Myc can induce DNA damage, increase reactive oxygen species, and mitigate p53 function: a mechanism for oncogene-induced genetic instability. Mol Cell 2002; 9(5):1031–1044.
40. Pusapati RV, Rounbehler RJ, Hong S, et al. ATM promotes apoptosis and suppresses tumorigenesis in response to Myc. Proc Natl Acad Sci U S A 2006; 103(5): 1446–1451.
41. Di Micco R, Fumagalli M, Cicalese A, et al. Oncogene-induced senescence is a DNA damage response triggered by DNA hyper-replication. Nature 2006; 444(7119):638–642.
42. Bartkova J, Horejsi Z, Koed K, et al. DNA damage response as a candidate anticancer barrier in early human tumorigenesis. Nature 2005; 434(7035):864–870.
43. Gorgoulis VG, Vassiliou LV, Karakaidos P, et al. Activation of the DNA damage checkpoint and genomic instability in human precancerous lesions. Nature 2005; 434(7035):907–913.
44. Bartkova J, Rezaei N, Liontos M, et al. Oncogene-induced senescence is part of the tumorigenesis barrier imposed by DNA damage checkpoints. Nature 2006; 444(7119):633–637.
45. Mallette FA, Gaumont-Leclerc MF, Ferbeyre G. The DNA damage signaling pathway is a critical mediator of oncogene-induced senescence. Genes Dev 2007; 21(1):43–48.
46. Lowe SW, Ruley HE, Jacks T, et al. p53-dependent apoptosis modulates the cytotoxicity of anticancer agents. Cell 1993; 74(6):957–967.
47. Lowe SW, Bodis S, McClatchey A, et al. p53 status and the efficacy of cancer therapy in vivo. Science 1994; 266(5186):807–810.
48. Schmitt CA. Senescence, apoptosis and therapy - cutting the lifelines of cancer. Nat Rev Cancer 2003; 3(4):286–295.
49. Chang BD, Broude EV, Dokmanovic M, et al. A senescence-like phenotype distinguishes tumor cells that undergo terminal proliferation arrest after exposure to anticancer agents. Cancer Res 1999; 59(15):3761–3767.
50. Chang BD, Xuan Y, Broude EV, et al. Role of p53 and p21waf1/cip1 in senescence-like terminal proliferation arrest induced in human tumor cells by chemotherapeutic drugs. Oncogene 1999; 18(34):4808–4818.
51. Han Z, Wei W, Dunaway S, et al. Role of p21 in apoptosis and senescence of human colon cancer cells treated with camptothecin. J Biol Chem 2002; 277(19): 17154–17160.
52. Schmitt CA, Lowe SW. Apoptosis and therapy. J Pathol 1999; 187(1):127–137.
53. Lee S, Schmitt CA. Mouse Models in Cancer Research. Germany: Wiley-VCH, 2006.

54. Palmiter RD. Transgenic mice—the early days. Int J Dev Biol 1998; 42(7):847–854.
55. Jonkers J, Berns A. Conditional mouse models of sporadic cancer. Nat Rev Cancer 2002; 2(4):251–265.
56. Johnson L, Mercer K, Greenbaum D, et al. Somatic activation of the K-ras oncogene causes early onset lung cancer in mice. Nature 2001; 410(6832):1111–1116.
57. Meuwissen R, Linn SC, van der Valk M, et al. Mouse model for lung tumorigenesis through Cre/lox controlled sporadic activation of the K-Ras oncogene. Oncogene 2001; 20(45):6551–6558.
58. Pear WS, Miller JP, Xu L, et al. Efficient and rapid induction of a chronic myelogenous leukemia-like myeloproliferative disease in mice receiving P210 bcr/abl-transduced bone marrow. Blood 1998; 92(10):3780–3792.
59. Hemann MT, Fridman JS, Zilfou JT, et al. An epi-allelic series of p53 hypomorphs created by stable RNAi produces distinct tumor phenotypes in vivo. Nat Genet 2003; 33(3):396–400.
60. Xue W, Zender L, Miething C, et al. Senescence and tumour clearance is triggered by p53 restoration in murine liver carcinomas. Nature 2007; 445(7128):656–660.
61. Nagano M, Brinster CJ, Orwig KE, et al. Transgenic mice produced by retroviral transduction of male germ-line stem cells. Proc Natl Acad Sci U S A 2001; 98(23): 13090–13095.
62. Carmell MA, Zhang L, Conklin DS, et al. Germline transmission of RNAi in mice. Nat Struct Biol 2003; 10(2):91–92.
63. Juweid ME, Cheson BD. Positron-emission tomography and assessment of cancer therapy. N Engl J Med 2006; 354(5):496–507.
64. Lyons SK. Advances in imaging mouse tumour models in vivo. J Pathol 2005; 205(2):194–205.
65. Lyons SK, Meuwissen R, Krimpenfort P, et al. The generation of a conditional reporter that enables bioluminescence imaging of Cre/loxP-dependent tumorigenesis in mice. Cancer Res 2003; 63(21):7042–7046.
66. Edinger M, Cao YA, Hornig YS, et al. Advancing animal models of neoplasia through in vivo bioluminescence imaging. Eur J Cancer 2002; 38(16):2128–2136.
67. Hoffman RM, Yang M. Dual-color, whole-body imaging in mice. Nat Biotechnol 2005; 23(7):790; author reply 791.
68. Vooijs M, Jonkers J, Lyons S, et al. Noninvasive imaging of spontaneous retinoblastoma pathway-dependent tumors in mice. Cancer Res 2002; 62(6):1862–1867.
69. Bearss DJ, Subler MA, Hundley JE, et al. Genetic determinants of response to chemotherapy in transgenic mouse mammary and salivary tumors. Oncogene 2000; 19(8):1114–1122.
70. Barrington RE, Subler MA, Rands E, et al. A farnesyltransferase inhibitor induces tumor regression in transgenic mice harboring multiple oncogenic mutations by mediating alterations in both cell cycle control and apoptosis. Mol Cell Biol 1998; 18(1):85–92.
71. Smith JB, Wickstrom E. Antisense c-myc and immunostimulatory oligonucleotide inhibition of tumorigenesis in a murine B-cell lymphoma transplant model. J Natl Cancer Inst 1998; 90(15):1146–1154.
72. Chin L, Tam A, Pomerantz J, et al. Essential role for oncogenic Ras in tumour maintenance. Nature 1999; 400(6743):468–472.
73. Felsher DW, Bishop JM. Reversible tumorigenesis by MYC in hematopoietic lineages. Mol Cell 1999; 4(2):199–207.

74. Adams JM, Harris AW, Pinkert CA, et al. The c-myc oncogene driven by immunoglobulin enhancers induces lymphoid malignancy in transgenic mice. Nature 1985; 318(6046):533–538.
75. Schmitt CA, Wallace-Brodeur RR, Rosenthal CT, et al. DNA damage responses and chemosensitivity in the Eμ-myc mouse lymphoma model. Cold Spring Harb Symp Quant Biol 2000; 65:499–510.
76. Schmitt CA, Lowe SW. Apoptosis and chemoresistance in transgenic cancer models. J Mol Med 2002; 80:137–146.
77. Eischen CM, Weber JD, Roussel MF, et al. Disruption of the ARF-Mdm2-p53 tumor suppressor pathway in Myc-induced lymphomagenesis. Genes Dev 1999; 13(20):2658–2669.
78. Schmitt CA, Fridman JS, Yang M, et al. A senescence program controlled by p53 and p16INK4a contributes to the outcome of cancer therapy. Cell 2002; 109: 335–346.
79. Hsu B, Marin MC, el-Naggar AK, et al. Evidence that c-myc mediated apoptosis does not require wild-type p53 during lymphomagenesis. Oncogene 1995; 11(1): 175–179.
80. Schmitt CA, Lowe SW. Bcl-2 mediates chemoresistance in matched pairs of primary Eμ-myc lymphomas in vivo. Blood Cells Mol Dis 2001; 27(1):206–216.
81. Waldman T, Zhang Y, Dillehay L, et al. Cell-cycle arrest versus cell death in cancer therapy. Nat Med 1997; 3(9):1034–1036.
82. Chang BD, Swift ME, Shen M, et al. Molecular determinants of terminal growth arrest induced in tumor cells by a chemotherapeutic agent. Proc Natl Acad Sci U S A 2002; 99(1):389–394.
83. Krtolica A, Parrinello S, Lockett S, et al. Senescent fibroblasts promote epithelial cell growth and tumorigenesis: a link between cancer and aging. Proc Natl Acad Sci U S A 2001; 98(21):12072–12077.
84. Sage J, Miller AL, Perez-Mancera PA, et al. Acute mutation of retinoblastoma gene function is sufficient for cell cycle re-entry. Nature 2003; 424(6945):223–228.
85. Bannister AJ, Schneider R, Kouzarides T. Histone methylation: dynamic or static? Cell 2002; 109(7):801–806.
86. Lachner M, O'Carroll D, Rea S, et al. Methylation of histone H3 lysine 9 creates a binding site for HP1 proteins. Nature 2001; 410(6824):116–120.
87. Ait-Si-Ali S, Guasconi V, Fritsch L, et al. A Suv39h-dependent mechanism for silencing S-phase genes in differentiating but not in cycling cells. EMBO J 2004; 23(3):605–615.
88. Shi Y, Lan F, Matson C, et al. Histone demethylation mediated by the nuclear amine oxidase homolog LSD1. Cell 2004; 119(7):941–953.
89. Metzger E, Wissmann M, Yin N, et al. LSD1 demethylates repressive histone marks to promote androgen-receptor-dependent transcription. Nature 2005; 437(7057): 436–439.
90. Yamane K, Toumazou C, Tsukada Y, et al. JHDM2A, a JmjC-containing H3K9 demethylase, facilitates transcription activation by androgen receptor. Cell 2006; 125(3):483–495.
91. Klose RJ, Kallin EM, Zhang Y. JmjC-domain-containing proteins and histone demethylation. Nat Rev Genet 2006; 7(9):715–727.
92. Shin S, Janknecht R. Diversity within the JMJD2 histone demethylase family. Biochem Biophys Res Commun 2007; 353(4):973–977.

93. Cloos PA, Christensen J, Agger K, et al. The putative oncogene GASC1 demethylates tri- and dimethylated lysine 9 on histone H3. Nature 2006; 442:307–311.
94. Klose RJ, Yamane K, Bae Y, et al. The transcriptional repressor JHDM3A demethylates trimethyl histone H3 lysine 9 and lysine 36. Nature 2006; 442(7100): 312–316.
95. Whetstine JR, Nottke A, Lan F, et al. Reversal of histone lysine trimethylation by the JMJD2 family of histone demethylases. Cell 2006; 125(3):467–481.
96. Lindemann RK, Newbold A, Whitecross KF, et al. Analysis of the apoptotic and therapeutic activities of histone deacetylase inhibitors by using a mouse model of B cell lymphoma. Proc Natl Acad Sci U S A 2007; 104(19):8071–8076.
97. Wendel HG, De Stanchina E, Fridman JS, et al. Survival signalling by Akt and eIF4E in oncogenesis and cancer therapy. Nature 2004; 428(6980):332–337.
98. Amaravadi RK, Yu D, Lum JJ, et al. Autophagy inhibition enhances therapy-induced apoptosis in a Myc-induced model of lymphoma. J Clin Invest 2007; 117(2): 326–336.
99. Jonkers J, Berns A. Oncogene addiction: sometimes a temporary slavery. Cancer Cell 2004; 6(6):535–538.
100. Ventura A, Kirsch DG, McLaughlin ME, et al. Restoration of p53 function leads to tumour regression in vivo. Nature 2007; 445(7128):661–665.
101. Martins CP, Brown-Swigart L, Evan GI. Modeling the therapeutic efficacy of p53 restoration in tumors. Cell 2006; 127(7):1323–1334.

17

Therapy-Induced Cellular Senescence: Clinical Relevance, Implications, and Applications

Daniel Y. Wu and Hui Wang

Department of Medicine, VA Puget Sound Health Care System; Department of Medicine, Division of Oncology, University of Washington; and Fred Hutchinson Cancer Research Center, Seattle, Washington, U.S.A.

Qin Wang

Department of Medicine, VA Puget Sound Health Care System, Seattle, Washington, U.S.A.

Peter C. Wu

Department of Surgery, University of Washington, Seattle, Washington, U.S.A.

Hubert J. Vesselle

Department of Nuclear Medicine, University of Washington, Seattle, Washington, U.S.A.

INTRODUCTION

Accelerated cellular senescence (ACS) is an emerging concept that implicates sustained, telomere-independent cell cycle arrest of immortalized cells in response to antineoplastic drugs, ionizing radiation, differentiating agents, oxidative stress, or the presence of selective oncogenic stimuli. The purpose of this chapter is to review the current understanding of ACS relevant to anticancer therapy. Recent findings from our laboratory suggest that therapy-induced cellular senescence exists in reversible and irreversible states. Possible molecular mechanisms mediating the conversion of reversible to irreversible ACS and their potential clinical implications will be discussed in this chapter. Advances in positron-emission tomography (PET) using novel metabolic tracers have allowed in vivo characterization of tumor biology. This technology may be used to determine the dominant therapy responses of tumors in patients during treatment. This chapter will also review the current progress in the field of in vivo imaging and discuss potential clinical significance of this technology.

IN VIVO ACCELERATED CELLULAR SENESCENCE AS A TUMOR RESPONSE TO TREATMENT

For a decade now, ACS has been clearly demonstrated in a variety of cancer tissue culture models following abbreviated courses of antineoplastic drugs at sublethal concentrations (1,2). These studies show that the senescence response can be elicited by a wide spectrum of agents suggesting that shared downstream pathways can be triggered by a variety of cellular damage and stress signals. A particularly surprising finding is that although ACS is clearly dependent on key players of the DNA damage and cell cycle checkpoint pathways, senescence can be induced in their absence. Individually or in combination p53, p21, p16, and pRB function may be dispensable in the ACS response (3–6). Particularly for epithelial solid tumors, this observation conforms to the decade long clinical experience on the equivocal importance of these tumor suppressor genes in patient outcome (7–11). Given our current understanding of p53/p21 and p16/RB pathways in tumorigenesis and physiological cellular senescence, it can be argued that both pathways are functionally inactivated in all solid tumors including the minority of the neoplasms that do not carry detectable loss or mutation of these genes (12,13). Therefore, in vivo occurrence of ACS as a tumor response should have been predicted despite the heterogeneity of genetic backgrounds in human cancer, especially at the sublethal concentrations of anticancer drugs that are conventionally used in clinical settings.

The occurrence of ACS in vivo has recently been demonstrated in nude mice xenograft, carcinogen-induced rat mammary tumor, and transgenic murine models (6,14–16). Of importance is the Eμ-*Myc* transgenic mice model, where the aberrant *Myc* expression results in a B-cell lymphoma that responds differentially to chemotherapy depending on the genetic background of the tumor (6). p53 was shown to be indispensable to chemotherapy response as Eμ-*Myc*/p53 null tumors were refractory to chemotherapy. $p16^{INK4a}$, $p19^{ARF}$, and Bcl-2 were all determinants of therapy response. Senescence response predominated in Eμ-*Myc*/$p19^{ARF-/-}$ tumors that also overexpress Bcl-2, presumably because of the anti-apoptotic effect by Bcl-2. Therapy-induced arrest in this situation was linked to $p16^{INK4a}$, when lost confers an additional chemotherapy resistance. These studies clearly show that both apoptosis and senescence contribute to outcome of cytotoxic therapy.

In vivo ACS in human cancer has been examined in a single retrospective study of archival tumor samples from patients with breast carcinoma obtained following cyclophosphamide, doxorubicin, and 5-FU therapy. Senescence-associated beta galactosidase (SA-β-gal) expression was found in 41% of samples derived from treated patients but only 10% in the untreated controls (17). SA-β-gal expression correlated with high expression level of p16 but inversely with the overexpression of p53 indicative of p53 mutations. It is noteworthy that 20% of tumor samples in p53 overexpressing samples were positive for ACS, confirming that while p53-dependent mechanisms promote ACS, p53-independent mechanisms also mediate ACS response to therapy. Our work in human lung cancer following neoadjuvant chemotherapy represents the only perspective demonstration of ACS as a physiological response to chemotherapy (5). The accrual of additional cases is ongoing. Thus far, SA-β-gal expression could be detected in nearly all cases (8/10) for which residual tumors were found at the time surgery (Table 1). Complete remission was induced by either neoadjuvant chemotherapy (carboplatin and paclitaxel) or chemoradiotherapy (CMT) in two cases (patients 1 and 8). Both patients are alive at the time of follow-up. The data is preliminary; however, the early trend suggests that detection of SA-β-gal on tumor tissue may portend to an adverse outcome. Additional cases and longer follow-up will be necessary to confirm these early observations.

DEFINING IRREVERSIBLE FROM REVERSIBLE ACCELERATED CELLULAR SENESCENCE AND ITS APPLICATION

Tumor cells in therapy-induced senescence appear to exist in two states with respect to their ability to reenter the cell cycle, reversible and irreversible senescence. We hypothesized that arrested cells in senescence gradually convert from a reversible to an irreversible state. Escape from ACS is thus possible only for the small number of cells remaining in the reversible state after prolonged cell cycle arrest. Recent findings in human mesenchymal stem cells and aging cardiac myocytes suggested that cells in senescence may be less susceptible to adenovirus infection as result of downregulation of the coxsackie-adenovirus receptor (CAR) (18,19). Utilizing this concept, we have recently developed an approach to distinguish the two states of therapy-induced cellular senescence. We marked H1299 cells during chemotherapy-induced senescence using adenovirus harboring a marker transgene, either the bacterial LacZ or the red fluorescence protein (RFP), and found that two subsets of senescent cell populations, marked and unmarked, clearly exist following chemotherapy (Fig. 1A). These two cell populations are morphologically indistinguishable under light microscopy and in their expression of SA-β-gal. Furthermore, propidium iodide exclusion studies confirmed that both cell populations retained their membrane integrity (data not shown). Following chemotherapy, the relative abundance of the marked cell population declines over time and is associated with an inverse rise of the unmarked cell population (Fig. 1B). When the marked

Table 1 Tumor Response and Clinical Outcome of Lung Cancer Patients following Neoadjuvant Therapy

Patient no.	TMN staging	Neoadjuvant therapy	Tumor pathology	SA-β-galactosidase		Follow-up
				Lung	Tumor	
1	T2N0M0	Carbo/taxol × 3	CR	+/–	NA	NED/46 mo
2	T3N1M0	Carbo/taxol × 3	Viable tumor	+/–	++++	†/14 mo
3	T3N0M0	Carbo/taxol × 3	Viable tumor	+/–	+++	†/30 mo
4	T2N2M0	CMT	Viable tumor	+/–	++	RD/36 mo
5	T2N2M0	CMT	Viable tumor	+	+++	†/27 mo
6	T3N2M0	CMT	Viable tumor	+/–	+	NED/28mo
7	T3N1M0	Carbo/taxol × 3	Viable tumor	–	+++	RD/12 mo
8	T3N2M0	CMT	CR	+/–	NA	NED/24 mo
9	T1N2M0	CMT	Viable tumor	+/–	++++	†/31mo
10	T2N2M0	CMT	Viable tumor	+/–	++	NED/10 mo
11	T2N2M0	CMT	Viable tumor	+/–	+++	†/17 mo
12	T3N0M0	CMT	Viable tumor	+/–	+	NED/12 mo

Abbreviations: TMN, tumor-nodes-metastasis; SA, senescence associated; CMT, combined modality therapy; +/–, scant/spotty positive; CR, complete response; NED, no evidence of disease; NA, not applicable; RD, recurrent disease/alive; †, deceased/recurrent disease.

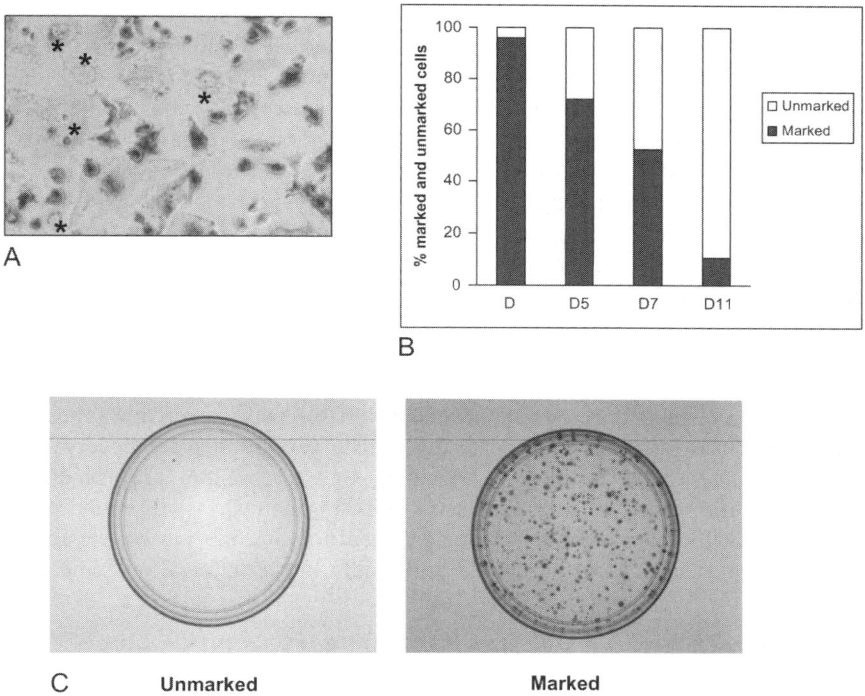

Figure 1 Reversible and irreversible states of ACS in H1299 cells. H1299 cells were treated with 60 nM camptothecin for three days and then rinsed and released into fresh media for recovery. On days 5, 7, and 11, the cells were infected with adenovirus-LacZ or adenovirus-RFP for three hours. Marker gene expression was determined after an overnight incubation by in situ X-gal assay or by FACS analysis. (A) In situ X-gal assay on H1299 cells in ACS (day 7) marked by adenovirus-LacZ after chemotherapy. * indicates unmarked senescence cells that failed to express marker bacterial β-galactosidase. (B) Population of marked cells and unmarked cell analyzed by FACS flow cytometer. Marked cells represent cells infected by adenovirus and expressing RFP, and unmarked cells represent cells uninfected by adenovirus. Untreated cell populations are designated as D0. Marked and unmarked cell populations on days 5, 7, and 11 are shown. (C) Colony formation assay of marked and unmarked cells. H1299 cells marked on in ACS (day 7) were sorted by FACS flow cytometer. 5×10^5 cells from each population were collected and replated into tissue culture plates. After incubation in growth medium for 10 days, colonies were fixed in 1% formaldehyde and stained with 2% trypan blue stain. *Abbreviations*: ACS, accelerated cellular senescence; RFP, red fluorescence protein; FACS, fluorescence-activated cell sorting (*The color version of this figure is provided on the DVD*).

and unmarked cells were sorted by fluorescence-activated cell sorting (FACS and replated into tissue cultures, colonies formed nearly exclusively from the marked cells suggesting that marked cells retain a significant ability to reenter the cell cycle and is therefore reversible in their arrest (Fig. 1C). In contrast, the unmarked cells proceed to cell death and are thus irreversibly arrested in cell cycle. The adenovirus marking correlates to the surface CAR expression of the senescent cells by gated FACS analysis (data not shown); therefore, either adenoviral marking or CAR expression could be used as potential strategies to distinguish reversible from irreversible cell populations. We have verified this application in several other tissue culture models.

These and other findings in our laboratory suggest that therapy-induced cellular senescence involves a further conversion of cells in sustained but reversible arrest to an irreversibile arrest that destines senescent cells to eventual death. This conversion may involve several possible mechanisms discussed below. The ability to distinguish reversible from irreversible senescent states using either adenoviral marking or based on CAR expression represents a powerful tool for studying early events in ACS, thereby making it possible to define the molecular differences between reversible and irreversible senescent cells and pathways that may regulate this conversion. Studies are underway to further define possible pathways regulating this conversion. Further understanding of the molecular mechanisms or signaling pathways will undoubtedly facilitate development of novel strategies to enforce the in vivo conversion to irreversible senescence, which should profoundly impact clinical outcome.

POSSIBLE MOLECULAR MECHANISMS OF IRREVERSIBLE ACCELERATED CELLULAR SENESCENCE

The conversion of a stressed cell from an arrested state that is potentially reversible to one that is irreversible may involve mechanisms shared between physiological and therapy-induced senescence. Epigenetic alterations resulting in chromatin remodeling have long been observed in aging of explanted epithelial cell in culture, where the overall CpG methylation decreases with increased population doublings (20,21). In normal human fibroblasts, this corresponds to reduced activity of DNA methyltransferase (22). Although global DNA methylation tends to decrease with cellular aging, paradoxical hypermethylation has been demonstrated to occur at selective CpG-rich promoters particularly at tumor suppressor genes, as precursor events to neoplastic transformation (23). It is presently unknown whether altered methylation pattern is responsible for therapy-induced cellular senescence; however, supporting evidences exist. For example, the methyltransferase inhibitor 5-azacitidine, recently approved by the FDA for treatment of myelodysplastic syndromes, can cause global hypomethylation and induce senescence phenotype in both neoplastic and primary cell lines (24). Furthermore, 5-azacitidine has been shown to interfere with hypermethylated human telomerase catalytic subunit (hTERT) promoter in telomerase positive tumor cell lines and induce replicative senescence after 15 population doublings (25).

Covalent modification of N-terminal tails of histone H3 and H4 also contributes to chromatin remodeling. A recent study showed that low dose doxorubicin induces acetylation of histone H3 at the *p21/WAF1* promoter and senescence in neuroblastoma cell line SKN-SH (26). Histone deacetylase-1 (HDAC1), frequently recruited to promoters by transcription repressor complexes, has been reported to be associated with CpG-binding proteins (27,28). At selected promoters, gene silencing appears to be initially mediated by histone deacetylation then followed by DNA methylation (29–31). SIR2 family of histone acetylases have been linked to aging, transcriptional silencing, chromatin remodeling, mitosis, and cell aging (32,33). Its human homologue SIRT1 found in complex with p53 may function to inhibit p53-mediated transcriptional activation of checkpoint genes and interfere with DNA damage–related arrest (34,35). HDAC inhibitors, presently in early clinical trial, have also been shown to induce senescence in both primary and neoplastic cell lines (36,37). Of interest, HDAC inhibitor suberoylanilide hydroxamic acid can cause mitotic defect and results in senescence in polypoidy HCT116 colon carcinoma cells by mechanisms apparently independently of p53, p21, and p16 (38). Therefore, both HDAC and methytransferase inhibitors may be excellent candidates for novel regimens in combination with conventional chemotherapeutic drugs to enforce cell cycle arrest in senescent tumors.

Another possible event leading to irreversible senescence may be occurring at telomeres of damaged cells. In a number of cell lines, we have found that irreversible senescence in cancer cells in ACS often accompanies massive loss of telomere length (data not shown). Moreover, cells that have bypassed senescence invariably partially or totally recovered the telomere loss. Our observations imply that uncapping of telomere ends is a frequent event in ACS. The ability of damaged cells to regain telomere length or recap, therefore, may determine whether the cells can regain proliferative capacity. Telomere uncapping as a result of DNA damage or by interfering with the TRF2 telomere–binding protein function has been shown to induce senescence (39,40). The molecular pathways mediating the senescence signal produced by telomere dysfunction have been proposed to involve ataxia telangiectasia mutated/ataxia telangiectasia Rad3-related (ATM/ATR), $p19^{ARF}$, and possibly p16 (41–43), although $p19^{ARF}$ and p16 are clearly dispensable since neither protein is expressed in H1299 lung carcinoma cells in senescence. Telomestatin, a potent G-quadruplex ligand that interacts with 3′ telomeric overhang at the end-cap structure, induces telomere uncapping associated with the loss of POT1 and TRF2 at these sites (44–46). Telomestatin triggers a delayed senescence in HT1080 fibrosarcoma cell line that could be partially blocked by overexpression of GFP-POT1. It is presently unknown if the cells that have bypassed irreversible senescence is inherently more resistant to telomere uncapping or have acquired a recap mechanism of the uncapped ends.

IN VIVO IMAGING

By and large, most human cancers are systemic diseases. Systemic failure following surgical therapy with curative intent occurs in a significant percentage of solid tumors even in those diagnosed at early stages. Taking stage IB non–small cell lung cancer (NSCLC) as an example, only 60% of those patients are expected to be alive at five years despite having no detectable lymph node or distant disease at the time of diagnosis (47). Unresectable NSCLC is rarely curable. Even with chemotherapy and radiation, less than 20% of the patients are expected to be alive at two years (48). Most solid tumors are also inherently chemoresistant. In NSCLC, first line chemotherapy produces only 25% to 40% tumor response (as defined by a reduction of the tumor on imaging). The response is mostly transient with tumor progression occurring one to six months following completion of therapy. The relative chemotherapy insensitivity in NSCLC and other solid tumors has been attributed to complex inherent and acquired apoptosis resistance mechanisms found in cancer cells (49). Given the heterogeneity of most solid tumors in their clinical evolution, it is likely that the treatment response of these cancers represents a mixture of apoptosis, transient arrest, reversible, and irreversible senescence. Developing noninvasive imaging technologies that enable clinicians to gain insights to specific tumor biology and treatment response will be important as more novel targeted drugs become available. For tumors demonstrating predominately senescence response, therapy directed at enforcing irreversibility should be considered.

Replicative arrest and reduced metabolic activity are obligatory features of cancer cells in senescence. In the past decade, the innovations in positron-emission tomography (PET) technology have enabled measurements of in vivo metabolic and proliferative activities of human cancers using radiolabeled tracers. The most widely used tracer is [^{18}F]-fluorodeoxyglucose (FDG), a glucose analog that has established clinical applications. Although FDG is not taken up specifically by cancer cells, the enhanced glycolytic rate in most malignant tumors increases FDG uptake and facilitates their detection using FDG-PET imaging (50,51). In NSCLC, FDG-PET has gained acceptance as a staging modality and the level of tracer uptake can provide prognostic information in early stages of the disease (52). [^{18}F]-3′-deoxy-3′-fluorothymidine (FLT) is a thymidine analog that has been used to measure tumor cell proliferation (53,54). FLT is more tumor specific than FDG. FLT uptake on PET imaging has been correlated with the Ki67 proliferation marker in human lung cancer (55). In practice, it is now possible to define both proliferative and metabolic responses to induction therapy in specific patient. An example of a patient with NSCLC is shown in Figure 2. In this patient, the post-therapy tumor FDG uptakes is significantly reduced but continued to exhibit tracer uptake suggesting the remaining of residual metabolically active, viable tumor. However, the FLT-tracer scan showed complete absence of signal indicating the lack of tumor cell proliferation. This type of PET response is highly suggestive of therapy-induced ACS. Along with well-designed clinicopathological studies, PET imaging information could be used to direct the development and the use of novel therapy to enforce senescence in clinical setting. Finally, additional biological information can be gained with other tracers, such as radiolabeled annexin V and NST-732, currently in active investigation (56). Both of these

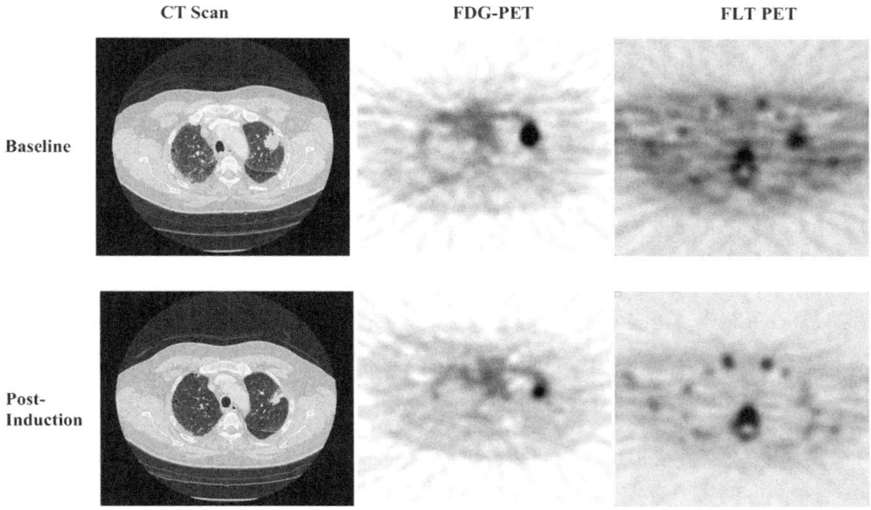

Figure 2 CT and PET scans of a lung cancer before and after induction therapy. CT and PET scans using either FDG or FLT tracers are carried out in a patient with T2 non–small cell carcinoma before or after induction chemotherapy and radiation. Arrow indicates the location of the primary tumor. *Abbreviations*: CT, computed tomography; PET, positron-emission tomography; FDG, fluorodeoxyglucose; FLT, fluorothymidine.

agents may be useful for assessing the extent of apoptosis occurring following therapy. Thus, it may be possible in the future to tailor therapy for individual patients based on dominant therapy responses occuring in vivo.

CONCLUSION

Over the last decade, our knowledge of ACS has been expanding at an amazing pace. Continued unfolding of the molecular mechanisms of therapy-induced senescence leading to irreversible ACS will promote the development of novel therapy based on rational targets. In vivo imaging potentially represents a strategy for patient risk stratification based on tumor therapy response and designing of clinical trials to tailor therapy based on biology.

REFERENCES

1. Roninson IB. Tumor cell senescence in cancer treatment. Cancer Res 2003; 63: 2705–2715.
2. Shay JW, Roninson IB. Hallmarks of senescence in carcinogenesis and cancer therapy. Oncogene 2004; 23:2919–2933.
3. Chang BD, Broude EV, Dokmanovic M, et al. A senescence-like phenotype distinguishes tumor cells that undergo terminal proliferation arrest after exposure to anticancer agents. Cancer Res 1999a; 59:3761–3767.

4. Chang BD, Xuan Y, Broude EV, et al. Role of p53 and p21waf1/cip1 in senescence-like terminal proliferation arrest induced in human tumor cells by chemotherapeutic drugs. Oncogene 1999b; 18:4808–4818.
5. Roberson RS, Kussick SJ, Vallieres E, et al. Escape from therapy-induced accelerated cellular senescence in p53-null lung cancer cells and in human lung cancers. Cancer Res 2005; 65:2795–2803.
6. Schmitt CA, Fridman JS, Yang M, et al. A senescence program controlled by p53 and p16INK4a contributes to the outcome of cancer therapy. Cell 2002; 109: 335–346.
7. Cuddihy AR, Bristow RG. The p53 protein family and radiation sensitivity: yes or no? Cancer Metastasis Rev 2004; 23:237–257.
8. Elledge RM, Allred DC. Prognostic and predictive value of p53 and p21 in breast cancer. Breast Cancer Res Treat 1998; 52:79–98.
9. Munro AJ, Lain S, Lane DP. P53 abnormalities and outcomes in colorectal cancer: a systematic review. Br J Cancer 2005; 92:434–444.
10. Thames HD, Petersen C, Petersen S, et al. Immunohistochemically detected p53 mutations in epithelial tumors and results of treatment with chemotherapy and radiotherapy. A treatment-specific overview of the clinical data. Strahlenther Onkol 2002; 178:411–421.
11. Zhu CQ, Shih W, Ling CH, et al. Immunohistochemical markers of prognosis in non-small cell lung cancer: a review and proposal for a multiphase approach to marker evaluation. J Clin Pathol 2006; 59:790–800.
12. Bond JA, Haughton MF, Rowson JM, et al. Control of replicative life span in human cells: barriers to clonal expansion intermediate between M1 senescence and M2 crisis. Mol Cell Biol 1999; 19:3103–3114.
13. Shay JW, Pereira-Smith OM, Wright WE. A role for both RB and p53 in the regulation of human cellular senescence. Exp Cell Res 1991; 196:33–39.
14. Christov KT, Shilkaitis AL, Kim ES, et al. Chemopreventive agents induce a senescence-like phenotype in rat mammary tumours. Eur J Cancer 2003; 39: 230–239.
15. Elmore LW, Rehder CW, Di X, et al. Adriamycin-induced senescence in breast tumor cells involves functional p53 and telomere dysfunction. J Biol Chem 2002; 277:35509–35515.
16. Roninson IB, Broude EV, Chang BD. If not apoptosis, then what? Treatment-induced senescence and mitotic catastrophe in tumor cells. Drug Resist Updat 2001; 4:303–313.
17. te Poele RH, Okorokov AL, Jardine L, et al. DNA damage is able to induce senescence in tumor cells in vitro and in vivo. Cancer Res 2002; 62:1876–1883.
18. Communal C, Huq F, Lebeche D, et al. Decreased efficiency of adenovirus-mediated gene transfer in aging cardiomyocytes. Circulation 2003; 107:1170–1175.
19. Hung SC, Lu CY, Shyue SK, et al. Lineage differentiation-associated loss of adenoviral susceptibility and Coxsackie-adenovirus receptor expression in human mesenchymal stem cells. Stem Cells 2004; 22:1321–1329.
20. Fairweather DS, Fox M, Margison GP. The in vitro lifespan of MRC-5 cells is shortened by 5-azacytidine-induced demethylation. Exp Cell Res 1987; 168: 153–159.
21. Wilson VL, Jones PA. DNA methylation decreases in aging but not in immortal cells. Science 1983; 220:1055–1057.

22. Vertino PM, Issa JP, Pereira-Smith OM, et al. Stabilization of DNA methyltransferase levels and CpG island hypermethylation precede SV40-induced immortalization of human fibroblasts. Cell Growth Differ 1994; 5:1395–1402.
23. Baylin SB, Herman JG, Graff JR, et al. Alterations in DNA methylation: a fundamental aspect of neoplasia. Adv Cancer Res 1998; 72:141–196.
24. Momparler RL. Pharmacology of 5-Aza-2′-deoxycytidine (decitabine). Semin Hematol 2005; 42:S9–S16.
25. Guilleret I, Benhattar J. Demethylation of the human telomerase catalytic subunit (hTERT) gene promoter reduced hTERT expression and telomerase activity and shortened telomeres. Exp Cell Res 2003; 289:326–334.
26. Rebbaa A, Zheng X, Chu F, et al. The role of histone acetylation versus DNA damage in drug-induced senescence and apoptosis. Cell Death Differ 2006; 13: 1960–1967.
27. Ng HH, Zhang Y, Hendrich B, et al. MBD2 is a transcriptional repressor belonging to the MeCP1 histone deacetylase complex. Nat Genet 1999; 23:58–61.
28. Razin A. CpG methylation, chromatin structure and gene silencing-a three-way connection. EMBO J 1998; 17:4905–4908.
29. Cameron EE, Bachman KE, Myohanen S, et al. Synergy of demethylation and histone deacetylase inhibition in the re-expression of genes silenced in cancer. Nat Genet 1999; 21:103–107.
30. Strunnikova M, Schagdarsurengin U, Kehlen A, et al. Chromatin inactivation precedes de novo DNA methylation during the progressive epigenetic silencing of the RASSF1A promoter. Mol Cell Biol 2005; 25:3923–3933.
31. Wischnewski F, Pantel K, Schwarzenbach H. Promoter demethylation and histone acetylation mediate gene expression of MAGE-A1, -A2, -A3, and -A12 in human cancer cells. Mol Cancer Res 2006; 4:339–349.
32. Blander G, Guarente L. The Sir2 family of protein deacetylases. Annu Rev Biochem 2004; 73:417–435.
33. Chang KT, Min KT. Regulation of lifespan by histone deacetylase. Ageing Res Rev 2002; 1:313–326.
34. Chua KF, Mostoslavsky R, Lombard DB, et al. Mammalian SIRT1 limits replicative life span in response to chronic genotoxic stress. Cell Metab 2005; 2:67–76.
35. Vaziri H, Dessain SK, Ng Eaton E, et al. hSIR2(SIRT1) functions as an NAD-dependent p53 deacetylase. Cell 2001; 107:149–159.
36. Ota H, Tokunaga E, Chang K, et al. Sirt1 inhibitor, Sirtinol, induces senescence-like growth arrest with attenuated Ras-MAPK signaling in human cancer cells. Oncogene 2006; 25:176–185.
37. Place RF, Noonan EJ, Giardina C. HDACs and the senescent phenotype of WI-38 cells. BMC Cell Biol 2005; 6:37.
38. Xu WS, Perez G, Ngo L, et al. Induction of polyploidy by histone deacetylase inhibitor: a pathway for antitumor effects. Cancer Res 2005; 65:7832–7839.
39. Blackburn EH, Chan S, Chang J, et al. Molecular manifestations and molecular determinants of telomere capping. Cold Spring Harb Symp Quant Biol 2000; 65: 253–263.
40. Li GZ, Eller MS, Firoozabadi R, et al. Evidence that exposure of the telomere 3′ overhang sequence induces senescence. Proc Natl Acad Sci U S A 2003; 100:527–531.
41. Ben-Porath I, Weinberg RA. The signals and pathways activating cellular senescence. Int J Biochem Cell Biol 2005; 37:961–976.

42. Incles CM, Schultes CM, Kempski H, et al. A G-quadruplex telomere targeting agent produces p16-associated senescence and chromosomal fusions in human prostate cancer cells. Mol Cancer Ther 2004; 3:1201–1206.
43. Tauchi T, Shin-Ya K, Sashida G, et al. Activity of a novel G-quadruplex-interactive telomerase inhibitor, telomestatin (SOT-095), against human leukemia cells: involvement of ATM-dependent DNA damage response pathways. Oncogene 2003; 22:5338–5347.
44. Gomez D, O'Donohue MF, Wenner T, et al. The G-quadruplex ligand telomestatin inhibits POT1 binding to telomeric sequences in vitro and induces GFP-POT1 dissociation from telomeres in human cells. Cancer Res 2006a; 66:6908–6912.
45. Gomez D, Wenner T, Brassart B, et al. Telomestatin-induced telomere uncapping is modulated by POT1 through G-overhang extension in HT1080 human tumor cells. J Biol Chem 2006b; 281:38721–38729.
46. Riou JF, Guittat L, Mailliet P, et al. Cell senescence and telomere shortening induced by a new series of specific G-quadruplex DNA ligands. Proc Natl Acad Sci U S A 2002; 99:2672–2677.
47. Schiller JH. Adjuvant systemic therapies in early-stage non-small-cell lung cancer. Clin Lung Cancer 2003; 5(suppl 1):S29–S35.
48. Bunn PA Jr. Chemotherapy for advanced non-small-cell lung cancer: who, what, when, why? J Clin Oncol 2002; 20:23S–33S.
49. Shivapurkar N, Reddy J, Chaudhary PM, et al. Apoptosis and lung cancer: a review. J Cell Biochem 2003; 88:885–898.
50. Ak I, Blokland JA, Pauwels EK, et al. The clinical value of 18F-FDG detection with a dual-head coincidence camera: a review. Eur J Nucl Med 2001; 28:763–778.
51. Avril N, Menzel M, Dose J, et al. Glucose metabolism of breast cancer assessed by 18F-FDG PET: histologic and immunohistochemical tissue analysis. J Nucl Med 2001; 42:9–16.
52. Higashi K, Ueda Y, Arisaka Y, et al. 18F-FDG uptake as a biologic prognostic factor for recurrence in patients with surgically resected non-small cell lung cancer. J Nucl Med 2002; 43:39–45.
53. Buck AK, Halter G, Schirrmeister H, et al. Imaging proliferation in lung tumors with PET: 18F-FLT versus 18F-FDG. J Nucl Med 2003; 44:1426–1431.
54. Rasey JS, Grierson JR, Wiens LW, et al. Validation of FLT uptake as a measure of thymidine kinase-1 activity in A549 carcinoma cells. J Nucl Med 2002; 43: 1210–1217.
55. Vesselle H, Grierson J, Muzi M, et al. In vivo validation of 3′deoxy-3′-[(18)F] fluorothymidine ([(18)F]FLT) as a proliferation imaging tracer in humans: correlation of [(18)F]FLT uptake by positron emission tomography with Ki-67 immunohistochemistry and flow cytometry in human lung tumors. Clin Cancer Res 2002; 8: 3315–3323.
56. Aloya R, Shirvan A, Grimberg H, et al. Molecular imaging of cell death in vivo by a novel small molecule probe. Apoptosis 2006; 11:2089–2101.

18

Mitotic Catastrophe in Cancer Therapy

Eugenia V. Broude, Jadranka Loncarek, Ikuo Wada, Kelly Cole, Christine Hanko, and Igor B. Roninson

Cancer Center, Ordway Research Institute, Albany, New York, U.S.A.

Mari Swift

Department of Molecular Genetics, University of Illinois at Chicago, Chicago, Illinois, U.S.A.

INTRODUCTION: CHECKPOINT DEFICIENCIES ARE THE ACHILLES HEEL OF TUMORS

Is it possible to name one general characteristic that distinguishes all the tumor cells from normal cells? Perhaps surprisingly, we believe that the answer to this question is "yes." Uncontrolled proliferation of tumor cells means that such cells are prone to progress through the cell cycle and divide under the conditions where normal cells become arrested in their cycle. Cell cycle arrest in response to physiological and damage signals is mediated by a variety of checkpoint mechanisms that become activated in different phases of the cell cycle. The process of carcinogenesis involves the inactivation of one or more cell cycle

This chapter refers to video presentations in the attached DVD: E. V. Broude, J. Loncarek, I. Wada, K. Cole, C. Hanko, and I. B. Roninson, Fluorescent and phase contrast video microscopy of mitotic catastrophe in irradiated tumor cells.

checkpoint mechanisms. As a result, tumor cells divide under inappropriate conditions, where normal cells would not. Almost all, and probably all the tumors are completely or partially deficient in at least some cell cycle checkpoints, including G1 checkpoints (which are regulated primarily by p53), G2 checkpoint (1), early mitosis checkpoint (2), and mitotic spindle checkpoint (3). Hence, cell cycle checkpoint deficiency is the most general property of all the tumor cells. While this property enables neoplastic transformation, it also provides the key weakness that allows selective killing of tumor cells by chemotherapeutic drugs and radiation.

Mitosis is by far the most complex of all the phases of the cell cycle and also the most prone to aberrations, which may lead either to cell death or to ploidy changes in daughter cells, an event with a great carcinogenic potential. Abnormal mitosis may occur when cells enter mitosis with damaged DNA, under the conditions where proper spindle formation is disrupted (e.g., in the presence of antimicrotubular agents), or with a deficiency in any of the multitude of proteins or protein complexes that carry out the intricate choreography of mitosis. Activation of cell cycle checkpoints provides cells with the time to repair the damage or to replenish the pool of necessary proteins, thereby preventing abnormal mitosis. In particular, the G2 checkpoint is the key to preventing cells treated with DNA-damaging agents from entering abnormal mitosis. The G2 checkpoint is regulated primarily through the pathways centered on the mitosis-initiating Cyclin B1-Cdc2 kinase complex (1), and it includes both p53-independent and p53-dependent mechanisms (4). At least one half of human tumors are p53-deficient, and some tumors also show mutations or altered expression of other components of the G2 checkpoint (1). Inhibition or knockout of several G2 checkpoint genes, such as ATM, ATR, Chk1, Chk2, Plk1, Pin1 or Mlh1 (5), as well as p53 and p53-inducible checkpoint regulators, p21 and 14-3-3-σ (4,6,7) were found to promote abnormal mitosis after DNA damage. MC was potentiated by the inhibition of Chk2 with a dominant mutant or a chemical inhibitor (8), or by depletion of Chk1 (9). Abnormal mitotic entry is also potentiated by chemical inhibitors of the G2 checkpoint, such as caffeine, okadaic acid, pentoxifylline, or UCN-01. On the other hand, the early mitotic checkpoint mediated by the protein CHFR, which is frequently inactivated in cancers, stops cells treated with microtubule-targeting agents at the onset of mitosis, prior to chromosome condensation, providing resistance to cell death induced by taxanes (2,10). Hence, the defects of cell cycle checkpoints in tumor cells make them more prone to undergo abnormal mitosis, compared with checkpoint-competent normal cells.

WHAT IS MITOTIC CATASTROPHE?

The propensity of tumor cells to undergo abnormal mitosis has been exploited, often unwittingly, in cancer therapy. As discussed below, all or almost all the agents in the cancer clinics induce abnormal mitosis, and this response is greatly enhanced by disabling cell cycle checkpoints. To be of therapeutic use, such abnormal mitosis should lead to cell death or irreversible growth arrest, the process termed mitotic catastrophe (MC). The above definition of MC (abnormal mitosis that leads to cell death or permanent cessation of cell division) is flexible and allows one to discuss the forms, mechanisms, and outcomes of this process in the context of therapeutic action. Many other definitions of MC have been used (or more often implied) in the literature, causing significant confusion. It would be helpful to go through some of these alternative definitions, in order to understand the nature of MC.

1. MC is sometimes equated with any form of abnormal mitosis. However, some abnormal mitotic events can result in cell survival with potentially tumor-promoting ploidy changes. An example of the latter is the failure of cytokinesis, which in some cases can lead to the formation of viable cells with double the original DNA content. This happens, for example, when the function of a protein required for cytokinesis is inhibited (11). Such polyploidization events are not catastrophic and should not be defined as MC. On the other hand, when the failure of cytokinesis is due to the failure of sister chromatid segregation (as described below for one type of radiation-induced MC), this event is usually lethal for the cell, fitting the definition of MC. It is important, therefore, to understand the underlying mechanisms for superficially similar events with different outcomes.
2. MC is sometimes defined as cell death that occurs during mitosis. Such mitotic cell death indeed occurs through apoptosis or necrosis (12). However, more often, cells that undergo abnormal mitosis die later in the interphase where they exit from unsuccessful mitosis (12). These outcomes are illustrated in phase-contrast time-lapse microscopy videos provided on the attached DVD (Broude EV, Loncarek J, Wada I, Cole K, Hanko C, and Roninson IB, Fluorescent and phase contrast video microscopy of mitotic catastrophe in irradiated tumor cells.). *Video 6* shows HCT116 (p21−/−) colon carcinoma cells filmed after ionizing radiation, where one cell (on the left) enters mitosis (rounds up) but reverts to the interphase without dividing. This cell then undergoes apoptosis while in the interphase. Concurrently, several other cells in the same field enter mitosis, and two of these cells undergo apoptosis while in mitosis. *Video 7* shows the fate of another cell from the same experiment. This large (apparently polyploid) cell enters mitosis but reverts to the interphase without dividing (restitution). This cell then dies by breaking up and spilling its contents (necrosis). *Video 8* shows an interesting example of abnormal mitosis that appears to be non-catastrophic upon short-term observation (which is usually used in the literature) but is in fact catastrophic when followed up for a longer term. Here, irradiated HT1080 fibrosarcoma cells carrying a transdominant p53 inhibitor GSE56 (7) were filmed from 4 to 72 hours after irradiation. A cell at the top of the field enters mitosis but reverts to the interphase without dividing. Following a full-length interphase, the cell (now micronucleated) enters another mitosis and nearly completes the telophase,

but the daughter cells remain connected and undergo restitution. Following another interphase, the large cell (with 4× the original DNA content) dies apparently through apoptosis.

Cells that underwent MC may also become senescent rather than die, as indicated by a substantial number of cells that become both micronucleated (a hallmark of MC, see below) and positive for senescence-associated β-galactosidase activity (SA-β-gal), a marker of senescence (13), after exposure to DNA-damaging agents (7,14). Even more strikingly, senescence is the primary effect of discodermolide, a drug that directly interferes with mitosis by hyperpolymerizing microtubules (15). Hence, death during mitosis is only one of the outcomes of MC.

3. MC was sometimes described as an early stage of apoptosis. This misperception occurs because MC in apoptosis-competent cells is frequently followed by apoptosis (16–18). Apoptosis, however, is not required for the lethal effect of MC, which can lead to cell death through both caspase-dependent and caspase-independent mechanisms (19). For example, Lock and Stribinskiene (20) showed that etoposide treatment induces both micronucleation and apoptosis in HeLa cells. When apoptosis was suppressed by transfecting cells with *BCL2*, many more drug-treated tumor cells were found to die through MC alone, and there was no significant change in the clonogenic survival in the presence of etoposide. In another example, apoptosis suppressor Bcl-X_L was found to block apoptosis induced by high-dose doxorubicin in hepatoma cells, but it did not block MC followed by non-apoptotic death in cells treated with a low dose of doxorubicin (21). A similar result from our study (22) is illustrated in Figure 1. In this experiment, a HeLa-derived cell line with tetracycline-regulated expression of the MDR1 (ABCB1) gene, which blocks apoptosis through a mechanism distinct from its well-known function as a multidrug transporter (23,24), was exposed to 9 Gy of radiation, with or without *MDR1* induction. At different time points after irradiation, we measured fractions of cells that display morphological markers of apoptosis, MC

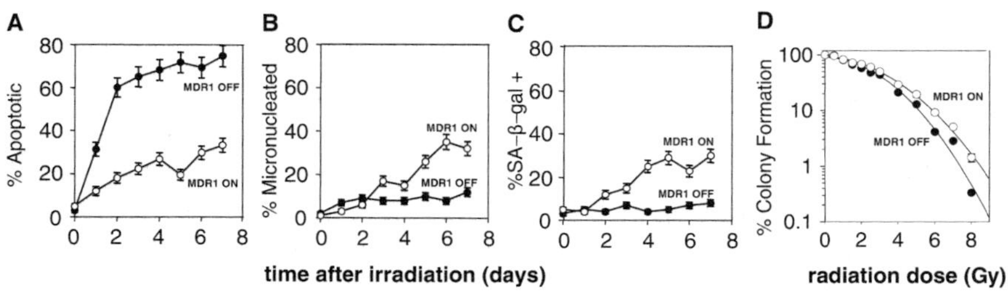

Figure 1 Effects of MDR1 expression in HeLa-derived HtTA cells on the induction of apoptosis (**A**), mitotic catastrophe (**B**), senescence (**C**), by 9 Gy of ionizing radiation, and clonogenic survival of different doses of radiation (**D**). *Source*: From Ref. 22.

(micronucleation, see below) or senescence (SA-β-gal staining). Without *MDR1*, apoptosis was the most prominent response to radiation in this cell line. Upon *MDR1* induction, apoptosis was greatly decreased (Fig. 1A), but this decrease was followed by a compensating increase in the fraction of cells that undergo MC (Fig. 1B) or senescence (Fig. 1C). Whereas the inhibition of apoptosis appears to provide radiation resistance if measured by the number of live cells in the short term, a criterion of resistance frequently used in the literature (not shown), it has only a minimal overall effect on radiation survival in the long term, as measured by the colony formation assay (Fig. 1D). Hence, cells that undergo MC may be dying *through* apoptosis but not *because* of it.

4. Micronucleation has been used by many authors (including ourselves) as a criterion for defining MC. Indeed, cells that reenter the interphase following abnormal mitosis develop multiple micronuclei that are completely or partially separated from each other. In the case of partial separation, the cell displays a large lobulated nucleus. The micronuclei arise through the formation of multiple nuclear envelopes around chromosome clusters at the end of abnormal mitosis. The formation of multiple micronuclei in cells that underwent abnormal mitosis is illustrated in Figure 2A. Chromosomes are randomly distributed among the resulting micronuclei, as illustrated in Figure 2B by fluorescence in situ hybridization of specific probes for chromosomes 18 and 21 with a multilobulated nucleus of a drug-treated tumor cell. Although micronucleation is a good sign that abnormal, and most probably catastrophic, mitosis has taken place, it does not occur when a cell dies during mitosis (as in *Video 6*). Therefore, micronucleation, while remaining the most easy to score marker of MC, reflects only a subset of MC events.

MITOTIC CATASTROPHE IS A GENERAL EFFECT OF DIFFERENT ANTICANCER AGENTS

With the above caveat, the frequency of micronucleated cells can still be used as a surrogate measure of MC in cells exposed to different agents. MC has long been known as the principal effect of ionizing radiation in vitro (25) and in vivo (26) and also identified as a prominent response to different anticancer drugs (20,27–29) or heat shock (30). In our survey of treatment responses in human tumor cells (14), equitoxic (ID_{85}) doses of all the tested anticancer agents induced MC in a high fraction (45–66%) of HT1080 fibrosarcoma cells. This included not only microtubule-targeting agents (taxol and vincristine) that perturb mitosis directly but also agents that act in the interphase, ultimately causing DNA damage, such as antimetabolite cytarabine, DNA replication inhibitor aphidicolin, topoisomerase II poisons doxorubicin and etoposide, and DNA-damaging agents cisplatin and ionizing radiation. MC (as measured by micronucleation) was the most uniform response to different agents, whereas the rate

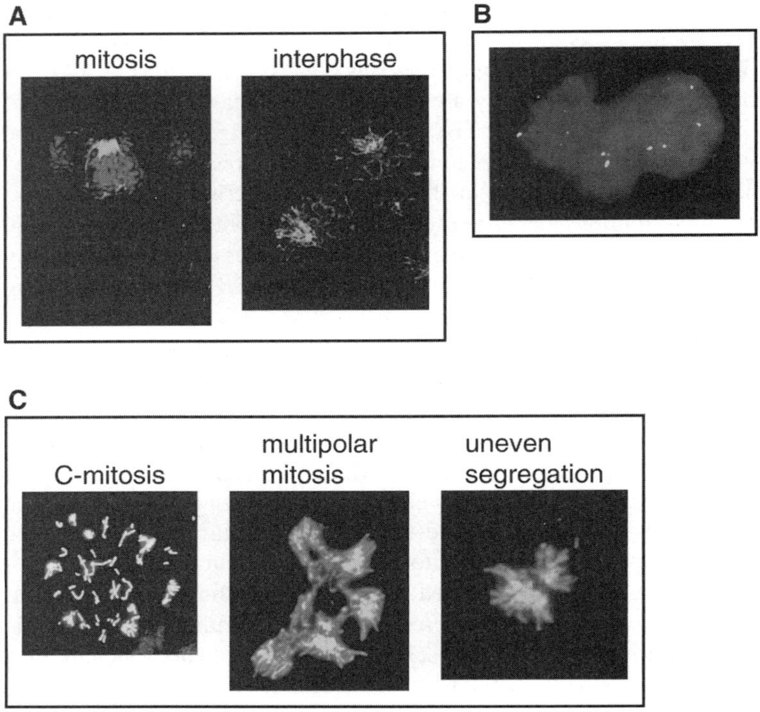

Figure 2 Morphological aspects of mitotic catastrophe. (**A**) Radiation-induced multiple micronuclei in HT1080 fibrosarcoma cells at the end of abnormal mitosis (*left*) and in interphase following abnormal mitosis. Cells were stained for DNA (propidium iodide) and microtubules (α-tubulin antibody) and photographed under a confocal microscope. (**B**) Fluorescence in situ hybridization analysis of chromosome distribution among the micronuclei in a partially fragmented nucleus of HT1080 cells treated with 20 nM doxorubicin for three days. Cells were hybridized with a mixture of fluorescent probes specific for chromosomes 18 and 21; nuclei were stained with DAPI. (**C**) Examples of abnormal mitotic figures in HT1080 cells recovering from p21-induced growth inhibition (DNA stained with DAPI). *Source*: From Refs. 14 (**B**) and 56 (**C**) (*The color version of this figure is provided on the DVD*).

of senescence (as measured by SA-β-gal staining) varied widely (high for DNA-damaging drugs and low for anti-microtubular agents) and apoptotic events (as measured by the frequency of cells with sub-G1 DNA contents) were rare in all cases. In the same study, only 2 of 14 solid tumor cell lines treated with moderate doses of doxorubicin showed predominantly apoptotic response, whereas the other 12 lines developed micronucleation, with or without significant apoptosis (14). MC is induced not only by the older drugs that specifically perturb the cell cycle but also by newer preclinical and clinical targeted therapeutic agents, such as HDAC inhibitors (31), geldanamycin (32), Notch-inhibiting gamma-secretase inhibitors (33), proteasome inhibitor bortezomib (34), genistein (35), E1B55K-defective adenovirus ONYX-015 (36), curcumin (37), and inhibitors of the PI3-kinase/Akt pathway (38). MC is induced by methylating agent temozolomide (39), and it also occurs in cells where DNA methylation is inhibited by the knockout of the enzyme DNMT1 (40). Hence, MC is the principal response of tumor cells to different anticancer agents.

MULTIPLE MECHANISMS OF TREATMENT-INDUCED MC

Why do drug-treated cells undergo MC? The answer to this question appears to be complex and different for different inducers of MC. The best-understood scenario is provided by microtubule-depolymerizing agents (e.g., Vinca alkaloids) or microtubule-hyperpolymerizing drugs (taxanes). These agents disrupt mitosis directly, since their primary target is the mitotic spindle. In the presence of taxol, chromosomes accumulate in an abnormal metaphase and eventually form random clusters and decondense (17). If taxol-treated tumor cells have retained a functional program of apoptosis, this program becomes triggered and executed, probably through the loss of anti-apoptotic protein survivin that localizes to centrosomes in mitotic cells (41). On the other hand, analysis of the cytotoxic effects of a taxane docetaxel in three breast carcinoma cell lines showed strong induction of MC but not of apoptosis (42). Microtubule-depolymerizing agents cause somewhat different mitotic abnormalities than taxol, where chromosomes often remain dispersed through the cell without forming a metaphase plate (C-mitosis); C-mitosis leads to the formation of multiple micronuclei and cell death (43,44). Preferential sensitivity of tumor cells to the induction of MC by microtubule-targeting drugs is likely to be due to the widespread deficiencies of tumor cells in the early mitosis checkpoint (2) and in the mitotic spindle checkpoint (3).

The mechanism of the induction of MC in cells that enter mitosis after the exposure to DNA-damaging agents presents a much more complicated puzzle. Several studies have described phenotypically different processes of abnormal mitosis in cells exposed to ionizing radiation. Thus, Ianzini and Mackey (45) described MC as starting with uneven chromatin condensation around nucleoli, which resembled premature chromosome condensation (PCC) originally observed in the fusions of mitotic cells with interphase cells in S or G2 (46). On the basis of this morphological similarity and on the observations of increased cellular levels of cyclin B1 and Cdc2 kinase in irradiated cells, these authors proposed that MC results from premature induction of mitosis before the completion of S or G2 (45,47,48). In agreement with this hypothesis, deliberate induction of premature mitosis by ectopic overexpression of the mitosis-initiating complex of Cyclin B1 and Cdc2 has been shown to result in PCC-like mitotic morphology and MC (49,50). On the other hand, uneven condensation could conceivably result not from PCC but from localized defects in the assembly of chromosome-condensing proteins (condensins and cohesins) in the damaged regions of DNA. In other studies, radiation-induced MC has been associated with overduplication of centrosomes, leading to multipolar mitosis (4,51–53), or with sustained metaphase arrest (54) and anaphase bridges (52,53).

In our ongoing studies on radiation induced MC in HT1080 fibrosarcoma cells, we have observed different pathways and outcomes of MC with apparently different causes. Several examples of radiation-induced MC are shown in *Videos 1 to 5*, which show real-time fluorescence video microscopy of mitosis in HT1080 cells that express histone H2B-GFP fusion protein (55) that produces green fluorescent chromatin. *Video 1* shows an example of normal mitosis in unirradiated cells, where chromosome condensation, alignment, segregation, and decondensation are easily discerned. In *Video 2*, HT1080 cells carrying p53 inhibitor GSE56 and histone H2B-GFP were irradiated in the presence of caffeine (G2 checkpoint inhibitor). These checkpoint-suppressed cells enter mitosis with a large amount of unrepaired DNA damage (not shown) and undergo a form of MC, characterized by incomplete chromosome condensation (compare with untreated cells in *Video 1*) and failure of chromosome segregation, leading to the formation of interphase cells with multiple micronuclei. Chromosomal morphology (not shown) indicates incomplete segregation of sister chromatids in these cells, suggesting that failure to decatenate newly replicated DNA could cause this form of MC.

On the other hand, HT1080 cells that enter mitosis after G1 or G2 checkpoint arrest, which allows them to repair much of the damage, are also prone to undergo several forms of MC, distinct from the above-described principal form observed in cells that enter mitosis with extensive DNA damage. In some cases, MC that follows checkpoint arrest can be due to chromosome bridges that result from chromosome fusion triggered by radiation-induced double-stranded DNA breaks. This is illustrated in *Video 3*, which shows a cell that entered mitosis after radiation-induced G2 checkpoint arrest. This cell shows proper chromosome condensation (unlike checkpoint-suppressed cells in *Video 2*), but it also fails to segregate its chromosomes in anaphase due to the formation of clearly visible chromatin bridges. The cell undergoes aberrant telophase; the incompletely separated daughter cells form multiple micronuclei and undergo apoptosis. A different form of MC is illustrated in *Video 4*, which shows the fate of another cell from the same experiment as in *Video 3*. Here, chromosomes fail to form the metaphase plate and undergo aberrant anaphase with uneven segregation and abortive telophase. Still another form of MC observed in the same experiment is shown in *Video 5*. In this irradiated cell, the chromosomes fail to form the metaphase plate and undergo aberrant anaphase with uneven segregation and abortive telophase. Condensed chromosomes fail to form metaphase and are randomly distributed for a prolonged period of time, producing the same morphology of "C-mitosis," as observed in cells treated with microtubule-depolymerizing agents. The state of "C-mitosis" often lasts for several hours, followed by chromatin decondensation (*Video 5*) and cell death.

Strikingly, the same forms of MC that are seen in cells that reenter the cycle after damage-induced checkpoint arrest are also observed in tumor cells that are recovering from cytostatic growth inhibition, suggesting that checkpoint arrest per se may be a cause of MC. A case in point is provided by our analysis of mitotic consequences of inducible expression of p21, the cyclin-dependent kinase inhibitor that mediates growth arrest produced by DNA damage (56). p21 expression from an inducible promoter in HT1080 cells leads to cell cycle arrest in both G1 and G2. Upon release from p21, essentially all the cells reentered the cycle, and G2-arrested cells underwent another round of DNA replication, doubling their ploidy. However, many of the cells that reentered cell cycle following p21-induced arrest died with features of MC. The extent of cell death and loss of clonogenicity after release from p21 were directly correlated with the duration of p21-induced growth arrest. Cell death after release from p21 was associated with multiple mitotic abnormalities, which paralleled those seen in cells damaged with drugs or radiation. Examples of abnormal mitotic figures in cells recovering from p21-induced growth arrest are shown in Figure 2C, including C-mitosis (failure of spindle formation or attachment), multipolar mitosis (overduplication of centrosomes), as well as uneven chromosome segregation.

One of the mechanisms for MC after p21-induced growth arrest can be explained by the finding that p21 induction is accompanied by rapid transcriptional inhibition of genes that are involved in all the stages of mitosis, including mitosis initiation, centrosome function, and spindle checkpoint control (56,57). The longer p21 induction is maintained, the more complete is the decay of p21-inhibited mitotic proteins. After release from p21, these proteins are asynchronously resynthesized, and the cells enter mitosis as soon as they regenerate sufficient pools of mitosis-initiating proteins (such as Cyclin B1 or Cdc2), although spindle checkpoint-control proteins (such as Mad2) are present at insufficient levels, leading to uneven chromosome segregation (Fig. 2C, right panel) (56). p21 also inhibits proteins that regulate centrosome separation and maturation (such as aurora-related and polo-like kinases) (56), and p21 can also block centrosome duplication at the G1/S boundary (58). Hence, the inhibition of mitosis-regulating proteins by the checkpoint protein p21 could be responsible for some of the forms of MC that occur in DNA-damaged tumor cells that enter mitosis after checkpoint arrest.

Another example of the complicated effect of the checkpoint mechanisms on MC comes from the study of Nitta et al. (59). In this study, p53-deficient cells that entered mitosis with DNA damage transiently arrested at metaphase for more than 10 hours without segregation of chromosomes, through the activation of mitotic spindle checkpoint, subsequently dying directly from metaphase. Suppression of spindle checkpoint function by RNAi knockdown of BubR1 or Mad2 led to escape from this form of MC, but the cells still underwent subsequent abnormal mitosis. The status of the mitotic spindle checkpoint is thus likely to be a determinant of the type of MC induced by DNA damage.

CHECKPOINT DEFICIENCIES POTENTIATE DRUG-INDUCED MC

As discussed above, cell cycle checkpoint deficiencies of tumor cells should make them prone to MC, the most general antitumor effect of different drugs with proven clinical efficacy. The selectivity of drug-induced MC for checkpoint-deficient cells can be demonstrated by the following example. We have compared the effects of doxorubicin on BJ-EN, a line of hTERT-immortalized normal BJ fibroblasts and BJ-ELB, an isogenic line that was transduced with both hTERT and the early region of SV40 (gifts of Dr. W. Hahn, Massachusetts General Hospital). The early region of SV40 encodes the LT protein that inhibits both p53 and Rb tumor suppressors, thus disabling most of the cellular checkpoints (60,61). We have found that three-day exposure of BJ-EN and BJ-ELB to 30 nM doxorubicin had approximately equal growth-inhibitory effect in both cell lines (as measured by the cell number). Cells were exposed to this drug dose for three days and then released for two days (cell death occurs primarily after release from the drug), and the fractions of cells showing different drug responses was determined by morphological criteria. The percentage of senescent (SA-β-gal positive) cells was very similar in BJ-EN and BJ-ELB (53% and 47%, respectively), and the apoptotic fraction was low in both cell lines, albeit higher in BJ-ELB (0.9% and 6.6%). On the other hand, the fraction of cells undergoing MC was much higher in BJ-ELB than in BJ-EN, as determined by the percentage of micronucleated cells (26.4% vs. 1.7%) and of apparently abnormal mitotic figures (92% vs. 40%). Thus, MC induced by a chemotherapeutic drug is much more common in checkpoint-deficient (cancerous or pre-cancerous) cells than in checkpoint-competent (normal or nearly normal) cells.

In summary, MC is the most general antitumor effect of different classes of anticancer agents, which is potentiated by cell cycle checkpoint deficiency, characteristic of tumor cells. MC is conceivably the principal determinant of tumor selectivity of all or almost all the agents with proven utility in cancer treatment. Therapy-induced MC occurs through several different pathways. It is important to determine in the future studies which of these pathways are invariably lethal, and which pathways allow for the recovery of some of the cells that underwent abnormal mitosis. Identification of compounds that induce or potentiate the most lethal forms of MC in tumor cells appears to be a justified and practical strategy in anticancer drug development.

ACKNOWLEDGMENTS

We apologize to all the colleagues whose relevant studies were not covered in this review due to space limitations. We thank former members of our laboratory who participated in the studies reviewed in this chapter, including Dr. Bey-Dih Chang, Dr. Adam Ruth, Tatiana Kalinichenko, Izolda Popova, and Dr. Claire Vivo; Dr. Yury Verlinsky and Mark Chmyra for FISH analysis; Dr. William Hahn and Dr. Burt Vogelstein for the gifts of cell lines used in these studies; Dr. Andrei Gudkov for p53 inhibitor GSE56; and Dr. Andrew Maniotis, Dr. Conly Rieder, Dr. Alexey Khodjakov, and Richard Cole for advice on setting up real-time video microscopy. Studies from our laboratory were supported by grants RO1 CA95727 and RO1 CA89636 from the National Cancer Institute.

REFERENCES

1. Stark GR, Taylor WR. Control of the G2/M transition. Mol Biotechnol 2006; 32:227–248.
2. Scolnick DM, Halazonetis TD. CHFR defines a mitotic stress checkpoint that delays entry into metaphase. Nature 2000; 406:430–435.
3. Cahill DP, Lengauer C, Yu J, et al. Mutations of mitotic checkpoint genes in human cancers [see comments]. Nature 1998; 392:300–303.
4. Bunz F, Dutriaux A, Lengauer C, et al. Requirement for p53 and p21 to sustain G2 arrest after DNA damage. Science 1998; 282:1497–1501.
5. Stewart ZA, Pietenpol JA. G2 checkpoints and anticancer therapy. In: Blagosklonny MV, ed. Cell Cycle Checkpoints and Cancer. Georgetown, TX: Landes Bioscience, 2001:155–178.
6. Chan TA, Hermeking H, Lengauer C, et al. 14-3-3Sigma is required to prevent mitotic catastrophe after DNA damage [see comments]. Nature 1999; 401:616–620.
7. Chang BD, Xuan Y, Broude EV, et al. Role of p53 and p21$^{waf1/cip1}$ in senescence-like terminal proliferation arrest induced in human tumor cells by chemotherapeutic drugs. Oncogene 1999; 18:4808–4818.
8. Castedo M, Perfettini JL, Roumier T, et al. The cell cycle checkpoint kinase Chk2 is a negative regulator of mitotic catastrophe. Oncogene 2004; 23:4353–4361.
9. Niida H, Tsuge S, Katsuno Y, et al. Depletion of Chk1 leads to premature activation of Cdc2-cyclin B and mitotic catastrophe. J Biol Chem 2005; 280:39246–39252.
10. Privette LM, Gonzalez ME, Ding L, et al. Altered expression of the early mitotic checkpoint protein, CHFR, in breast cancers: implications for tumor suppression. Cancer Res 2007; 67:6064–6074.
11. Jiang W, Jimenez G, Wells NJ, et al. PRC1: a human mitotic spindle-associated CDK substrate protein required for cytokinesis. Mol Cell 1998; 2:877–885.
12. Chu K, Teele N, Dewey MW, et al. Computerized video time lapse study of cell cycle delay and arrest, mitotic catastrophe, apoptosis and clonogenic survival in irradiated 14-3-3sigma and CDKN1A (p21) knockout cell lines. Radiat Res 2004; 162:270–286.
13. Dimri GP, Lee X, Basile G, et al. A biomarker that identifies senescent human cells in culture and in aging skin in vivo. Proc Natl Acad Sci U S A 1995; 92: 9363–9367.
14. Chang BD, Broude EV, Dokmanovic M, et al. A senescence-like phenotype distinguishes tumor cells that undergo terminal proliferation arrest after exposure to anticancer agents. Cancer Res 1999; 59:3761–3767.
15. Klein LE, Freeze BS, Smith AB III, et al. The microtubule stabilizing agent discodermolide is a potent inducer of accelerated cell senescence. Cell Cycle 2005; 4:501–507.

16. Demarcq C, Bunch RT, Creswell D, et al. The role of cell cycle progression in cisplatin-induced apoptosis in Chinese hamster ovary cells. Cell Growth Differ 1994; 5:983–993.
17. Jordan MA, Wendell K, Gardiner S, et al. Mitotic block induced in HeLa cells by low concentrations of paclitaxel (Taxol) results in abnormal mitotic exit and apoptotic cell death. Cancer Res 1996; 56:816–825.
18. Waldman T, Lengauer C, Kinzler KW, et al. Uncoupling of S phase and mitosis induced by anticancer agents in cells lacking p21. Nature 1996; 381:713–716.
19. Mansilla S, Priebe W, Portugal J. Mitotic catastrophe results in cell death by caspase-dependent and caspase-independent mechanisms. Cell Cycle 2006; 5:53–60.
20. Lock RB, Stribinskiene L. Dual modes of death induced by etoposide in human epithelial tumor cells allow Bcl-2 to inhibit apoptosis without affecting clonogenic survival. Cancer Res 1996; 56:4006–4012.
21. Park SS, Kim MA, Eom YW, et al. Bcl-xL blocks high dose doxorubicin-induced apoptosis but not low dose doxorubicin-induced cell death through mitotic catastrophe. Biochem Biophys Res Commun 2007; 363:1044–1049.
22. Ruth AC, Roninson IB. Effects of the multidrug transporter P-glycoprotein on cellular responses to ionizing radiation. Cancer Res 2000; 60:2576–2578.
23. Robinson LJ, Roberts WK, Ling TT, et al. Human MDR 1 protein overexpression delays the apoptotic cascade in Chinese hamster ovary fibroblasts. Biochemistry 1997; 36:11169–11178.
24. Smyth MJ, Krasovskis E, Sutton VR, et al. The drug efflux protein, P-glycoprotein, additionally protects drug-resistant tumor cells from multiple forms of caspase-dependent apoptosis. Proc Natl Acad Sci U S A 1998; 95:7024–7029.
25. Jonathan EC, Bernhard EJ, McKenna WG. How does radiation kill cells? Curr Opin Chem Biol 1999; 3:77–83.
26. Falkvoll KH. The occurrence of apoptosis, abnormal mitoses, cells dying in mitosis and micronuclei in a human melanoma xenograft exposed to single dose irradiation. Strahlenther Onkol 1990; 166:487–492.
27. Tounekti O, Pron G, Belehradek J Jr, et al. Bleomycin, an apoptosis-mimetic drug that induces two types of cell death depending on the number of molecules internalized. Cancer Res 1993; 53:5462–5469.
28. Torres K, Horwitz SB. Mechanisms of Taxol-induced cell death are concentration dependent. Cancer Res 1998; 58:3620–3626.
29. Taylor BF, McNeely SC, Miller HL, et al. p53 suppression of arsenite-induced mitotic catastrophe is mediated by p21CIP1/WAF1. J Pharmacol Exp Ther 2006; 318:142–151.
30. Nakahata K, Miyakoda M, Suzuki K, et al. Heat shock induces centrosomal dysfunction, and causes non-apoptotic mitotic catastrophe in human tumour cells. Int J Hyperthermia 2002; 18:332–343.
31. Magnaghi-Jaulin L, Eot-Houllier G, Fulcrand G, et al. Histone deacetylase inhibitors induce premature sister chromatid separation and override the mitotic spindle assembly checkpoint. Cancer Res 2007; 67:6360–6367.
32. Nomura M, Nomura N, Newcomb EW, et al. Geldanamycin induces mitotic catastrophe and subsequent apoptosis in human glioma cells. J Cell Physiol 2004; 201:374–384.
33. Curry CL, Reed LL, Broude E, et al. Notch inhibition in Kaposi's sarcoma tumor cells leads to mitotic catastrophe through nuclear factor-kappaB signaling. Mol Cancer Ther 2007; 6:1983–1992.

34. Strauss SJ, Higginbottom K, Juliger S, et al. The proteasome inhibitor bortezomib acts independently of p53 and induces cell death via apoptosis and mitotic catastrophe in B-cell lymphoma cell lines. Cancer Res 2007; 67:2783–2790.
35. Tominaga Y, Wang A, Wang RH, et al. Genistein inhibits Brca1 mutant tumor growth through activation of DNA damage checkpoints, cell cycle arrest, and mitotic catastrophe. Cell Death Differ 2007; 14:472–479.
36. Cherubini G, Petouchoff T, Grossi M, et al. E1B55K-deleted adenovirus (ONYX-015) overrides G1/S and G2/M checkpoints and causes mitotic catastrophe and endoreduplication in p53-proficient normal cells. Cell Cycle 2006; 5:2244–2252.
37. Wolanin K, Magalska A, Mosieniak G, et al. Curcumin affects components of the chromosomal passenger complex and induces mitotic catastrophe in apoptosis-resistant Bcr-Abl-expressing cells. Mol Cancer Res 2006; 4:457–469.
38. Hemstrom TH, Sandstrom M, Zhivotovsky B. Inhibitors of the PI3-kinase/Akt pathway induce mitotic catastrophe in non-small cell lung cancer cells. Int J Cancer 2006; 119:1028–1038.
39. Hirose Y, Katayama M, Mirzoeva OK, et al. Akt activation suppresses Chk2-mediated, methylating agent-induced G2 arrest and protects from temozolomide-induced mitotic catastrophe and cellular senescence. Cancer Res 2005; 65: 4861–4869.
40. Chen T, Hevi S, Gay F, et al. Complete inactivation of DNMT1 leads to mitotic catastrophe in human cancer cells. Nat Genet 2007; 39:391–396.
41. Jiang X, Wilford C, Duensing S, et al. Participation of survivin in mitotic and apoptotic activities of normal and tumor-derived cells. J Cell Biochem 2001; 83: 342–354.
42. Morse DL, Gray H, Payne CM, et al. Docetaxel induces cell death through mitotic catastrophe in human breast cancer cells. Mol Cancer Ther 2005; 4:1495–1504.
43. Jordan MA, Wilson L. The use and action of drugs in analyzing mitosis. Methods Cell Biol 1999; 61:267–295.
44. Therman E, Kuhn EM. Mitotic modifications and aberrations in cancer. Crit Rev Oncog 1989; 1:293–305.
45. Ianzini F, Mackey MA. Spontaneous premature chromosome condensation and mitotic catastrophe following irradiation of HeLa S3 cells. Int J Radiat Biol 1997; 72:409–421.
46. Johnson RT, Rao PN. Mammalian cell fusion: induction of premature chromosome condensation in interphase nuclei. Nature 1970; 226:717–722.
47. Mackey MA, Anolik SL, Roti Roti JL. Cellular mechanisms associated with the lack of chronic thermotolerance expression in HeLa S3 cells. Cancer Res 1992; 52: 1101–1106.
48. Mackey MA, Zhang XF, Hunt CR, et al. Uncoupling of M-phase kinase activation from the completion of S-phase by heat shock. Cancer Res 1996; 56:1770–1774.
49. Heald R, McLoughlin M, McKeon F. Human wee1 maintains mitotic timing by protecting the nucleus from cytoplasmically activated Cdc2 kinase. Cell 1993; 74:463–474.
50. Jin P, Hardy S, Morgan DO. Nuclear localization of cyclin B1 controls mitotic entry after DNA damage. J Cell Biol 1998; 141:875–885.
51. Dodson H, Wheatley SP, Morrison CG. Involvement of centrosome amplification in radiation-induced mitotic catastrophe. Cell Cycle 2007; 6:364–370.

52. Eriksson D, Lofroth PO, Johansson L, et al. Cell cycle disturbances and mitotic catastrophes in HeLa Hep2 cells following 2.5 to 10 Gy of ionizing radiation. Clin Cancer Res 2007; 13:5501s–5508s.
53. Sato N, Mizumoto K, Nakamura M, et al. Radiation-induced centrosome over-duplication and multiple mitotic spindles in human tumor cells. Exp Cell Res 2000; 255:321–326.
54. Kodym E, Kodym R, Choy H, et al. Sustained metaphase arrest in response to ionizing radiation in a non-small cell lung cancer cell line. Radiat Res 2008; 169: 46–58.
55. Kanda T, Sullivan KF, Wahl GM. Histone-GFP fusion protein enables sensitive analysis of chromosome dynamics in living mammalian cells. Curr Biol 1998; 8:377–385.
56. Chang BD, Broude EV, Fang J, et al. $p21^{Waf1/Cip1/Sdi1}$ -induced growth arrest is associated with depletion of mitosis-control proteins and leads to abnormal mitosis and endoreduplication in recovering cells. Oncogene 2000; 19:2165–2170.
57. Chang BD, Watanabe K, Broude EV, et al. Effects of $p21^{Waf1/Cip1/Sdi1}$ on cellular gene expression: implications for carcinogenesis, senescence, and age-related diseases. Proc Natl Acad Sci U S A 2000; 97:4291–4296.
58. Matsumoto Y, Hayashi K, Nishida E. Cyclin-dependent kinase 2 (Cdk2) is required for centrosome duplication in mammalian cells. Curr Biol 1999; 9:429–432.
59. Nitta M, Kobayashi O, Honda S, et al. Spindle checkpoint function is required for mitotic catastrophe induced by DNA-damaging agents. Oncogene 2004; 23: 6548–6558.
60. Morales CP, Holt SE, Ouellette M, et al. Absence of cancer-associated changes in human fibroblasts immortalized with telomerase. Nat Genet 1999; 21:115–118.
61. Jiang XR, Jimenez G, Chang E, et al. Telomerase expression in human somatic cells does not induce changes associated with a transformed phenotype. Nat Genet 1999; 21:111–114.

19

How Do Cells Die After Irradiation? Time-Lapse Studies of Cells in Culture

William C. Dewey

Department of Radiation Oncology, University of California San Francisco, San Francisco, California, U.S.A.

INTRODUCTION

In the past, time-lapse studies have demonstrated that HeLa cells (1,2) and mouse L cells (3) undergo postmitotic death after the cells divided one or more times; the death processes were associated with trapping in mitosis, fusion of daughter cells, giant cell formation, and death during interphase that was classified as pyknosis or necrosis. When these studies were conducted, apoptosis had not been identified. Also, with an aerial photographic camera that was used for visualizing a large number of cells in order to construct pedigrees (3), individual cells could not be adequately visualized to discern morphological alterations that could be related to the ultimate fate of the cells. Attempts have been made to develop time-lapse systems using multiple fields (4,5) at sufficient magnification so that progeny from one or two cells in each field at the time of irradiation could be followed as a colony of approximately 50 cells developed. Retrieval of the information was very complicated and time consuming, however. For information in this chapter, we have used computerized video time lapse (CVTL) (6) for analyzing (7) the fates of individual cells and their progeny at 200× in about 50 microscopic fields that are followed for as long as four to six days.

SUMMARIES OF OUR CVTL STUDIES

CVTL is Needed for Quantification of Cell Death by Apoptosis and Senescence

Following individual cells by CVTL is essential for quantifying the modes of death (6). For example, by CVTL analysis, the loss of clonogenic survival of x-irradiated (9.5 or 2.5 Gy) REC:myc cells (expressing c-myc oncogene and wild-type p53) was attributed almost entirely to the cells dying by apoptosis, with almost all the apoptosis occurring after the progeny had divided one to four times. In contrast, the loss of clonogenic survival of x-irradiated REC:ras cells (expressing c-Ha-ras oncogene and wild-type p53) was attributed to two processes. After 9.5 Gy, approximately 60% of the nonclonogenic REC:ras cells died by apoptosis, and the other 40% underwent a senescent-type process in which some of the cells and their progeny stopped dividing but remained as viable cells throughout 140 hours of observation. Scoring cell death in whole populations of cells gave erroneous results since both clonogenic and nonclonogenic cells were dividing as nonclonogenic cells were apoptosing or senescing over a period of many days. For example, after 9.5 Gy, which causes clonogenic reproductive cell death in 99% of both types of cells, the cumulative percentage of the cells scored as dead in the whole population at 60 to 80 hours after irradiation, when the *maximum* amount of cumulative apoptosis occurred, was approximately 60% for REC:myc cells compared with only approximately 30% for REC:ras cells.

Quantifying Apoptosis When REC:MYC Cells were Irradiated in Different Phases of the Cell Cycle

After x-irradiation, the cells died almost exclusively by postmitotic apoptosis during interphase after one or more divisions (8); i.e., the cells rounded, followed by vigorous membrane blebbing. When blebbing and cell movement stopped, the cell was classified as being dead, although the membrane did not rupture until a few hours later (the apoptotic process is illustrated in Fig. 3 in Ref. 8). CVTL analysis provided information on the variation in rates and amounts of apoptosis for cells irradiated in different phases of the cell cycle. Cells irradiated with 4 Gy in G1 divided one to six times and survived 40 to 120 hours before apoptosing, compared with only one to two times and 5 to 40 hours for cells irradiated in late S/G2. Most importantly, cells irradiated in late S or G2 were more radiosensitive than cells irradiated in G1 for both loss of clonogenic survival and the time of death and number of divisions completed after irradiation.

After the nonclonogenic cells divided and yielded progeny entering the first generation after irradiation with 4 Gy, 60% of the progeny either had micronuclei or were sisters of cells that had micronuclei, compared with none of the progeny of clonogenic cells having micronuclei in generation one (Fig. 10 in Ref. 8). However, another 20% of the nonclonogenic cells had progeny with micronuclei appearing first in generation 2 or 3. As a result, 80% of the nonclonogenic cells had progeny with micronuclei. Thus, cell death was attributed to postmitotic apoptosis during interphase associated with formation of micronuclei, which have been attributed to chromosomal aberrations (9). These relationships support the evidence that postmitotic cell death is caused by chromosomal aberrations (10).

Quantifying Apoptosis and Mitotic Death in X-Irradiated Human Lymphoid Cells

A CVTL analysis (11) demonstrated that x-irradiated human lymphoid cells apoptosed either premitotically without attempting to divide or during an abortive mitosis. After 4 Gy, the lymphoid cells (L5178Y-S and MOLT-4) died primarily by apoptosis during a prolonged abortive mitosis (called mitotic death) that is distinct from mitotic catastrophe. This prolonged abortive mitosis often persisted for 10 to 40 hours (Figs. 5 and 6 in Ref. 11) and usually occurred during the first mitosis after irradiation, but in about 10% of the cases, an abortive mitosis occurred in the second mitosis after an apparently normal mitosis. Lymphoid ST4 cells behaved differently in that after 2.5 or 4 Gy, 100% of the cells apoptosed premitotically during interphase (Figs. 2 and 6 in Ref. 11), but after 1 Gy, about 90% of ST4 cells died premitotically during interphase, with 10% producing lineages that died postmitotically during interphase (Figs. 2 and 3 in Ref. 11).

Quantifying Necrosis in X-Irradiated Human Bladder Carcinoma Cells

A CVTL study of x-irradiated EJ 30 human bladder carcinoma cells showed that the cells die primarily by postmitotic necrosis during interphase after they have divided several times (12). Necrosis was observed to occur from the outside inward as the cell membrane ruptured and the cell literally exploded, with the cytoplasmic contents released and a condensed nuclear body persisting for several hours (see Fig. 1 in Ref. 12). These cells that undergo very little mitotic catastrophe contain both an oncogenically activated HRAS gene and a nonexpressed and nonfunctional p53 because of a mutation in exon 5. A typical pedigree is shown in Figure 1. Note that several giant cells were observed, and in a subsequent study of the fates of giant cells (13), 8 of 102 giant cells observed beyond five days after irradiation (6 Gy) divided four or five times, with two of the giant cell pedigrees producing two normal-sized daughter cells. These results suggest that a small fraction of giant cells might be potentially clonogenic.

An interesting observation was that most nonclonogenic cells irradiated in mid-S phase (9–12 hours after mitosis) died by the second generation, while those irradiated either before or after this short period in mid-S phase had cell deaths occurring over one to nine generations postirradiation (Figs. 5 and 6 in Ref. 12). The nonclonogenic cells irradiated in mid-S phase also experienced the

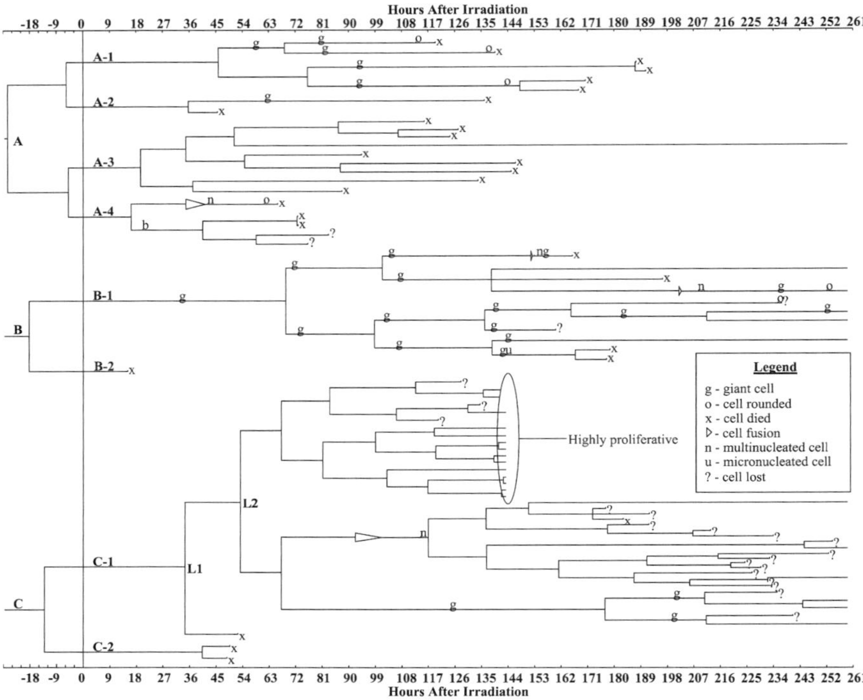

Figure 1 Pedigree diagrams illustrating the fate of EJ30 cells after irradiation (6 Gy). The figure was taken from Ref. 12. The root cells A, B, and C were allowed to divide one to two times before their progenies were irradiated with 6 Gy during different parts of the cell cycle. The irradiated cells A-1, A-2 (age = 5.7 hours), A-3, A-4 (age = 5 hours), B-1, B-2 (age = 18.3 hours), and C-2 (age = 13.2 hours) were all nonclonogenic. The significant cellular events associated with nonclonogenicity, namely cell death, cell fusion, and giant cell formation, are marked on the diagrams according to the times of occurrence. These events can occur in the irradiated cells or in their progenies over several generations. The cumulative effect is that colony formation was prevented. In contrast, clonogenic cells proliferate rapidly after irradiation with occasional nonclonogenic events. The irradiated cell C1 (age = 13.2 hours) produced a clonogenic pedigree with two *lethal-sectoring* events. *Lethal sectoring is defined as a division that produces only one daughter cell that forms a clonogenic subcolony.* After the first division (L1), one daughter died, while the other kept dividing. Then, after the second division (L2), the lower daughter had progeny with long generation times and aberrant events (such as cell fusion, giant cell formation, and cell death) that did not occur in its sister's progeny.

longest average division delay before their first divisions, and clonogenic cells (11/12 cells) divided soon after irradiation than the average nonclonogenic cells derived from the same phase of the cell cycle (see Fig. 8 in Ref. 12). Similarly, time-lapse studies of HeLa and Chinese hamster V79 cells showed that division delay was the greatest for irradiation in mid-S phase (14), and time-lapse studies of V79 (15) and mouse L cells (16) showed that clonogenic cells had less division delay than nonclonogenic cells. Thus, our CVTL study and time-lapse studies in the literature suggest that an increase in division delay is associated more with an increase in damage than with increased time for repair. The early death and long division delay observed for nonclonogenic cells irradiated in mid-S phase could possibly result from an increase in damage induced during the transition from the replication of euchromatin to the replication of heterochromatin (17).

Quantifying Apoptosis and Mitotic Catastrophe in X-Irradiated Human Colorectal Carcinoma Cell Lines

CVTL microscopy was used to observed cellular events induced by ionizing radiation (10–12 Gy) in the nonclonogenic cells of the wild-type HCT116 colorectal carcinoma cell line and its three isogenic derivative lines in which $p21^{WAF/CIF}$ (CDKN1A), 14-3-3σ, or both checkpoint genes had been knocked out (7). The postirradiation cellular events that occurred during or at the end of the irradiated generation (generation 0) are illustrated in Figure 2. Cells that fused after mitosis or failed to complete mitosis were classified together as cells that had undergone mitotic catastrophe. Seventeen percent of the wild-type cells and 34% to 47% of the knockout cells underwent mitotic catastrophe to enter generation 1 with a 4N content of DNA, i.e., the same DNA content as cells arrested in G2 at the end of generation 0 (quantified in Fig. 3 in Ref. 7). Radiation caused a transient delay in generation 0 (called division delay and quantified in Fig. 3 in Ref. 7) before the cells divided or underwent mitotic catastrophe (quantified in Fig. 4 in Ref. 7). Compared with the transient division delay for wild-type cells that express p21 and 14-3-3σ, knocking out p21 reduced transient division delay the most for cells irradiated in G1 (from ~15 to ~3–5 hours), while knocking out 14-3-3σ reduced transient division delay the most for cells irradiated in late S and G2 (from ~18 to 3–4 hours). Note that the CVTL analysis clearly distinguished between the transient division delay consisting of less than 20 hours compared with an arrest in generation 0 that lasted as long as 96 hours. Data quantified in Fig. 3 in Ref. 7 showed that 27% of wild-type cells and 17% of 14-3-3σ–/– cells were arrested at 96 hours in generation 0, compared with less than 1% for p21–/– and double-knockout cells. Therefore, expression of p21 is necessary for the prolonged delay or arrest in generation 0. Furthermore, while 14-3-3σ appears to play a role in the transient G2 checkpoint, the p21 protein plays a crucial role in generation 1, greatly inhibiting progression into subsequent generations of both diploid cells and polyploid cells produced by mitotic catastrophe (Figs. 3 and 4). Thus, in p21-deficient cell lines, a series of mitotic catastrophe events occurred to produce highly polyploid progeny cells during generations 3 and 4.

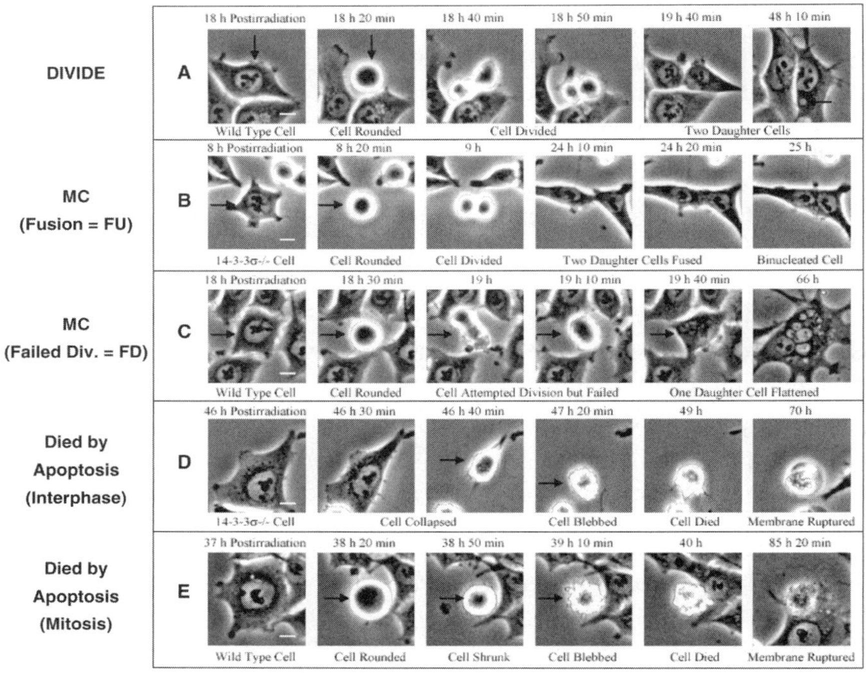

Figure 2 Montages of postirradiation cellular events that occurred in HCT116 cells during or at the end of the irradiated generation (generation 0). The figure was taken from Ref. 7. The same types of events occurred during subsequent generations. Three different outcomes from mitosis were observed, as illustrated in the top three montages (*panels A–C*). Panel A: A wild-type cell divided, and its daughter cells remained separate. Also, note the presence of a micronucleus (*arrow*) in one of the daughter cells. Panel B: A 14-3-3σ–/– cell divided to produce progeny cells that subsequently fused to form a binucleated cell. The time interval from division to fusion was highly variable (7.1 $\pm$ 10.7 hours SD for wild-type cells, 4.4 $\pm$ 5.3 hours for p21 –/– cells, 12.9 $\pm$ 11.3 hours for 14-3-3σ–/– cells, and 6.4 $\pm$ 7.4 hours for double-knockout cells). Panel C: A wild-type cell failed in the process of division. The cell rounded up at 18 hour 30 minutes, attempted division at 19 hours, then failed to complete the process and flattened at 19 hour 40 minutes as one cell. In the image at 66 hours, note the fragmentation of the nucleus into multiple micronuclei that resulted from the failed division. The last two scenarios are grouped together as MC events that led to potential nuclear fragmentation and polyploidy. Two modes of death are illustrated in the bottom two montages (*panels D and E*). Apoptosis during interphase was the most common mode of death. Panel D: A 14-3-3σ–/– cell in interphase collapsed rapidly at 46 hour 40 minutes after irradiation, blebbed at 47 hour 20 minutes, and then died (i.e., stopped moving) at 49 hours. The membrane of the dead cell did not rupture until 70 hours. Panel E: A much less common mode of death was apoptosis during mitosis, or abortive mitosis. A wild-type cell rounded up at 38 hour 20 minutes after irradiation as the cell entered mitosis, shrank at 38 hour 50 minutes, blebbed at 39 hour 10 minutes, and then died (stopped moving) at 40 hours. The cell membrane did not rupture until 85 hour 20 minutes. Images were captured at 200×. White bars represent 10 μm. *Abbreviations*: SD, standard deviation; MC, mitotic catastrophe.

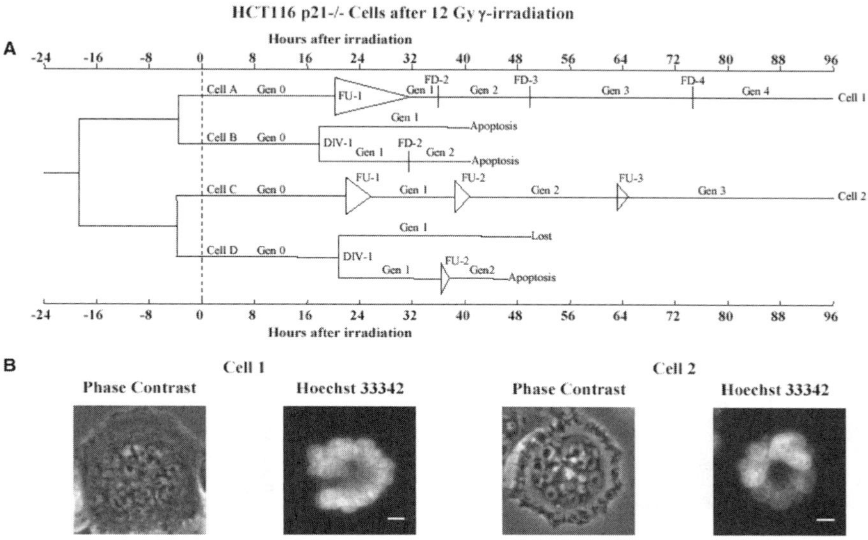

Figure 3 Lack of p21 expression resulted in a series of mitotic catastrophes. The figure was taken from Ref. 7. The pedigree in panel A illustrates the occurrences of MC events over time for HCT116 p21–/– cells. The pedigree shows four related cells of generation zero (Gen 0) that were irradiated and then underwent a series of mitotic events (labeled as DIV for division, FU for fusion after division, and FD for failed division, as well as the number of mitoses since irradiation) to produce progeny cells of later generations (Gen 1, Gen 2, etc.). Cell A and Cell B were both irradiated at an age of 3.7 hours. Cell A was observed to divide once followed by fusion of the daughter cells (FU-1) and then to undergo three more rounds of failed divisions (FD-2, FD-3, and FD-4) to form Cell 1. Cell B divided (DIV-1) with one daughter cell undergoing one more failed division (FD-2) before apoptosis and the other dying by apoptosis without further attempt at division. Cell C and Cell D were both irradiated at an age of 3.8 hours. Cell C was observed to divide three times, each time followed by a fusion (FU-1, FU-2, and FU-3) to form Cell 2. Note that FU-3 represents a tripolar division followed by fusion, which points to the accumulation of multiple centrosomes. Cell D divided (DIV-1) to produce one daughter cell that divided followed by fusion (FU-2) before dying by apoptosis, while the other was lost from the field without further division. Note that Cell 1 was a fourth-generation progeny of the irradiated Cell A, while Cell 2 was a third-generation progeny of the irradiated Cell C. Because the series of mitotic catastrophes in each cell's lineage prevented the production of more than one progeny, Cell 1 should have a putative ploidy of 32N–64N, while Cell 2 should have a ploidy of 16N–32N. Montage B shows the phase-contrast and Hoechst 33342 fluorescence images of Cell 1 and Cell 2 from the pedigree; both cells were multimicronucleated. Images were captured at 200×. White bars represent 10 μm. *Abbreviation*: MC, mitotic catastrophe.

Most importantly, for the cells entering generation 1, polyploid progeny produced by mitotic catastrophe did not die more rapidly than the progeny of dividing cells (see Fig. 10B, C in Ref. 7). In both cases, death was almost exclusively apoptotic, for which the time of death was identified as loss of cell movement, i.e.,

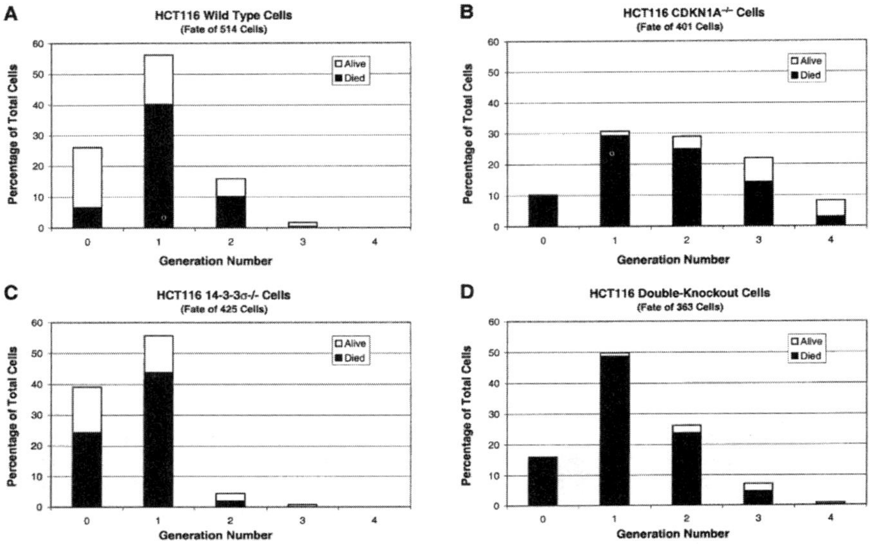

Figure 4 Cell fates. The figure was taken from Ref. 7. The fates of the cells were scored as whether they died or were still alive by the end of the 96-hour observation period. The open bars represent the percentage of cells from different generations that survived to the end of the 96-hour postirradiation period. Solid bars show the percentage of cells that died in different generations after irradiation. Generation 0 includes cells that were irradiated. Generation 1 and above include progeny cells that were derived from irradiated cells through one or more rounds of mitosis. The distribution of cell fates over generations, plotted for each cell line, illustrates the following: (*i*) the wild-type and 14-3-3σ-/– cell lines (*panels A and C*) had surviving cells from earlier generations (i.e., 0–2), while p21-/– and double-knockout cell lines (*panels B and D*) had surviving cells predominantly from late generations (i.e., 2–4) because of the proliferation. Many of the cells not expressing p21 had successive mitotic catastrophe events in their lineages (Fig. 3); for example, 50% of the p21-/– cells in generation 4 had four successive MC events in their lineages (Fig. 9 from Ref. 7). (2) There were more cell deaths in generation 0 in the 14-3-3σ-deficient cell lines (*panel C*) than in the wild-type cell line (*panel A*) (3); 14-3-3σ-deficiency appeared to counter the proliferation unleashed by p21-deficiency, with more cell deaths occurring in generation 1 in the double-knockout cell line (*panel D*) than in the p21-deficient cell line (*panel B*). Total numbers of cell fates analyzed are 517 wild-type cells, 403 p21-/– cells, 425 14-3-3σ-/– cells, and 363 double-knockout cells. *Abbreviation*: MC, mitotic catastrophe.

metabolic activity. Thus, mitotic catastrophe itself is not a direct mode of death. Instead, postmitotic apoptosis during interphase of both uninucleated and polyploid cells was the primary mode of death observed in all of the four cell types.

Knocking out either p21 or 14-3-3σ increased the amount of cell death at 96 hours, from 52% to approximately 70%, with even a greater increase to 90% when both genes were knocked out (see Fig. 10A in Ref. 7). Thus, in addition to

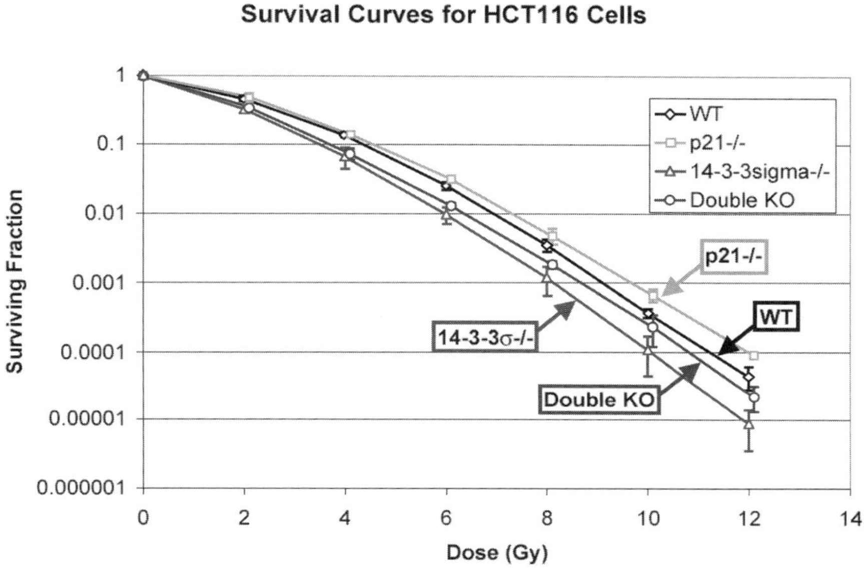

Figure 5 Clonogenic survival. The figure was taken from Ref. 7. Results from two experiments were averaged for each point. The average plating efficiencies for unirradiated cells were: 55 ± 2% SD for wild-type cells, 46 ± 0.1% for p21–/– cells, 38 ± 4% for 14-3-3σ–/– cells, and 32 ± 5% for double-knockout cells. Error bars represent SDs. The data were fit with a model (computer program written by Norm Albright), which assumes that $\ln S = -(\alpha D + \beta D^2)$. The ratios of the doses at an isosurvival level of 0.01 were p21–/–/wild type = 1.05 ± 0.02 SEM, p21 –/–/double knockout = 1.13 ± 0.03, wild type/14-3-3σ–/– = 1.13 ± 0.03, wild type/double knockout = 1.08 ± 0.02, and double knockout/14-3-3σ–/– = 1.04 ± 0.03. On the basis of these ratios, ± 2 SEM being different from 1.00 (after small corrections for differences in cellular multiplicity at the time of irradiation), all survival curves, except for double knockout compared with 14-3-3σ–/–, were considered to be significantly different from each other. *Abbreviations*: SD, standard deviation; SEM, standard error of mean (*The color version of this figure is provided on the DVD*).

effects of p21 and 14-3-3σ expression on transient cell cycle delay, the p21 protein has both an antiproliferative and antiapoptotic function, while 14-3-3σ has only an antiapoptotic function. Finally, the large alterations in the amounts of cell death did not correlate overall with the small alterations in clonogenic survival (Fig. 5). However, knocking out p21 resulted in a decrease in arrested cells (senescent?) and a small increase in survival,[a] and knocking out 14-3-3σ resulted in an increase in apoptosis and a small decrease in clonogenic survival.

[a]The radiation response of a p53 knockout HCT116 isogenic derivative cell line (G. Prieur–Carrillo unpublished data) was almost identical to that of the p21 knockout cell line shown: with multiple mitotic catastrophe events in the pedigree in Figure 3, for cells progressing through several generations in Figure 4 (panel B), and for clonogenic survival in Figure 5. This might be expected since p21 is downstream of p53.

FUTURE STUDIES

The use of CVTL combined with other techniques may be used to explore mechanisms responsible for different radiation responses observed in different cell lines that express different genes. As discussed in our CVTL publications, these differences and an understanding of the mechanisms involved may have clinical implications. Hypotheses related to differences in gene expression are presented, on the basis of our findings and the literature, but these hypotheses need to be extended. Most importantly, the mechanisms responsible for the different responses and differences in radiosensitivity when cells are irradiated in different phases of the cell cycle may be explored by using CVTL combined with other techniques. The power of observing and recording the events in *individual* cells by CVTL is illustrated in the movies and pedigrees included in our DVD. Note that the history for each cell is recorded prior to its death, senescence, or continued division to produce a colony. For example, in the field 15 pedigree for irradiated HCT116 p21 –/– cells, cell no.2 underwent two divisions prior to irradiation of eight of its progeny cells; one of the progeny cells irradiated in generation 0 was 15-2222 (bottom line on the pedigree); the phase contrast picture showed that this cell appeared to be normal at the time of irradiation. By 96 hours after irradiation, this cell and its progeny had gone through three generations with a failed division (marked with H and coded with 8) at the end of each generation. The final multimicronucleated cell in generation 3 at 96 hours was identified as 15-2222-888; both the phase image and fluorescence image are shown for this and other cells. The challenge for future studies is to identify molecular signals, possibly with GFP proteins, at various times after irradiation so that the ultimate fate of the cell can be traced back in time to identify particular signals and events in the cell's ancestors.

SUMMARY

1. The amount of apoptosis that occurs in irradiated populations can be quantified. The proliferation of irradiated cells, in particular clonogenic cells, complicates the determination of the cumulative fraction of the cell population that has apoptosed when the cultures as a whole are analyzed. Also, quantification of individual cells arrested in G2 can be distinguished from cells that have entered G1 after undergoing mitotic catastrophe, i.e., failing to complete division or fusing after division. Since both cell types have a 4N content of DNA, they cannot be distinguished from one another by standard flow cytometry based on DNA content.
2. To examine mechanisms of cell death, one should determine first how the *cellular endpoints of radiosensitivity*, cell cycle checkpoints, and modes of death depend on the *age of the cell at the time of irradiation*. Cell age is determined by observing individual cells growing asynchronously for one to two generations before they are irradiated.

3. *Different cell types expressing different genes* can be compared with one another as they are *irradiated at the same age*, e.g., in G1, S, or G2.
4. Once the *cellular endpoints are quantified* and compared between *different cell types within a given age category*, a logical and systematic approach can be taken in investigating fundamental *molecular and biochemical pathways* responsible for the differences observed in radiosensitivity, checkpoints, and modes of death.
5. *Within each cell type*, observed *differences in the cellular endpoints* as cells are irradiated in *different phases of the cell cycle* and can be systematically investigated at the *molecular level*. Utilization of GFP proteins should be most useful, as the histories of events that precede the different fates of the cells are determined.
6. Ultimately, determine the cell cycle phase irradiated and study the histories of events in nonclonogenic cells compared with clonogenic cells (also identify colonies with lethal sectoring).
7. In considering the points above, we wish to emphasize that events related to mitosis can be identified by video time-lapse system (CVTL). These events discussed and *illustrated in the DVD provided* are: (*i*) premitotic death (apoptosis and necrosis during interphase), (*ii*) delay and/or prolonged arrest (senescence?) in the cell cycle, (*iii*) mitotic death (apoptosis and necrosis *during* an abortive mitosis), (*iv*) death after an abnormal mitosis—mitotic catastrophe leading to binucleated and multinucleated cells that result from failed divisions or after cells divide followed by fusion of the daughters, and (*v*) postmitotic death (apoptosis, necrosis, and senescence after cells divide 1 or more times).

ACKNOWLEDGMENTS

The CVTL research conducted in the laboratory of W.C. Dewey was supported in part by NIH Cancer Institute grants CA 31808 and CA 85610 to W.C.D and CA 52713 and CA 61019 to C. Clifton Ling, a collaborator. Appreciation is expressed to Dr. Bert Vogelstein for providing the HCT116 isogenic derivative cell lines. I wish to thank several individuals who made major scientific and technical contributions: N. Albright, K. Chu, M.W. Dewey, B. Endlich, H.B. Forrester, C. King, E.A. Leonhardt, J. Lindquist, C.C. Ling, G. Pieur-Carrillo, N. Teele, M. Trinh, and C.A. Vidair.

REFERENCES

1. Hurwitz C, Tolmach LJ. Time lapse cinemicrographic studies of X-irradiated HeLa S3 cells I. Cell progression and cell disintegration. Biophys J 1969; 9:607–633.
2. Hurwitz C, Tolmach LJ. Time-lapse cinemicrograhic studies of x-irradiated HeLa S3 cells II. cell fusion. Biophys J 1969; 9:1131–1143.

3. Thompson LH, Suit HD. Proliferation kinetics of X-irradiated mouse L cells studied with time-lapse photography. II. Int J Radiat Biol 1969; 15:347–362.
4. Heye RR, Kiebler EW, Arnzen RJ, et al. Multiplexed time-lapse photomicrography of cultured cells. J Microsc 1981; 125:41–50.
5. Kallman RF, Blevins N, Coyne MA, et al. Novel instrumentation for multifield time-lapse cinemicrography. Comput Biomed Res 1990; 23:115–129.
6. Forrester HB, Vidair CA, Albright N, et al. Using computerized video time lapse for quantifying cell death of X-irradiated rat embryo cells transfected with c-myc or c-Ha-ras. Cancer Res 1999; 59:931–939.
7. Chu K, Teele N, Dewey MW, et al. Computerized video time lapse study of cell cycle delay and arrest, mitotic catastrophe, apoptosis and clonogenic survival in irradiated 14-3-3sigma and CDKN1A (p21) knockout cell lines. Radiat Res 2004; 162:270–286.
8. Forrester HB, Albright N, Ling CC, et al. Computerized video time lapse (CVTL) analysis of apoptosis of REC:Myc cells x-irradiated in different phases of the cell cycle. Radiat Res 2000; 154:625–639.
9. Heddle JA, Carrano AV. The DNA content of micronuclei induced in mouse bone marrow by gamma-irradiation: Evidence that micronuclei arise from acentric chromosomal fragments. Mutat Res 1977; 44:63–69.
10. Revell SH. Relationship between chromosome damage and cell death. In: Ishihara T, Sasaki MS, eds. Radiation-Induced Chromosome Damage in Man. New York, NY: Liss, 1983:215–233.
11. Endlich B, Radford IR, Forrester HB, et al. Computerized video time-lapse microscopy studies of ionizing radiation-induced rapid-interphase and mitosis-related apoptosis in lymphoid cell lines. Radiat Res 2000; 153:36–48.
12. Chu K, Leonhardt EA, Trinh M, et al. Computerized video time lapse (CVTL) analysis of cell death kinetics in human bladder carcinoma cells (EJ30) x-irradiated in different phases of the cell cycle. Radiat Res 2002; 158:667–677.
13. Prieur-Carrillo G, Chu K, Lindqvjist J, et al. Computerized video time lapse (CVTL) analysis of the formation and fate of giant cells in x-irradiated human bladder carcinoma EJ30 cells. Radiat Res 2003; 159:705–712.
14. Froese G. Division delay in HeLa cells and Chinese hamster cells A time lapse study. Int J Radiat Biol 1966; 10:353–367.
15. Froese G, Cormack DV. A correlation between division delay and loss of colony-forming ability in Chinese hamster cells irradiated *in vitro.* Int J Radiat Biol 1968; 14:589–592.
16. Thompson LH, Suit HD. Proliferation kinetics of X-irradiated mouse L cells studied with time-lapse photography: I. Experimental methods and data analysis. Int J Radiat Biol 1967; 13:391–397.
17. Okeefe RT, Henderson SC, Spector DL. Dynamic organization of DNA replication in mammalian cell nuclei—spatially and temporally defined replication of chromosome-specific alpha-satellite DNA sequences. J Cell Biol 1992; 116:1095–1110.

20

Modes of Cell Death by Anticancer Agents: The Crucial Importance of Dose

J. Martin Brown
Division of Radiation and Cancer Biology, Stanford University Medical Center, Stanford, California, U.S.A.

DOSE RESPONSE CURVES FOR DIFFERENT MODES OF CELL DEATH

As detailed throughout this book and elsewhere (1,2), there are several different ways in which cells can die after injury, including that produced by conventional anticancer therapy. In many cases, cells manifest more than one mode of cell death following a given treatment (3). The mix of the different modes of cell death depends critically on the type of cell, its genetic makeup, and the agent used. For example, B-lymphocytes are particularly susceptible to dying by apoptosis, whereas cells derived from epithelial or connective tissues are more likely to experience mitotic death or be permanently arrested following DNA damage (1,4–7). What is perhaps less appreciated is the importance of the dose delivered to the cells in determining the mode of cell death. This is well illustrated in Figure 1, which shows the cumulative percentage of apoptosis in the human TK6 B-lymphocyte cell line as a function of radiation dose. Though this cell line is particularly susceptible to death by apoptosis, it is even more sensitive to cell inactivation measured by clonogenic survival probably because of the sensitivity of this cell line to mitotic death as a result of a deficiency in DNA double strand break repair (8). As shown in Figure 1A, a dose of 5 Gy kills approximately 50% of the cells by

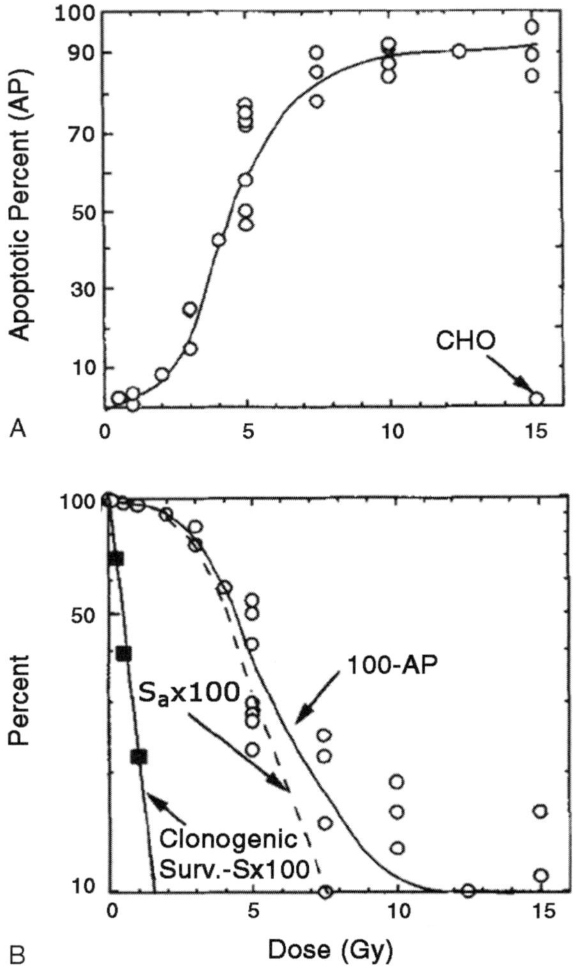

Figure 1 (**A**) Percentage of apoptotic cells as a function of radiation dose for the TK6 human B-lymphocyte cell line scored 24 hours after irradiation. CHO cells, which are not susceptible to apoptosis, are also shown. (**B**) Data in (**A**) were converted into percentage of the total population susceptible to apoptosis [assumed to be 90% from the plateau in (**A**)]. The dashed line labeled $S_a \times 100$ plots the survival of the apoptotic susceptible population. *Abbreviation*: CHO, Chinese hamster ovary. *Source*: From Ref. 4.

apoptosis. However, Figure 1B shows that few, if any, cells survive this dose when measured by clonogenic survival. At a lower dose (e.g., 2 Gy), few cells (<10%) die of apoptosis, but more than 90% die by clonogenic survival. Thus, at high doses (>5 Gy), most cells die of apoptosis, but at lower doses, another form of cell death predominates.

Figure 1B shows the two different dose response curves for clonogenic survival (which integrates all forms of cell death and growth arrest) and apoptosis, and shows a major difference between the sensitivity of the two forms of death. It is clear that these cells are much more sensitive to killing by the form of cell death associated with poor DNA repair than by apoptosis. However, the data also raise the question of how could a high level of apoptosis be measured at high doses if another form of cell death was most important in killing the cells. The answer lies in the fact that the kinetics of different modes of cell death are very different with primary apoptosis being a rapid way in which cells die after treatment. Thus, not only is it important to be aware of the different dose response curves for different modes of cell death, it is also important to be aware of their different kinetics.

These concepts are illustrated in Figure 2 for two hypothetical modes of cell death, mode A and mode B. For the purposes of this example, we assume that

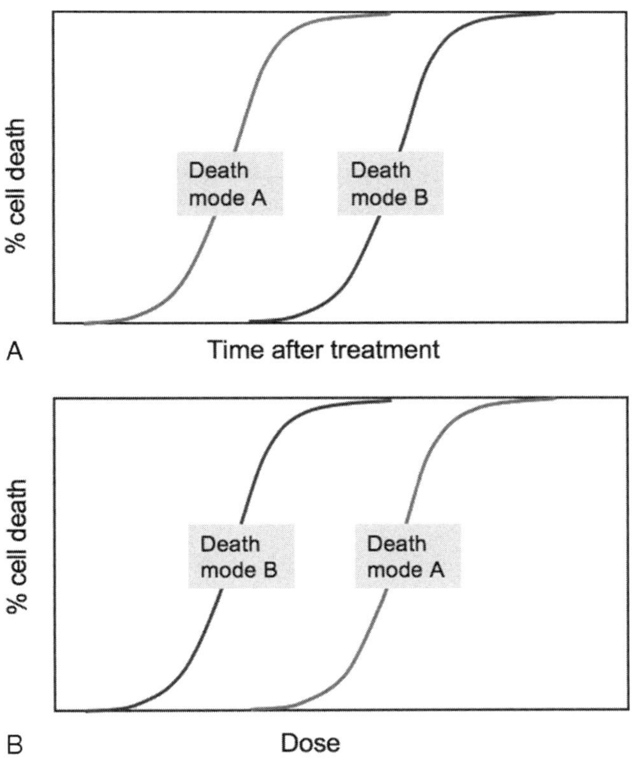

Figure 2 Illustrates different kinetics (**A**) and dose response (**B**) for cells undergoing two different forms of cell death, mode A and mode B. Death by mode A is assumed to occur rapidly after treatment but be more resistant than mode B as a function of the dose delivered (*The color version of this figure is provided on the DVD*).

mode A occurs rapidly after treatment compared to mode B (Fig. 2A), but is more resistant than mode B in terms of inductions as a function of dose. Thus, depending on the dose given, the investigator could conclude that cells die of mode A or mode B: at low doses, he/she would conclude that the cells were dying of death mode B, whereas at higher doses, he/she would conclude that death mode A was more important. As we illustrated with the data in Figure 1, death mode A is often primary (early) apoptosis, whereas death mode B is often mitotic death or permanent arrest.

FALLACIES FROM LOOKING ONLY UNDER ONE LAMPPOST

The above considerations suggest that an investigator could come to different conclusions as to the most important mode of cell killing depending on the dose he/she used. However, this is not what has led to the majority of inaccurate conclusions in the literature. Rather, most of the problems have stemmed from what has been a widespread assumption that cells die following cancer therapy by apoptosis (9–12). Thus, an investigator who wished to determine whether a particular genetic mutation or condition sensitized or protected against cell death by anticancer therapy would adjust the conditions of the experiment—notably, time of assay and dose of agent given—in order to produce apoptosis in the cell line used. Typically, the assay would be performed at an early time after treatment (1–2 days) with doses sufficient to cause apoptosis in this time scale. In Figure 3, we have reproduced the curves from Figure 2 but have shown that if death mode A were apoptosis, then changing levels of the apoptosis inhibitor Bcl-2 could markedly change the sensitivity of the cells to apoptosis but would not change the sensitivity to mitotic death. However, as the investigator has chosen apoptosis at an early time after treatment for the assay, he/she would not be aware of the fact that overall cell killing, which would be dominated by mitotic death, had not changed. Thus, the conclusion would be that changing levels of Bcl-2 would change overall cell killing. As a number of investigators have shown, such changes in Bcl-2 can have marked changes on the levels of apoptosis without changing overall cell killing as measured by clonogenic survival (13–15).

CONSIDERATIONS OF DOSE RESOLVE CONFLICTS IN THE LITERATURE

There are few areas in science in which the literature is more conflicting than that describing the effects of various genetic mutations on the response of tumor cells to anticancer therapy. The fact that this is a crucial area of importance to cancer patients and their physicians makes this confusion more than one of academic interest. Faced with such contradiction, hundreds of studies have been performed with clinical material to try to tease out the influence of various mutations or protein levels on therapy response. Unfortunately, these have not

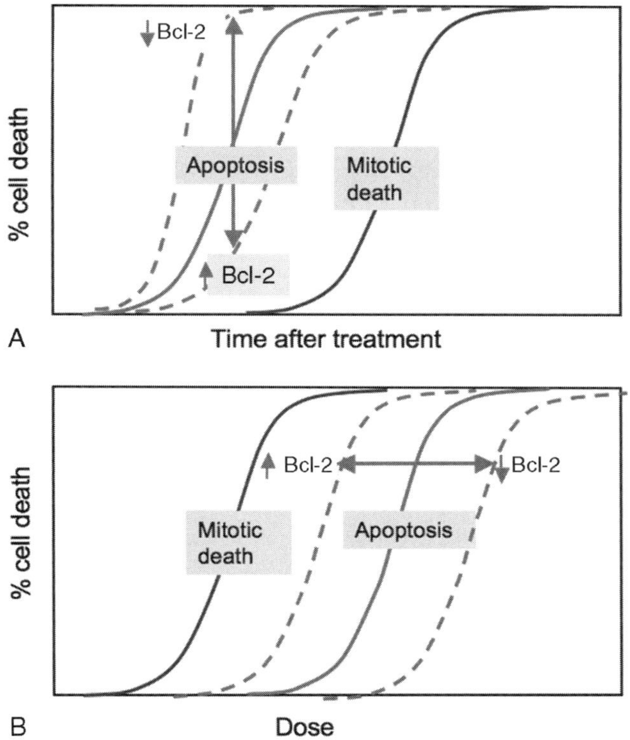

Figure 3 Illustrates the fact that if Bcl-2 overexpression or suppression decreases or increases the sensitivity of cells to apoptosis (**B**), this will produce a decrease or increase in the level of apoptosis at any time it is measured (**A**), but will not affect the sensitivity for mitotic death, assuming this is more sensitive than apoptotic death (**B**) (*The color version of this figure is provided on the DVD*).

led to definitive conclusions (16,17). We believe that the considerations of dose, time of assay, and the assay used described in the above section go a long way toward resolving many of these controversies.

To illustrate this, we show data from work of Zamble and colleague (18) (Fig. 4). These authors investigated the effect of p53 on the response of mouse testicular teratocarcinoma cells to cisplatin. They found that cisplatin treatment resulted in rapid apoptosis in p53 wild-type cells but not in $p53^{-/-}$ teratocarcinoma cells. In the latter case, cisplatin exposure produced prolonged cell-cycle arrest with accompanying high levels of p21 protein. When they performed clonogenic assays on the cells, they found that the p53 mutation did not confer resistance to cisplatin. The data for apoptosis is shown in Figure 4A, which demonstrates that continuous exposure to 10 μM cisplatin produces a p53-dependent apoptosis, with apoptosis being much lower in the cell line with

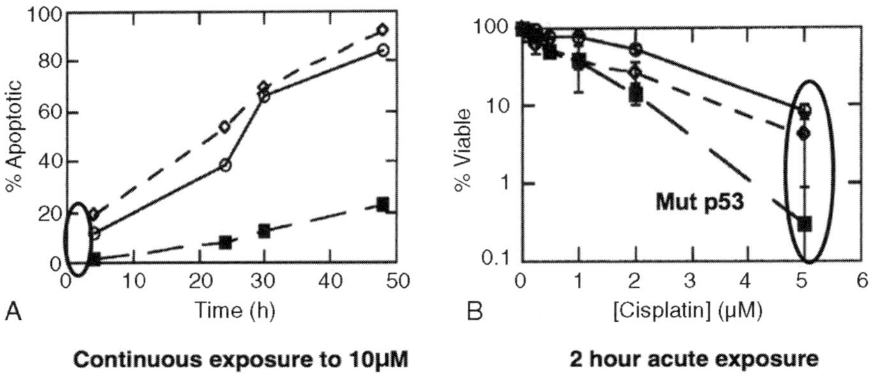

Figure 4 (**A**) Percentage of cells undergoing apoptosis as a function of time of continuous exposure to 10 μM cisplatin. The p53-mutated cells (■) are highly resistant to the induction of apoptosis. (**B**) Clonogenic survival of the three cell lines following a two-hour acute exposure to different concentrations of cisplatin. The p53-mutated cell line is more sensitive to this assay. The dose that produces 90% to 99% cell kill (5 μM) is outlined by the oval and is reproduced by the oval (**A**), demonstrating that this dose produces little or no apoptosis in any of the cell lines. *Source*: From Ref. 18.

mutated p53. Figure 4B shows the clonogenic assay for the three cell lines following a two-hour acute exposure to cisplatin and shows much less difference between the cell lines with, if anything, the mutated p53 cell line being more sensitive to cisplatin. The authors concluded that cisplatin inhibits cell proliferation in these cells by two mechanisms: a p53-dependent apoptosis and a p53-independent cell-cycle arrest. However, what is not emphasized in the publication is the very different sensitivities of the cells to the loss of clonogenic survival and to the induction of apoptosis. This can be seen by referring to the red ovals in Figure 4A, B. In the case of clonogenic survival, an exposure of two hours to 5 μM cisplatin kills 90% to 99% of the cells. This exposure is equivalent to a one-hour exposure to 10 μM (same concentration × time, C × T), which can be seen in Figure 4A and produces little or no apoptosis. Thus, the doses being used to measure apoptosis are superlethal for clonogenic survival, with no cells surviving such treatment, irrespective of the genetic makeup. This illustrates what we believe to be a general phenomenon: since for the majority of cell types, the induction of apoptosis requires higher doses than are required to inactivate cells by other cell death or permanent arrest modes, investigators who have used apoptosis as a surrogate for cell death have generally used superlethal doses of the agent—in effect, doses that kill all cells irrespective of genetic makeup. It is clear that this could lead to inaccurate conclusions as to the genetic determinates of cell killing.

But is the above an isolated example? Is it correct that investigators have in general used higher doses to measure cell killing by apoptosis than are needed to produce cell death by clonogenic survival? This question was addressed in a literature search by Berndtsson and colleagues (19) by examining 100 randomly selected recent publications where cisplatin had been used to measure apoptosis in tumor cell lines. They found that the mean cisplatin concentration used was continuous exposure to a concentration of 52 μM with apoptosis measured at 24 hours. As cisplatin is relatively stable in vitro (20), it is reasonable to assume a $C \times T$ area under the curve calculation to compare acute exposure with chronic exposure. Thus, if we assume a mean 10-hour exposure in the apoptosis studies, $C \times T$ would be approximately 500 μM h for an average apoptosis-inducing study. This is far higher than that needed to kill most of the cells by clonogenic survival as can be seen from Fig. 4B. Typically, values of 10 to 30 μM hr in the literature are needed to kill 90% of tumor cells by cisplatin when assayed by clonogenic survival. An additional comparison for growth arrest would be the mean GI50 concentration of the NCI 60 cell panel, which, for cisplatin, is 0.83 μM. As this is a two-day assay with a half-life of cisplatin of 16 hours in vitro (20), it would give a $C \times T$ of approximately 20 μM hr, which agrees with the results of clonogenic survival experiments. Thus, these comparisons suggest a factor of approximately of 20- to 50-fold in doses used to measure apoptosis and cell killing for cisplatin.

SUMMARY

Cells die or arrest permanently following cancer treatment in different ways, depending on the type of cell, their genetic makeup, the agent used, and the dose delivered to the cells. The kinetics of these different modes of cell killing differ, with some being rapid (e.g., primary apoptosis) and some delayed (e.g., mitotic death, which always follows mitosis). In many instances in the literature, investigators have chosen to measure apoptosis as a measure of cell killing and have adjusted the doses and times of assay to fit this endpoint. Unfortunately, they have not appreciated that one or more other forms of cell death might occur later and, importantly, at lower doses. This potentially invalidates any conclusions about genes or conditions that affect cell killing if these conclusions are based on doses that are relevant for apoptosis but "overkill" for another form of death. It is recommended that, if possible, investigators should determine the dose response curve for clonogenic survival following any treatment (as this integrates all forms of death), and not use doses above the range in which a substantial fraction of the cells can form colonies after treatment.

CONCLUSION

The dominant paradigm over the past 10 to 15 years that cells die of anticancer therapy by apoptosis has led investigators to study apoptosis as a surrogate for cell killing. They have therefore adjusted the doses and timing of the assay used to obtain measurable levels of apoptosis. However, it is now clear that at least for some agents, these doses are at least an order of magnitude greater than those needed to kill cells by other mechanisms or produce permanent growth arrest. Consequently, conclusions reached as to the genetic factors that affect apoptosis are unlikely to apply to these more sensitive modes of cell death. This in itself has created tremendous confusion in the literature as to the determinants of tumor sensitivity to anticancer drugs. As clonogenic survival is a means of integrating all forms of cell death for treatment in vitro, we recommend that investigators determine the dose response curve for clonogenic survival following any treatment, and not use doses above the range in which a substantial fraction of the cells can form colonies after treatment. If these doses do not induce apoptosis, then it is unlikely that apoptosis is an important form of cell killing after these treatments.

REFERENCES

1. Brown JM, Attardi LD. The role of apoptosis in cancer development and treatment response. Nat Rev Cancer 2005; 5(3):231–237.
2. Okada H, Mak TW. Pathways of apoptotic and non-apoptotic death in tumour cells. Nat Rev Cancer 2004; 4(8):592–603.
3. Ruth AC, Roninson IB. Effects of the multidrug transporter P-glycoprotein on cellular responses to ionizing radiation. Cancer Res 2000; 60(10):2576–2578.
4. Dewey WC, Ling CC, Meyn RE. Radiation-induced apoptosis: relevance to radiotherapy. Int J Radiat Oncol Biol Phys 1995; 33(4):781–796.
5. Chang BD, Broude EV, Dokmanovic M, et al. A senescence-like phenotype distinguishes tumor cells that undergo terminal proliferation arrest after exposure to anticancer agents. Cancer Res 1999; 59(15):3761–3767.
6. Roninson IB, Broude EV, Chang BD. If not apoptosis, then what? Treatment-induced senescence and mitotic catastrophe in tumor cells. Drug Resist Updat 2001; 4(5):303–313.
7. Mirzayans R, Scott A, Cameron M, et al. Induction of accelerated senescence by gamma radiation in human solid tumor-derived cell lines expressing wild-type TP53. Radiat Res 2005; 163(1):53–62.
8. Evans HH, Ricanati M, Horng MF, et al. DNA double-strand break rejoining deficiency in TK6 and other human B-lymphoblast cell lines. Radiat Res 1993; 134(3):307–315.
9. Weinberg RA. One Renegade Cell: How Cancer Begins. Perseus Publishing, 1999.
10. Herr I, Debatin KM. Cellular stress response and apoptosis in cancer therapy. Blood 2001; 98(9):2603–2614.
11. Johnstone RW, Ruefli AA, Lowe SW. Apoptosis: a link between cancer genetics and chemotherapy. Cell 2002; 108(2):153–164.
12. Kim R, Tanabe K, Uchida Y, et al. Current status of the molecular mechanisms of anticancer drug-induced apoptosis. The contribution of molecular-level analysis to cancer chemotherapy. Cancer Chemother Pharmacol 2002; 50(5):343–352.
13. Kyprianou N, King ED, Bradbury D, et al. bcl-2 over-expression delays radiation-induced apoptosis without affecting the clonogenic survival of human prostate cancer cells. Int J Cancer 1997; 70(3):341–348.

14. Lock RB, Stribinskiene L. Dual modes of death induced by etoposide in human epithelial tumor cells allow Bcl-2 to inhibit apoptosis without affecting clonogenic survival. Cancer Res 1996; 56(17):4006–4012.
15. Yin DX, Schimke RT. BCL-2 expression delays drug-induced apoptosis but does not increase clonogenic survival after drug treatment in HeLa cells. Cancer Res 1995; 55(21):4922–4928.
16. Brown JM, Wilson G. Apoptosis genes and resistance to cancer therapy: what do the experimental and clinical data tell us? Cancer Biol Ther 2003; 2(5):477–490.
17. Wilson GD. What do the clinical data tell us about the role of apoptosis in sensitivity to cancer therapy? In: Roninson IB, Brown JM, Bredesen DE, eds. Beyond Apoptosis: Cellular Outcomes of Cancer Therapy. New York: Informa Health Care, 2007.
18. Zamble DB, Jacks T, Lippard SJ. p53-Dependent and -independent responses to cisplatin in mouse testicular teratocarcinoma cells. Proc Natl Acad Sci U S A 1998; 95(11):6163–6168.
19. Berndtsson M, Hagg M, Panaretakis T, et al. Acute apoptosis by cisplatin requires induction of reactive oxygen species but is not associated with damage to nuclear DNA. Int J Cancer 2007; 120(1):175–180.
20. Doz F, Berens ME, Dougherty DV, et al. Comparison of the cytotoxic activities of cisplatin and carboplatin against glioma cell lines at pharmacologically relevant drug exposures. J Neurooncol 1991; 11(1):27–35.

Index